Consultancy and Innovat

T0228244

Consultancy and Innovation links two important aspects of European economic development in the past thirty years: the pace of technical and management innovation and the growing significance of technical and business consultancy. The rapid expansion of specialist consultancy, evident in every country, has affected the nature and rate of innovation among all its clients.

Written by national experts in the field, the book includes detailed studies of consultancy activities or 'knowledge-intensive services' (KIS) in eight EU countries – France, Germany, The Netherlands, the UK, Italy, Greece, Portugal and Spain. They have all experienced rapid consultancy growth since 1980. Although there are many similarities in trends between countries, the most critical differences affecting consultancy influence are the conditions governing the quality of consultancy–client interaction. In spite of their growth, innovation and economic development policies continue to neglect the contribution of consultancies to processes of European modernization.

The role of consultancies also needs to be judged in the context of the developing 'knowledge economy'. In general the corporate managements of large firms determine the exploitation of scientific and commercial knowledge. Increasingly they rely on commercial consultancies to help them interpret and respond to change. These commercial consultancies are therefore integral to change. One of their most important characteristics in this role is growing internationalization, opening up national and regional economies to global influences on business change. However, while consultancy expertise increasingly influences innovative change, it is also polarizing economic opportunities across Europe, as it concentrates in the existing prosperous regions.

Peter Wood is Professor and Head of the Geography Department at University College London. As well as working for the European Commission he has co-directed a series of Economic and Social Research Council (ESRC)-sponsored studies on the role of consultancies and their use by large client firms.

Routledge Studies in International Business and the World Economy

Consultancy and Innovation

The business service revolution
in Europe

Edited by Peter Wood

Routledge
Taylor & Francis Group

LONDON AND NEW YORK

First published 2002
by Routledge
2 Park Square, Milton Park, Abingdon, Oxfordshire OX14 4RN

Simultaneously published in the USA and Canada
by Routledge
711 Third Avenue, New York, NY 10017

First issued in paperback 2015

Routledge is an imprint of the Taylor and Francis Group, an informa business

Typeset in Galliard by Bookcraft Ltd, Stroud, Gloucestershire

British Library Cataloguing in Publication Data
A catalogue record for this book is available from the British Library

Library of Congress Cataloging in Publication Data
  Consultancy and innovation: the business service revolution in
Europe/edited by Peter Wood.
     p.   cm.
  Includes bibliographical references and index.
  1. Business consultants – European Union countries.
  2. Knowledge management – European Union countries.
  I. Wood, Peter

  HD69.C6 C6355 2001
  004'.094–dc21   2001019662

ISBN13: 978-1-138-86728-4 (pbk)
ISBN13: 978-1-84142-005-9 (hbk)

Contents

Illustrations

Contributors

Isabel André is Assistant Professor at the Department of Geography, University of Lisbon, and a member of the Geoideia economic and planning consultancy team. Her work concerns labour-market analysis, gender and local development.

Lucia Cavola is senior partner of Iter – Centro Ricerche e Servizi, a research centre based in Naples, which carries out studies on economic, social and policy issues in the South of Italy for the central and local governments. Her work focuses on the study of entrepreneurship, technological innovation and business services as a leverage for development.

Juan R. Cuadrado-Roura is Professor of Applied Economics at University of Alcalá. Director of the Service Industries Research Laboratory (Servilab), Spain. Former president of the European Regional Science Association. Author of many books and articles on both regional science and service activities.

Pavlos-Marinos Delladetsima is Assistant Professor in Geography at the University of the Aegean, specializing in urban geography, planning, and planning methods and land policy. He has coordinated and participated in many research programmes in Europe and Greece concerning land use and real estate development, local economic development, decision support systems and regional service economies.

Hélène Farcy received her PhD from the University of Lille I, on the role of producer services in the French urban hierarchy. She is currently working as an *ingénieur d'études* at the Faculty of Economics and Sociology in the University.

Paulo Areosa Feio is Lecturer at the Department of Geography, University of Lisbon. He specializes in the study of innovation and SMEs. Publications include *Território e Competitividade*, Edições Colibri, 1997.

João Ferrão is senior researcher at the Institute of Social Sciences, University of Lisbon, and a member of the Geoideia economic and planning consultancy team. His main interests are in producer services, innovation and regional development policies.

Camal Gallouj is Associate Professor of Economics and head of the *licence-maîtrise* in Labour Economics at the Faculty of Economics and Sociology at the University of Lille I. He specializes in service economy research.

Alekos G. Kotsambopoulos is an expert in economic analysis and employment generation strategies, and consultant on many research programmes and projects in Greece, Europe and the Balkan countries. Special concerns include project evaluation and appraisal, industrial restructuring policies and service sector development.

Jan Lambooy is Professor of Geography at Utrecht University and in the Faculty of Economics and Econometrics of the University of Amsterdam. He is an economic geographer with interests in innovation, small–medium enterprises and regional development.

Walter Manshanden is a regional economist, working as a senior researcher with the SEO (Foundation for Economic Research) in the Faculty of Economics and Econometrics, University of Amsterdam. He works mainly on the regional economy of the metropolitan area of Amsterdam.

Flavia Martinelli is Professor of Analysis of Territorial Systems in the Planning Programme of the University of Reggio Calabria. She has published several works on regional development issues and policies, in Italy and internationally, with a focus on industrial structures, services and tourism, especially in relation to the Italian 'southern question'.

Frank Moulaert is Professor of Economics and Sociology and head of Masters Programme in Industrial and Labour Economics at the University of Lille I. Recent books include: *Globalization and Integrated Area Development in European Cities* (with others) Oxford: Oxford University Press, 2000.

Luis Rubalcaba-Bermejo is Titular Professor of Applied Economics at the University of Alcalá. General Secretary of the Service Industries Research Laboratory (Servilab), Spain. Consultant for the European Commission since 1989 on service activities, business services in particular. Member of the RESER (Réseau Européen Services et Espace – Services and Space European Network) Council.

Simone Strambach is university researcher and teacher in the Department of Economic and Social Geography at the University of Stuttgart. Her major research areas are national and regional innovation systems, and regional policy development, as they reflect the evolution and functioning of inter-organizational networks, learning processes and their spatial implications.

Chris van der Vegt is an economist and senior researcher with the SEO (Foundation for Economic Research). His major research areas are national and regional innovation systems, and regional policy development, as they reflect the evolution and functioning of inter-organizational networks, learning processes and their spatial implications.

Peter Wood is Professor and Head of Department of Geography at University College London. His broad research interests are in regional economic development, especially in the regional impacts of business service growth on employment and the provision of technical and managerial expertise to other sectors. He has co-directed major Economic and Social Research Council (ESRC) projects on business service developments in various UK regions, the use of consultancies by large clients, and the regional conditions supporting the growth of business service growth, especially their exports.

Introduction

Knowledge-intensive services, consultancy and innovation

Peter Wood

Innovation has become the clarion call of modern economic management. Traditionally, it is associated with technological change and the development of new or increasingly complex manufactured goods. Technology is seen as the embodiment of human ingenuity in transforming the material world, with the exploitation of inventions being one of the most important bases for capitalist growth in an industrial age.

The revival of interest in these processes during the past quarter-century reflects the acceleration of technological change based especially on micro-electronics, information and computer technology, new materials and biotechnology. In the longer historical context, comparisons have been made with earlier phases of innovation, suggesting that we are moving towards a new, 'Fifth Kondratieff' wave of technology-induced growth.[1] Before this, during the long post-war period of relative technological stability, the advanced industrial economies were dominated by mass production, especially of consumer goods such as automobiles and electrical gadgetry; improved transport and communications based on electronics developments; and the widespread adoption of oil-based synthetic materials. Now the innovative focus has shifted especially to micro-electronics and computer networks, radically transforming communications and work patterns across all aspects of life. The former market conditions and price relationships that, at least notionally, linked supply and demand for materials, capital, labour and products, are being changed. Competition between old and new forms of production has intensified, exemplified by the impacts of 'lean' manufacturing methods on automobile production or of computer networks on banking (Womack *et al.* 1990; Harrison 1994). Serving the diversity of customer preferences now dominates business strategies. The new technologies enable capital investors to pursue their goals in new ways, with new social and political consequences. The stable consensus between corporate, state and labour union power, which dominated political debate in the 1950s and 1960s, and delivered social goals such the welfare state and full employment, has also had to be recast (Piore and Sabel 1984; Moulaert and Swyngedouw 1989; Harvey 1989). Thus, the more or less predictable rules of the established economic system have been overturned. Radical technological change brings new, inherently unpredictable market relationships, new forms of business opportunity and requires new methods of state involvement.

These conditions have prompted a return to earlier ideas, especially those of J.A. Schumpeter, who emphasized the importance of new technology in driving capitalist investment, growth and profit (Schumpeter 1939). The accelerating development of new materials, processes and products has created clusters of technology, supporting new composite markets in information technology, the media, construction or capital circulation, and directed towards personal consumption, leisure, mobility, health, and education and training provision. The result is not just high-technology factories but also high-tech offices, homes, TV studios, warehouses, superstores, theme parks, airports, banks, hospitals, security systems and databases. The culture of innovation has promoted new ways of organizing the global economy, new forms of capital exchange, and new ways of working and consuming, in a sort of 'hyper-Schumpeterian' world. The roles of economic actors, and of market relationships, appear to be constantly transforming. The tendency towards disequilibium is now more evident than the achievement of any short-term stability. Capitalism appears even more dedicated to exploiting the outcomes of such 'creative destruction'.

Of course, it is not always easy to separate the reality of current trends from the hype of those who benefit from promoting them. While change is all around us, perhaps much more remains the same. Innovation is a complex process with many outcomes which are seldom uncontroversially beneficial. Innovativeness is also not confined to the technological realm. In the 1970s it became increasingly obvious that economic success required, even more, the effective organization of production, with associated financing, human resources development and marketing. The success of the Japanese economy seemed to be based less on traditional scientific inventiveness, and more on the effective delivery of established technologies. The 1980s were also dominated by a search for new, more effective forms of organization, management and human resources development to serve these ends. Evidently, 'late industrial' competitiveness depended on adaptability across all the activities which support the production of goods and services. Manufacturing technology and production may be important means to this end but not the only keys to success.

In the broadest sense, innovativeness appears now to encompass the social ability, not just to initiate change, but also to adapt to myriad forms of change whose sources have become much wider. International and global trade, capital and information exchanges have opened up developed economies to commercial 'best practice' anywhere in the world. Multinational companies are key agents in such change, often deploying their global command of capital resources. But small to medium-sized companies have also come to be regarded as sources of particular forms of technical and organizational innovativeness, especially at the sub-national, regional level. Further, in some key markets, such as defence and health spending, government agencies dominate the pace of change. Even consumers, individually and collectively, have been drawn into the culture of change, being made constantly aware through the media of new global possibilities in arenas such as fashion, popular culture, the treatment of disease or environmental change.

Knowledge-intensive services

The culture of change has itself encouraged the emergence of new economic sectors orientated towards supporting change. Again, these might conventionally be associated with 'high-technology' production. Equally, and more pervasively, they consist of new and growing types of services, promoting new ways of doing things. These include such diverse activities as television production companies, new types of financial intermediary, contract cleaning corporations and 'bucket-shop' travel agencies. When the provision of knowledge about change is their purpose, these services may generically be described as the 'knowledge-intensive services' (KIS).

KIS are significant because they offer specialist knowledge to other organizations in a rapidly changing, increasingly uncertain, and internationally orientated economic environment. The terminology of their definition, or at least the confusion surrounding it, needs to be clarified for the purposes of this book. First, they are profit-making, private sector companies or partnerships. They do not include non-commercial information or advisory agencies supported, for example, by government or trade organizations. Second, KIS are conventionally defined as orientated towards organizations rather than consumer clients, and so exclude knowledge-intensive consumer services such as educational, information or cultural activities. They are therefore sometimes regarded as synonymous with 'producer services' or 'business services', or the more specifically termed 'knowledge-intensive business services' (KIBS). Each of these phrases, however, has a different connotation. 'Producer services' is a term originally coined to emphasize that not all services were simply consumer orientated, since some supported manufacturing and other material production. 'Business services' are more generally directed towards any private business organization customers. Either might include manual and material-handling activities such as contract cleaning or transport. Such activities would ideally not be part of KIBS, which are largely synonymous with KIS. The latter term is adopted here, mainly because such activities often serve government, statutory and voluntary bodies as well as private business. In summary, KIS are thus private sector firms providing knowledge-based services to other business and non-business organizations.

Such niceties of definition tend to be blunted by measurement practice and convention. Virtually all studies drawing on secondary data sources, including those in this volume, must accept definitions which, at best, only approximate to any formal definitions of producer, business or knowledge-intensive services. There is no standard approach to establishing such proxies, as Marshall (1988) and many other commentators have noted. As a result, the 'KIS' of one study are likely to be subtly different from that of another, since practical definitions are based on different adaptations of established classification systems, related to the International Standard Industrial Classification or the more recent European Union standard NACE (Nomenclature des Activités Communauté Européen).

All such systems, originally directed towards classifying manufacturing units of production on the basis of technology, are inherently unable to differentiate service activities effectively. The essential economic role of services is defined not simply by technology, but also by the markets they serve, for example among consumer or business clients. The distinction between consumer and producer markets is important for manufactured products, but the worth of services is much more directly market defined. For example, the difference between activities serving consumer, business or public sector clients is much more significant for banking, passenger transport, telecommunications, legal or accountancy services than the different production processes they employ. Some conventionally defined sectors, such as retailing, appear largely consumer orientated while others primarily serve private and public organizations, including advertising, computer services and 'business services'. The last conventionally combines very varied activities, including much (although not all) business consultancy and, as we have seen, catering, cleaning and security services, and even materials-handling activities such as transport and repair and maintenance. These official classifications thus offer a poor basis for identifying KIS accurately. Instead, measures of more aggregated 'business services' are usually employed, unavoidably including activities that also serve consumer markets, and often both private and public sectors (definitional issues are fully reviewed in Daniels 1985; Marshall 1988; Illeris 1989; Rubalcaba-Bermejo 1999).

In practice, therefore, fine distinctions between producer, business and knowledge-intensive services remain no more than notional. They cannot be properly measured and are often, in effect, treated as the same. The theoretical distinctions are nevertheless important because they have practical implications, especially in an increasingly service-dominated economy. Theoretically, the utility of any service is primarily defined by the market it serves (Hill 1977; Riddle 1986). Practically, it is also important in characterizing firms or organizations, whether their activities serve consumer or organizational markets. Theoretically, expertise is increasingly important in a knowledge-based or 'learning' economy (Lundvall and Johnson 1994). Practically, sources of key knowledge, both within and between organizations, are vital, for example in managing innovation. Neither market orientation nor knowledge intensity, the two defining characteristics of KIS, is consistently reflected in available secondary data.[2] The resulting imprecisions of measurement that dominate this field have therefore to be accepted. They nevertheless serve to demonstrate the continuing obsolescence of much contemporary monitoring of economic processes and change.

Consultancy

The notion of 'knowledge intensity' also needs to be elaborated. Obviously, any expert economic activity, including much manual labour, depends on specialist knowledge. KIS customers, however, pay primarily for information, usually associated with the knowledge required for its exploitation. This may be delivered in material form, as a report or computer disc, as a database, by word of mouth, or by any of these combined. Whatever the material form, the value of KIS depends on the utility of the knowledge conveyed to the client. This in turn may depend, for example, on the pace of market developments indicated, the means suggested for recruiting and training key staff, the arrangements suggested for obtaining capital, the scope of proposed computer or database developments, or the technical quality of recommended production changes.

If such 'knowledge intensity' is central to the role of KIS, even more important is the active process that conveys knowledge and enables it to be used by clients for their own purposes. This is the 'service' content of KIS, which consists of interactions with clients in knowledge exchange, a process commonly termed business 'consultancy'. Knowledge, like information, is of little value if it is not applied to a specific outcome. KIS consultancy must therefore apply *expertise* to knowledge, in ways that are related to the needs and complementary expertise employed by the client. Any consultancy process may be based on relatively routine information processing, or uniquely adapted to particular circumstances. The knowledge conveyed may be tacit, that is implicitly understood in the context of regular exchange, or explicitly codified. Often it may combine both modes. Consultancy is, however, essentially a process of exchange, whose most sophisticated and influential forms require close collaboration and often a sustained period of interaction with client expertise. Such exchanges are more likely to support significant KIS-influenced changes among clients than more routine interactions directed towards predetermined client outcomes. This form of consultancy is most likely to support KIS-induced client innovation. Therefore, as far as data sources allow, it forms the focus of this book.

Activities customarily regarded as knowledge-intensive services, based on various styles of consultancy, offer expertise on:

1   *Management and administration:* including management consultancy, legal, accounting, financial strategy and fiscal advice, mergers, and take-overs and restructuring.
2   *Production:* including architectural and engineering consultancy, distribution logistics, operational leasing, repair and maintenance, and quality control.
3   *Research:* including contract research, testing and specialist advice.
4   *Human resources:* including training, recruitment, job evaluation and temporary workers.

5   *Information and communications:* including data banks, information services, software services, technical computing, systems design and implementation, advanced telecommunications and express mailing.
6   *Marketing:* including advertising, sales, promotion, market research, direct marketing, public relations, export promotion, and fairs and exhibitions (adapted from Vogler-Ludwig *et al.* 1993).

These activities are also sometimes described as 'professional services'. This is because they depend on the skills and accreditation of individuals, and the establishment of client trust, supported by statutory or voluntary regulation. The modern growth of consultancy can be seen in some sectors of the traditional professions, especially law and accountancy. It can also be found in older-established commercial activities, such as marketing, advertising, personnel management and technical consultancy, which aspire to the trust-based status of the professions. These have been augmented in recent years by relatively new specialisms associated with computing, management advice and research. In all these, professionalization is being combined with active commercialization, including attempts to codify and standardize products. More radically, the consultancy professions are also subject to active corporate consolidation and internationalization. KIS developments also reflect both the professionalization of business management more generally, and the emergence of distinct market strategies by consultancy firms themselves (Tordoir 1994).

European perspectives

As later chapters demonstrate, the modern competitive emphasis on innovativeness across Europe appears to have been paralleled by the rapid growth of KIS, and of more intensive client–consultancy exchanges, over the past twenty or thirty years. The central questions addressed in this book concern how these two trends might be associated and, in particular, whether the growth of KIS has significantly affected processes of innovation and adaptability to change. The answers to these questions clearly depend on the relationship between aggregate knowledge-intensive services, usually measured in terms of employment, and the presence of active consultancy-type exchange with clients. The latter is clearly greater for some measured activities, such as management or computer consultancy, than it is for others, including data processing or commissioned advertising. In what follows, the term 'consultancy' refers to the process of active client–KIS expertise exchange, and agencies that engage in it; 'KIS' is used to encompass all private sector activities providing knowledge-based services to business and non-business organizations.

The significance of KIS in Europe arises from national and, more important, regional variability in both innovativeness and KIS development. Thus analyses of the European 'core' economic regions such as south-east England, the Paris Basin, Lombardy or Baden-Württemberg may not be relevant to other regions, or even nations. A comparison of international evidence is thus essential. Such an approach has been adopted in several recent studies of business services (see Illeris 1996; Rubalcaba-Bermejo 1999). In general, however, these studies have not made the link between either KIS development or consultancy activity, and wider innovativeness in client sectors. For this volume, these questions have been addressed by national specialists in eight European countries: France, Germany, The Netherlands, the United Kingdom, Italy, Greece, Portugal and Spain.

For most countries it is relatively easy to demonstrate the expansion and regional distribution of KIS and to identify key macro-economic trends that have favoured these developments. The trends are dominated by national economic conditions and forms of business regulation. As well as showing differences between states, these trends generally contrast the experience of various industrial nations of the 'North' with the rapidly transforming economies of southern Europe. The benefits that consultancy might bring to clients can also be identified in the form of specific knowledge and procedures. These have been enhanced, at least partly, by the increasingly international spread of key consultancy companies and networks, drawing in new sources of knowledge and experience. Benefits may also be focused in strong regional concentrations, reinforcing local knowledge systems of exchange. The achievement of this potential, however, appears uneven, varying between different countries, types of consultancy, client sectors and firm sizes, and regions.

The evidence presented also includes increasingly consistent survey results about the use of consultancies in many countries. Consultancies offer both technical and organizational expertise, and support strategically important private and public sector functions in a very wide range of circumstances and conditions. There is a similarly growing body of evidence, however, that traditional sectors, small businesses, and poorer, more peripheral regions have only limited access to consultancy. At the regional and micro-economic levels, therefore, there appears to be an association between innovative ability, the presence of KIS and the use of consultancy. Of course, this still does not confirm whether, and in what conditions, the intervention of consultancy as a business process may actually promote technical or organizational innovation by clients, or how far uneven access to KIS may be a basis for uneven regional, or even national, development.

KIS–client interaction

One conclusion that can be drawn from this book is that its central question cannot be answered consistently across Europe from currently available evidence. It can only present circumstantial evidence to support the hypothesis that the universal, if uneven, growth of KIS (especially of consultancy) has exerted a significant effect on rates of technical and organizational change among clients. Some might argue that growing consultancy influence is a symptom of the inability of client managements to respond for themselves to uncertainty. Gullible and inadequate managers are being persuaded to seek and adopt outside advice by the tactics of opportunists feeding off their insecurity, often with disastrous results. Such views tend to be based on anecdotal evidence of allegedly poor consultancy practice. The diversity of such practice, its rapid growth and wide penetration of modern organizational culture make it easy to find such cases, as well as cases of poor management in general. This scale and diversity also suggest, however, that a significant proportion of clients benefit from consultancy services, many even employing them on a routine basis.

What is clear from the survey evidence is that the impact of consultancy depends on how its expertise is matched to the needs of clients and how this interaction is managed. At its most basic level, much KIS activity is routine. Day-to-day functions are 'externalized' or outsourced through the provision of standard (including statutory) services such as accountancy or tax advice. These activities appear simply to transfer work from client employees to those of a specialist contractor. Even in these circumstances there may be innovative spin-offs. Benefits may arise from the economies of scale and specialist expertise offered by the supplier. Client resources may also be released, allowing them to be directed towards other activities. Outsourcing contracts usually include provisions for improving service efficiency over time. The innovative benefits of employing KIS therefore depend on the client's response to the situation, and may be positive even in the simplest outsourcing cases.

Much of the KIS interaction with clients assumed in this study, however, goes beyond such basic externalization. In general, clients seek out consultancies to do jobs which they believe cannot be done by in-house staff. This may simply compensate for gaps in client expertise, for example by providing auditing, market research or computer systems advice. In other cases, client staff possess considerable expertise, which consultancies can complement with more specialist skills, for example supporting market research and advertising campaigns, or recruiting and training new staff. Sometimes clients employ consultancies to reinforce the highest level of in-house expertise, such as in strategic reviews, complex project planning, technical development and testing programmes, new logistical plans or to support major construction projects. The innovative influence of consultancy may be significant in any of these contexts, although probably in different ways. For technical innovation, it may impinge directly on client core activities, reinforcing client strengths in product or computer systems development. On the other hand, outside specialists may selectively add to in-house expertise by supporting the organizational, training or marketing innovation required to deliver new technologies. Perhaps the least likely context for innovativeness is when KIS firms simply fill gaps in client capabilities, since this requires no other adjustment in established client procedures or products, whether or not these are innovation orientated.

These generalizations are hypothetical, although based on at least one established study (Wood 1996). They nevertheless emphasize how the nature of the client activity and response, based on the capabilities of their staff, is critical to the technical or organizational impact of consultancies. To be successful, good consultancy requires good clients. Tordoir characterizes three styles of consultancy as:

1 *Selling:* Professional expertise sold at an agreed price. This may take the form of a delivered report or training course, using methods developed in advance, although with some degree of customization and possible follow-up.
2 *Jobbing:* Carrying out a project to the order of client, for example economic analysis, architectural design, project engineers.
3 *Sparring:* The outcome is determined through consultant–client deliberations, requiring interaction over a sustained period. This may include consultant involvement in project implementation (Tordoir 1994).

These correspond to different forms of contractual relationship underpinning client–consultancy interaction. 'Selling' or 'jobbing' may influence client capacities to innovate, depending on the client's purpose in commissioning the work. The success of the 'sparring' relationship, however, is the most significant basis for a distinct consultancy influence on client behaviour. This is because the consultancy process and client competencies must be mutually adapted. These processes are not mechanistic or universally predictable, especially across the range of economic conditions encompassed in the European Union. They have hardly been researched, partly because of the recentness of the KIS phenomenon, but mainly because of the complexity of the context of change within which they operate. KIS may undoubtedly be active agents of change, and their growth has been one of the most marked economic phenomena of recent decades. However, their role in innovation is not easy to separate from the wider business systems of which, increasingly, they form an integral part.

Purpose of the book: basis of the KISINN study

This book is a revised and updated version of the analysis and conclusions of the Thematic Network, 'Knowledge-Intensive Services and Innovation' (KISINN), which met over an eighteen-month period in 1996–97 under the auspices of the European Union Framework IV Programme of Research (Targeted Socio-Economic Research (TSER) Area 1). Its purpose was to review the strategic role of private sector KIS in the transmission and application of technical and management innovation in Europe. It included nine participating research agencies, one each from Belgium, France, Germany, Greece, The Netherlands, Italy, Spain, Portugal and the United Kingdom.

The Network was not intended to undertake new research, but to draw on existing international research knowledge and experience. Its programme was organized around four international workshops attended by Network members and policy experts. Towards the end of the project, there was also a national workshop in each country. These involved public and private sector agencies, and aimed to examine and comment on the interim conclusions of the first three meetings. They were set within a continuous programme of consultation and report preparation, to encourage the exchange of ideas and research evidence. An international Policy Panel was engaged in this process, including representatives of innovation-supporting public agencies, KIS consultancies and their clients. Interviews were also held with leading KIS firms in each country.

The first part of the Final Report, completed in September 1997, synthesized the earlier interim reports from the workshops. The aim was a policy-orientated analysis of best practice in the uses of consultancies to support economic innovation across the widely varied conditions found in the European Union. Part I of this book, Chapters 1–3, is based on this analysis, extensively adapted and rewritten for publication here. Part Two of the Final Report consisted of nine country reports, charting the diversity of KIS activities and their impacts across Europe, in both participating countries and representative regions. The major part of this book, Part II, includes eight country-based chapters, excluding Belgium, based on this work, again extensively adapted and updated for publication here. A common approach was agreed, through which patterns and recent trends in KIS supply were analysed for each country, including regional variations; the national patterns of demand explaining these trends; the economic, business, regulatory and institutional conditions governing the interaction of KIS supply and demand; and evidence for the role of consultancies in national and regional processes of technical and organizational innovation. Contributors were encouraged to emphasize the themes that they believed to be most significant for each country, however, rather than adopt a highly specified common approach. The diversity of outcomes is reviewed in the concluding chapter, which also reviews the innovation and regional policy implications of the growing influence of consultancies across the EU.

Chapter 1 begins with a summary of current theoretical understanding of the processes that support economic and industrial innovation at both micro- and macro-levels. Modern conceptions of these processes within the 'learning organization' emphasize how key expertise may be developed within and between organizations, and combined to create specific innovative circumstances. KIS as sources of expertise may enhance innovative potential, although the conditions under which this occurs are not fully researched. KIS growth marks a significant change in the commercial management of expertise, as KIS have become increasingly implicated in competitive global technical and organizational transformations, as well as state responses to these transformations.

The nature and scope of innovation increasingly dominate economic thinking and policy, and the conditions under which innovative activity might interact with KIS expertise have thus broadened in recent decades. The incidence of both across Europe vary, nationally and regionally within each country. Established approaches to the measurement of innovation have been overtaken by these developments. Such approaches are orientated towards technical change in manufacturing production rather than to the organizational, marketing and other service changes that are required to deliver novel outputs. There has also been a change in the role of policy in supporting innovation, summarized in the 1996 European Commission *Green Paper on Innovation*. Traditional policies directed at individual firms and key sectors have been augmented by a new emphasis on supportive regulation, information provision, training and inter-agency coordination. The value of supporting regional and national innovation systems to stimulate wider social and economic adaptability to change is increasingly recognized. In this context, the role of facilitating agencies such as KIS may grow.

Chapter 2 reviews the scale of growth of KIS in Europe, despite definitional and measurement problems. The conditions that have encouraged this growth are summarized, including demands for specialist expertise, and the improved supply response of the KIS sector itself. One of the most significant changes in the innovative environment created by the growth of KIS is the internationalization of expertise exchange, through the rise of trade, consultant mobility and, especially, global consultancies. These have, in turn, created adaptive responses in nationally and regionally based KIS across Europe.

Chapter 3 turns to the innovative contributions of KIS, which are dependent on sustained processes of client–KIS interaction and an often mutual learning process through consultancy. The innovative qualities of consultancies are summarized, including their distinctive modes of operation and the diversification they offer in sources of key forms of expertise. Consultancy exerts a catalytic influence on wider processes of change, including innovation. An increasingly significant dimension of a 'KIS perspective' is its international scope. The emerging system of links between international, national and local levels of consultancy development shadow corporate management structures. The implications of this system are reviewed for both strategic change and technology transfer. Consultancy-influenced innovation favours some regions over others, because of both their own uneven patterns of concentration into urban core regions, and their influence on client outcomes, often reinforcing established patterns of inequality.

Part II focuses on national and regional experiences in both innovation practice and KIS influence. The eight country chapters (Chapters 4–11) are linked by a common concern for the role of KIS activity in economic development and its implications for innovation. The major variation in experience reflects differences between the more and less industrially developed northern and southern countries of the EU. Put simply, innovation policies in the former are dominated by the need to sustain international competitiveness through economic restructuring. This requires support for new forms of production, including new technologies and materials; new forms of economic organization, both within and between firms; and a more liberal basis for state influence in relation to more complex, service-dominated economies. In the South, equally significant developments are creating a more competitive technical and organizational culture. Market liberalization over the past twenty years has sought to exploit labour availability, growing consumer demand, and key growth sectors including tourism, the media, trade or agriculture. Success has also often depended on a combination of inward and public investment.

There are also common characteristics across Europe. One is the overall growth of KIS, as a symptom of these other economic changes. Another is a preoccupation with supporting indigenous innovativeness and economic growth. This is often regarded as synonymous with the success of small–medium enterprises (SMEs), or effective spin-off from local universities and research institutions. National and regional innovation policies increasingly employ quasi-public intermediary agencies to support the networking of expertise. Thus, commercial consultancies have been engaged in only a marginal and controlled way. However, conditions in all parts of Europe are very varied, requiring local adaptation of agencies and methods, and also a degree of realism about what can be achieved when diverse countries and regions aspire to similar goals of innovation-orientated production.

Chapter 12 summarizes these national variations, and reviews the lessons for European innovation policies from the detailed evidence and experience described earlier. It examines the means of obtaining greater effective involvement of consultancies in SME and regional development policies, to counter the economic polarization inherent in their wider development and encourage a complementary role for them in relation to public agencies.

The study on which this book was based and its preparation have depended on the support of many individuals, especially the authors and their colleagues in eight institutions across Europe. I am particularly grateful for the efficient and enthusiastic support of Natalie Foos, the KISINN Network administrator and Kate Dracup-Jones, whose financial management skills were vital. The work was also greatly helped by the encouragement of the European Commission staff supporting the TSER Area of the Framework IV Programme, especially Ronan O'Brien. The preparation of the manuscript has also been much facilitated by the assistance at various stages by Simon Maxwell and Kersty Hobson.

Notes

1   Kondratieff waves, named after the Russian economist who first noted them (see Kondratieff 1925), are roughly half-century phases of development since the 1780s, which Schumpeter (1939) associated with the economic impacts of major new groups of technology (Freeman and Soete 1997: 18).

2   Ingenious attempts to overcome these problems include those of Hepworth (1989), adapting occupational and industrial data, and a few studies employing national or (in the US) regional input–output tables, which indicate inter-sector market exchanges (cf. Chapter 6).

References

Daniels, P.W. (1985) *Service Industries: A Geographical Appraisal*, London: Methuen.

European Commission (1996) *Green Paper on Innovation (Bulletin of the European Union*, Supplement 5/95), Brussels: European Commission.

Freeman, C. and Soete, L. (1997) *The Economics of Industrial Innovation*, 3rd edn, London: Pinter.

Harrison, B. (1994) *Lean and Mean: The Changing Landscape of Corporate Power in the Age of Flexibility*, New York: Basic Books.

Harvey, D. (1989) *The Condition of Post-Modernity*, London: Basil Blackwell.

Hepworth, M. (1989) *Geography of the Information Economy*, London: Belhaven Press.

Hill, T.P. (1977) 'On goods and services', *Review of Income and Wealth* 23: 315–38.

Illeris, S. (1989) *Services and Regions in Europe*, Aldershot: Avebury.

—— (1996) *The Service Economy: A Geographical Approach*, Chichester: Wiley.

Kondratieff, N. (1925) 'The long wave in economic life' (English translation), *Review of Economic Statistics* 17: 105–15.

Lundvall, B. and Johnson, B. (1994) 'The learning economy', *Journal of Industry Studies* 1: 23–41.

Marshall, J.N. (1988) *Services and Uneven Development*, Oxford: Oxford University Press.

Moulaert, F. and Swyngedouw, E. (1989) 'A regulation approach to the geography of flexible production systems', *Environment and Planning D: Society and Space* 7: 327–45.

Piore, M. and Sabel, C. (1984) *The Second Industrial Divide*, New York: Basic Books.

Riddle, D. (1986) *Service-led Growth: The Role of the Service Sector in World Development*, New York: Praeger.

Rubalcaba-Bermejo, L. (1999) *Business Services in European Industry: Growth, Employment and Competitiveness*, Brussels and Luxemburg: European Commission, D-G III Industry.

Schumpeter, J.A. (1939) *Business Cycles: A Theoretical, Historical and Statistical Analysis of the Capitalist Process*, 2 vols, New York: McGraw-Hill.

Tordoir, P.P. (1994) 'Transactions of professional business services and spatial systems', *Tijdschrift voor Economische en Sociaale Geografie* 85: 322–32.

Vogler-Ludwig, K., Hofmann, H. and Varloou, P. (1993) 'Business services', *European Economy: Social Europe, Report and Studies* 3: 381–400.

Womack, J.P., Jones, D.T. and Roos, D. (1990) *The Machine that Changed the World*, New York: Rawson/Macmillan.

Wood, P.A. (1996) 'Business services, the management of change and regional development in the UK: a corporate client perspective', *Transactions, Institute of British Geographers* NS 21: 649–65.

Part I

Business consultancy growth and innovation

A European perspective

1 Innovation and the growth of business consultancy

Parallel but associated trends?

Peter Wood

The 1996 European Commission *Green Paper on Innovation* (European Commission 1996) was published shortly before the work on which this study is based was commissioned. In 1994, the European Commission White Paper, *Growth, Competitiveness, Employment*, had also been published (European Commission 1994). This placed significant emphasis on the impacts of research and technological development on employment, incomes, economic organization and competitiveness. Many of the constituent governments of the EU presented similar analyses during the 1990s, as will be evident in later chapters. The Green Paper adopted a very broad definition of innovation: 'the successful production, assimilation and exploitation of novelty in the economic and social spheres' (European Commission 1996: 9). It nevertheless particularly emphasized Europe's failure to exploit its traditional scientific and technological ingenuity, and the inadequacies of investment and coordination in research and development (R&D) activities. Priority was therefore given to scientific and technical skills, and the leading role of technology in driving economic change (ibid.: 12–13). As was pointed out, the term 'innovation' is used to describe both a *process*, transforming ideas for new products, services, or ways of creating them into reality, and the *results* of this process, marked by the successful establishment of the novelty in the market. The process of innovation is nowadays understood to be more complex than was often perceived in the past, and its market impact also involves a much wider array of changes than those primarily dependent on new technologies alone.

Changing perspectives on innovation and the firm: the learning organization

For any business to survive in modern competitive conditions, it is not enough for current patterns of production to be successful. Support for process and product innovation in anticipation of changing market conditions must now be a core organizational function. The creation of new products, services and processes requires continuous self-criticism and active organizational learning (Tushman and Nadler 1986). Innovation takes many forms, acting simultaneously within the same firm and affecting various processes and products through overlapping time periods. Much is *incremental*, building on established methods and ideas. This depends not only on technical R&D for success but also changes in organization, human resources and marketing. Many important innovations are *synthetic*, exploiting new combinations of established ideas and technologies. The successful development and application of entirely new ideas, technologies or products, termed *discontinuous* innovation, is rarer and more risky, although it may eventually reap major rewards.

The modern, 'neo-Schumpeterian' emphasis on continuous change through both product and process innovation implies sustained organizational learning (Freeman and Soete 1997; Freeman 1994; Lundvall 1988). The balance of learning priorities, however, may change over time, for example during the so-called 'product life cycle', moving from initiation, through growth, to final market saturation and decline. Within firms, *product innovation* is most obviously important in the early, product development phase of the cycle, as well as later, when there is intensified competition in crowded markets. While an established market is growing, the quality of *process innovation* dominates, reflecting investment strategies and management changes. Different forms of learning behaviour may be associated with each stage of the cycle (Tushman and Nadler 1986).

Much analysis and commentary on effective innovation has focused on the conditions that encourage it *within* organizations (Rothwell 1994). These include a high quality of technical and managerial staff, effective systems of communication, and flexible teamwork directed towards problem solving. There must also be a capacity to monitor and respond to key changes in the external market and technical environment. Innovation-orientated forms of job design, education and training are required, with appropriate career development and incentive systems. Wider organizational values and norms are important, including the attitudes to risk set by senior management. Recognized processes for solving problems and resolving conflicts will often be needed. These must operate within informal interactive communication networks which link, for example, R&D with marketing functions, innovators with managers, internal with external intelligence, and planned development strategies with opportunities which may arise at any time.

The incremental nature of much business innovation derives from day-to-day practice and much modern organizational restructuring aims to foster the continuous evaluation of change, even in routine functions. Where major synthetic or discontinuous innovations are planned, however, special arrangements may be necessary, such as earmarked investment in R&D or the establishment of specialist business units. For the most innovative programmes, separate licensing, joint venture or venture capital arrangements may be set up with distinct targets and performance requirements, separate from those of parent organizations (Tushman and Nadler 1986: 84).

Recognition of this learning-based scenario has challenged at least two traditional assumptions about innovation: (1) that technology-based R&D

Table 1.1 Share of innovation expenditure on R&D and non-R&D costs, 1992 (%)

		R&D	*Non-R&D*
Belgium	Small companies	40	60
	Large companies	64	36
Germany	Small companies*	17	83
	Large companies	41	59
Greece	Small companies	45	55
	Large companies	44	56
Italy	Small companies	30	70
	Large companies	61	39
Netherlands	Small companies*	58	42
	Large companies	61	39
Spain	Small companies	28	72
	Large companies	47	53

Source: European Commission 1996: Table 26.

Notes
Small companies: 49 employees or fewer; * 50–249 employees.
Large companies: 500 employees or more.
Non-R&D costs include patent costs, product design, trial production, training and market analysis.

functions are the primary base for innovation; and (2) that innovative functions can be successfully sustained by the expertise resources of individual organizations alone. A growing volume of evidence on the corporate context of R&D since the 1970s has established that R&D spending is only selectively necessary and never sufficient for successful innovation. Table 1.1 indicates that in Europe only about half of innovation expenditure is on R&D. The proportion appears to be low in Germany and among small companies in Italy and Spain. More detailed estimates for The Netherlands confirm that just over half of total manufacturing and service innovation expenditure went on product-related R&D. This proportion fell to about one-quarter, however, when related investment in fixed assets was included: 35 per cent in manufacturing (close to separate Italian estimates) and only 19 per cent in services (Brouwer and Kleinknecht 1996). Whatever the general significance of these figures, processes such as outside patent acquisition, product design and development, trial production, training, tooling and market analysis, are clearly important for innovation. Similarly, many important *inter-organizational* relations have been shown to foster innovation, especially with clients and suppliers, other commercial collaborators, and research-orientated bodies such as universities (Teece 1989; Marceau 1994). This inter-organizational, 'systems' perspectives will be returned to below.

New, learning-based models of the innovation process have been developed in recent years. Traditional thinking has been dominated by linear models of events, moving sequentially from initial scientific invention, to research, development, innovation, diffusion, manufacture and finally to sales; or from basic science, to engineering and manufacturing, to marketing. Innovation was seen as generated by basic science or technology, followed by a necessary time lag during which commercial exploitation opportunities might emerge, and demand might 'catch up'. Such 'technology–push' innovation processes might be augmented by 'need–pull' pressures to improve products from the market. By the 1980s, these approaches had been linked in Rothwell and Zegveld's 'coupling model' (1985). The stages between invention and the market were nevertheless still conceptualized as functionally distinct, rather than as part of a complex, continuous and interactive set of processes. Dodgson and Bessant (1996) believe that much innovation policy is still implicitly based on such linear models.

Since the early 1980s, more interactive models have been proposed and have themselves evolved towards more complex and universal forms. These interpretations emphasize the simultaneous and overlapping nature of the various phases of innovation. One of the most influential is the 'chain-link' model of Kline and Rosenberg (Kline and Rosenberg 1986; Rothwell 1992). Others, such as Foray's (1993) 'recombination model' and Henderson and Clark's (1990) model of 'architectural innovation', emphasize the incremental and synthetic nature of much innovation, based on recombining known characteristics of products by adding small differences. Among the common attributes of these approaches is the lowering of barriers through informal group work, not only within organizations but also between them (Gadrey *et al.* 1995).

Rothwell (1994) has argued that we are entering a 'fifth generation' in the development of innovation processes. This is characterized by the 'learning organization', in which technology strategies are embedded within wider organizational changes, and continuous innovation is integral to various forms of 'lean' and 'flexible' production. These involve new forms of work organization and training, devolved managerial responsibility, responsiveness to customer and market demands, and continuous quality control. Strategic inter-firm links are also increasingly significant in promoting innovation. These extend vertically, to customers and suppliers, and horizontally, through collaboration in networks or joint ventures. These developments are all supported by increasing levels of information exchange, often exploiting modern information and communications technology (Dodgson and Rothwell 1994).

System perspectives

Complementing these deepening micro-economic insights, considerable emphasis is now placed on the more general social and political context of innovation, including inter-organizational perspectives, through the national, technological and regional systems within which individual firms operate (Marceau 1994; Carlsson 1994; Lundvall 1992; Nelson 1993; Dodgson and Bessant 1996). These are particularly significant in relation to the development of policies to support innovativeness. The influences affecting national innovativeness have been summarized by Porter (1990) in terms of the quality of interacting factor inputs, and of demand, key support sectors and management behaviour. Innovation systems encompass the prevailing business and social relations that support financial, regulatory, educational and research institutions, and public policy (Freeman and Soete 1997: 371–432). In key industrial clusters, based on specific technologies such as electronics, pharmaceuticals, biotechnology or defence, important interactions occur between agencies with common commitments to development (Freeman and Soete 1997: 21–187; Dodgson and Rothwell 1994: 145–257; Hagedoorn and Schakenraad 1990). There has also been an extended debate about the supply and demand conditions at the regional level that favour innovation, including their significance for national economic competitiveness (Cooke and Morgan 1994; Braczyk *et al.* 1998).

In the light of these relationships, the interactive, 'networked' nature and functioning of business firms have been widely discussed (Rothwell 1994: 43–4). Key contributions to innovation are as likely to come from outside as within them, and the boundaries of firms are increasingly difficult to delineate. As Granstrand *et al.* (1992) point out, modern commercial interactions now encompass many forms of contractual relationship other than the two embodied in classic transaction cost analysis: those tying employees to firms and those governing market exchanges between firms. Conglomerate relationships, strategic alliances, partnerships, production agreements, outsourcing and marketing networks are only some of the alternatives. They also include consultancy–client relationships.

Organizational learning is a collective task, combining formally codified knowledge with the tacit and informal understanding arising from familiar routines and practices within firms. New knowledge arises from the interaction of diverse and complementary expertise (Lundvall 1988), but it is becoming less possible for individual organizations to command this diversity alone. In modern conditions of rapid economic change, organizations must be open and responsive to outside knowledge. Therefore, while relationships and obligations within firms must remain strong, to innovate in modern conditions also requires organizations that sustain outside sources of knowledge.

Such relationships may include formal contractual links to partly owned subsidiaries, joint ventures or network collaborators (Kleinknecht and Reijnen 1992; Dodgson 1992). Even multinationals in Europe must develop various forms of strategic collaboration, since key innovations do not come simply from within the particular core technologies that they may command, whether in telecommunications, micro-electronics, computers, robotics, new materials or biotechnology. Increasingly they also come from opportunities in areas of overlap between these (van Tulder and Junne 1988: 16). More generally, new knowledge may arise from customer–supplier interchange at any stage of the production and delivery chain, as well as from subcontractors; or from the mobility of skilled technical and management labour; or the adaptation of established technical and commercial knowledge to new circumstances. Again, the use of consultancies adds to this list of sources.

Corporations are thus becoming more 'multi-technological', acquiring ideas through the whole range of in-house, contractual, market and informal relations. Within these, cost pressures are encouraging externalization. Granstrand *et. al* (1992), after studying technology-acquisition strategies in the electrical and telecommunications sectors in Japan, Sweden and the US, observe that, even as more resources are being expended on internal R&D, technology externalization is growing. Growing complexity requires more external technology scanning, and a close complementarity of in-house and external innovation acquisition.

The complexity of some of the most innovative production systems, with multiple inputs and highly customized outputs, including chemical processing plant, aerospace or military production, imposes elaborate collaborative requirements. Such systems are usually closely regulated and dominated by a few major firms and trade associations. In the case of the flight simulator industry, Miller *et al.* note that:

> technological innovation was not viewed by industry experts as the most important inducer of industrial transformation. Technology was seen as a facilitator; a necessary but insufficient condition for change. Regulatory/ institutional turning points were viewed as the most important turning points ... Without these, new technology breakthroughs could not have been exploited by the industry.
>
> (Miller *et al.* 1995: 384)

In this case, such turning points included the certification of advanced simulation-based training in the US and the development, through mergers, of key specialist companies responding to market expansion in the 1970s and 1980s. Even for complex high-technology products, where strong in-house support and protection are required, many innovations depend on external knowledge acquisition, imitation and incremental improvement. Innovating companies need inputs of specialist computer systems, instrumentation and software, and personnel training. They often also depend on public agency supervision and approval as new technologies emerge. These interactions must be adapted through effective communication and coordination, often in the form of interdisciplinary teams.

Similar processes, supported by regulatory changes and new combinations of technology, characterize much service innovation. Complex 'production systems', involving radically new forms of employment, business organization and technology, have created major new markets. These include 'value added' information and communications services; many facets of the media; global and domestic financial services; mass food retailing; and international tourism. In terms of profit, they are among the most successful innovation systems of modern times, linking formerly separate functions and places to potential markets, especially through new communications and information technology. Realizing technological potential therefore requires a myriad of organizational, workplace and regulatory adaptations, often coordinated by service agencies. Radical changes in public services, such as health care, are also associated with advances in information and computer technology and pharmaceuticals, but they depend on much more than these technologies alone for their impact.

In summary, learning organizations are supported by interactive exchanges with customers, suppliers and other collaborators within national, sectoral and regional innovative systems. They depend fundamentally on how expertise, based on complementary specialist skills, is developed and combined. Dodgson (1992) emphasizes the importance of human resources management for the success of inter-firm collaboration in innovation. Such resources relate not only to technological, still less scientific expertise. They include management, marketing, training, logistical and marketing skills. Perhaps, above all, there is nowadays a premium on integrative and risk-taking experience. Some firms are more successful than others in marrying technological knowledge and applications expertise to appropriate financial and organizational skills and market intelligence. A few small to medium-sized firms may achieve this for a time through the skills and flexibility of individuals. Large organizations must plan to achieve it, but only a few succeed in creating an innovative culture. Both depend on the quality of human inputs, capable of promoting innovation across many functions, and adapting to dynamic social processes. This is the essence of modern entrepreneurship.

As the 1996 European Commission Green Paper stated, the innovation process:

> is not a linear process, with clearly-delimited sequences and automatic follow-on, but rather a system of interactions, of comings and goings between different functions and players whose experience, knowledge and know-how are mutually reinforcing and cumulative. This is why more and more importance is attached in practice to mechanisms for interaction within the firm [...], as well as to *the networks* linking the firm to its environment (other firms, support services, centres of expertise, research laboratories, etc).
>
> (European Commission 1996: 12, emphasis in original)

This is the context in which KIS companies may support, transmit or even originate technical, personnel and organizational change and innovation.

The impact of consultancy growth

Innovation processes require individual firm performance to be linked to the wider economic system. One of the most startling features of this system across Europe over the past twenty years has been the growing use, and thus presumably influence, of consultancy (see Chapter 2). Consultancies offer many types of specialist knowledge to complement or reinforce that of clients, in association with more or less sustained consultancy expertise. Their growth presumably reflects the same complex commercial changes, including internationalization and increasing uncertainty, which place such a premium on innovation. This does not mean, of course, that innovation processes depend on consultancies, or that consultancies are inherently innovative. The processes of innovation and the activities of consultancies may nevertheless be associated in a growing variety of conditions and circumstances, which will be examined in Chapter 3.

These conditions and circumstances are implicit in the above summary of current understanding of innovation processes. Innovativeness depends on flexible organizations, orientated towards learning and change, able to collaborate with other organizations at any stage of production within sectoral, regional, national or even international innovation systems. The development of potentially innovative skills, including their increased specialization, appears also to lead organizations increasingly to engage outside expertise, augmenting, complementing or reinforcing in-house strengths (Wood 1996). Both service and manufacturing products have become increasingly complex, requiring integrative expertise and experience, which combine technical and organizational processes. Innovative organizations might therefore be expected increasingly to need the particular forms of expertise that KIS can provide, including specialist knowledge and a range of organizational and technical experience.

Leaving aside wider interpretations of innovation, Table 1.2 offers evidence that consultancies are significant even for technology acquisition. While there were other significant vectors of innovation, including the purchase of equipment, the recruitment of skilled employees, and service inputs from other enterprises, the table shows that technical consultancies were among the six most mentioned sources of technological acquisition, particularly in Germany and among large firms in Italy and the UK.

In addition to offering specialist advice and varied project experience, consultancy use may offer further support for the innovative potential of clients by encouraging the wider interchange of experts. It has been noted by van Tulder and Junne (1988: 152–5), for example, that there is lower mobility of multinational managerial staff in Europe than in either the US, where inter-organizational movement is much higher, or Japan, which has a tradition of high intra-organizational movement. These authors suggest that this may be a significant constraint on the diffusion of experience and internal company flexibility. Consultancy interaction with client staff, and indeed the movement

Table 1.2 Acquisition of technology from outside domestic sources, large and small companies, 1992

		Percentages of innovative enterprises indicating acquisition						
		Belgium	France	Germany	Italy	Neths	UK	Spain
Right to use others' inventions	Small	8	11	10	7	6	27	33
	Large	8	18	26	11	5	42	23
Result of R&D contracted out	Small	9	40	15	6	18	16	n.d.
	Large	45	57	38	22	48	37	n.d.
Use of consultancy	Small	10	11	61	29	34	30	42
	Large	21	13	68	52	24	53	28
Purchase of equipment	Small	43	32	72	66	33	62	n.d.
	Large	24	35	53	64	11	68	n.d.
Communication with specialist services from other enterprises	Small	22	n.d.	82	21	36	54	n.d.
	Large	13	n.d.	66	31	23	63	n.d.
Hiring of skilled employees	Small	48	31	45	33	15	54	n.d.
	Large	36	40	81	45	27	84	n.d.

Source: European Commission 1996: Table 25a.

Notes
Small companies: 49 employees or fewer.
Large companies: 500 employees or more.
n.d.: data not available.

of experts between consultancies and other sectors, thus offer alternative channels for such diffusion. Consultancy is also pervasive in the US, but the more static context of expertise interchange in Europe suggests that KIS growth may be a more significant basis for innovative exchange there. These effects will be discussed further in Chapter 3.

The potentially positive significance of consultancy for innovative sectors and innovation systems, however, must be set against the negative impact of its growth on less innovative activities, or those otherwise unable to exploit such expertise. As this book will demonstrate, KIS growth and influence are nationally, regionally and sectorally highly concentrated. Small–medium enterprises (SMEs) appear to face particular difficulties in benefiting from consultancy-based development. Such problems have begun to be recognized in innovation policies, which in some cases now incorporate a consultancy contribution in support of technology transfer (Chapter 12). One example is the early 1990s' advanced manufacturing technology programme of the UK Enterprise Initiative. This demonstrated the multiple roles of consultants as intermediaries, especially for training and the reduction of the associated uncertainty for SMEs (Bessant and Rush 1995, 1996).

A recurrent theme in the country chapters of this volume (Chapters 4–11) is the uneven influence of KIS, both through their own patterns of growth, and the responses to them of different clients and areas. This is evident between European countries, between their regions, and between different business sectors and size segments. Generally, KIS impacts are concentrated in the corporate business and public sectors, and in the fastest growing markets and regions. Outside these, especially among SMEs and more peripheral regions, KIS growth has had much less effect, either directly or indirectly. If their influence is shown to be significant, therefore, it may require a more robust policy approach than has emerged so far to compensate for their uneven impact.

The changing context of innovation and consultancy

Another important change in the context of innovation, which may increase the influence of consultancies, is that it has become a more globally competitive phenomenon. In the past, the complexity of innovation as a social, technical, economic and learning process was perhaps best demonstrated at the regional level and, by implication, best managed there, at least for SMEs. At this level, competitiveness could be fostered within the regional 'production milieu' of industrial clusters, based on rapid information flows and the diverse and effective diffusion of ideas (Porter 1990: 151). Economies of scale and scope could be sustained, supported by an appropriate regulatory environment, enabling close collaboration between public and private investment (Amin and Goddard 1986; Aydalot and Keeble 1988; Benko and Dunford 1991).

Today, these conditions must be projected into the national and even global economic arena. While exploiting 'local' strengths, national and regional innovation systems must also be responsive to global change (Amin and Thrift 1992, 1994, 1995). New international and national divisions of labour have emerged, favouring nations and regions where corporate control functions are concentrated, or where globally marketable manufacturing or service activities can thrive. This has supported the success, for example, of the design-led consumer/craft industries of the 'Third Italy', the financial services of the City of London, and the engineering and automobile industries of Baden-Württemberg. Such regions undoubtedly contain local KIS activities, undertaking complementary technical, financial, trading and training functions, as part of the local social division of labour. Nowadays, however, their traditional knowledge and experience is not enough. It must be augmented by wider consultancy knowledge and experience of national and global conditions, increasing the need for outside advice even in traditional regionally based innovation systems.

The global success stories of such regions over the past twenty or thirty years should not detract from other circumstances in which innovation may succeed, especially when it is supported by less locationally tied corporate or even public resources. Any firm with sufficient managerial flexibility and capital may now exploit international markets across extended geographical space. From a global perspective, the dominant recent innovations can be seen to originate not simply from specific sectors, technical inventions, or from established regional systems, but from virtually universal transformations in business organization and control. These affect working methods and training, logistical processes and marketing. They are commonly directed towards achieving greater sensitivity to market needs, and flexibility in responding to change (Coombes 1994; Bessant 1994). They have transformed sectors as diverse as automobiles, electronics, fast food, retail banking and mass-produced clothing. Japanese and US success in the 1980s and 1990s can be ascribed largely to such corporate organizational adaptation. Multiple technical opportunities have thus been exploited through effective business strategy, especially in relation to information and computer technologies. Such transformations may have significant implications for the organization or success of regional systems, but they are not dependent on them.

A recent study of the international transfer of key organizational innovations argues that they not only add growth potential to technical innovation, but are also often prerequisites for firm growth, future innovativeness and international expansion (IMIT 1996). Social agents, including top management, business associations and also consultants, have played important roles in diffusing such organizational innovations. These include 'continuous improvement' methods, globalization strategies, and technology extension to SMEs. Lundvall argued in the late 1980s that in the current period of rapid change, 'social innovations might become more important for the wealth of nations than technical innovations', including institutional change strengthening the competence and the power of final users (Lundvall 1988).

The rapidity of consultancy growth has especially reflected and facilitated these organizational and social forms of innovation. The IMIT study emphasizes the organizational complexity of innovation, and acknowledges the potential contribution of consultancies to organizational transformation. It also contrasts the ease of diffusion of organizational innovations by consultancies in North America with the perceived economic, legal and cultural barriers to such diffusion across European countries and regions. In view of this, new agencies are proposed at EU and national levels to encourage the exchange of organizational innovation experience within Europe. It is argued that such mediation would offer significant net benefits, especially when judged in relation to the huge policy resources already directed towards technological development.

In emphasizing the barriers to consultancy activity in Europe, this account may undervalue the rapidity of its growth during the past two decades. Global, often US-based, consultancies have been expanding rapidly and many nationally based firms have developed Europe-wide markets (Chapter 3). It is true that much consultancy remains localized and relatively routine, facing significant regulatory and cultural barriers to a Continent-wide market for their services. For strategically significant expertise, however, the consultancy sector has shown an impressive capacity to overcome these barriers in Europe since 1980, in association with wider processes of organizational transformation. It is doubtful whether public agencies could be so effective in disseminating commercial best practice except possibly to more marginalized regions, sectors and business-size segments of the economy.

Some types of consultancy are directly involved in technical innovation, for example through the design and testing of specialist machinery, the adaptation of computer systems, or the assessment of the market or financial prospects of alternative products. However, consultancy activities show how there is more to organizational change than simply supporting technical innovation. Very often, especially for IT and computing innovation, consultancies address both the technical and organizational contexts of change, challenging the very dichotomy between them. The social complexity of modern innovation, the global competitive context that drives it, and the interdependence of technical and organizational innovation are all attributes of innovation systems favouring the involvement of consultancy activities.

Measurement

The importance of recognizing the influence of both technical and organizational innovation is especially evident in current debates over the measurement of innovation itself. Traditionally, this focuses on expenditure and employment in research and development, patent applications, and investment in the production of new goods by manufacturing firms: in other words, on measurable innovation-orientated inputs, rather than their effectiveness through outputs. Unfortunately, much input effort, especially on R&D and product innovation, never achieves any economic return or wider social impact, for reasons implicit in Rothwell's complex 'learning model', described above. Strong basic R&D must not only be coordinated with other input factors, including financing, complementary staff skills, patent strategy, speed of development, risk-taking attitudes and joint ventures with other firms. The wider management competencies enabling such coordination must also support the ability to link technical possibilities to market needs (Freeman and Soete 1997: 210–18).

These measurement approaches also generally neglect non-manufacturing innovation. We have seen that, quite apart from the significance of service support for manufacturing innovation, many modern innovations originate from service activities (Gershuny and Miles 1983; de Wit 1990; Miles 1993, 1994). Retailing, banking and tourism have also led demand for new communications, data-processing and transport technologies, and pioneered associated new organizational forms. Commentators on service developments, such as Marshall and Wood (1995), argue that nowadays our view of economic change should be 'service informed', recognizing the interdependence of manufacturing and service functions, with the latter often taking the lead in promoting innovation, especially through the market delivery of new technologies.

The manufacturing–service distinction thus appears increasingly irrelevant to modern practice. Much innovative manufacturing needs to be customized and 'service-like', while many conventional services are becoming more 'industrialized', mass produced, and dependent on manufactured technology. Much 'value added' in modern manufacturing also derives from innovative design, quality and marketing, which only partly depend on technical innovation. The value of R&D spending is itself a 'service' input and, if unsuccessful, is a cost on innovation rather than a basis for it. Once a product or process is established, material processing now seldom accounts for more than 15–25 per cent of costs. Innovative success depends on 'service' inputs. Perhaps we should now regard material products as no more than the physical embodiments of the capital and labour service skills assembled to produce them; 'tools for the production of final services' (Coombes and Miles 1998; Quinn and Dorley 1988; Gershuny and Miles 1983). All companies, even conventional manufacturers, may thus primarily be seen as providers of services by alternative means.

The significance of manufacturing-orientated studies of technological innovation is thus limited, even without their neglect of non-technical, organizational and marketing innovation. A common metric is needed for manufacturing and service innovation. Kleinknecht and his associates have addressed these basic issues, developing alternative measures to those of technologically-related inputs (Brouwer and Kleinknecht 1996; Kleinknecht 1998). These also emphasize the creative application of *available* technologies, and the *results* of innovation, based on market introduction and diffusion. Their work for the Organization for Economic Cooperation and Development (OECD) and the EU has analysed announcements of new products and services in trade and technical journals, and surveyed the shares of total company sales taken by innovative products for both manufacturing and service firms. A survey of R&D and patent spending for Dutch business showed, as expected, that service firms are less engaged than manufacturers in technical innovation. On the other hand, from a market impact perspective, the sectors hardly differed in their shares of sales in new products.

Measurement methods are thus beginning to acknowledge the implications of the broad EU definition of innovation referred to at the beginning of this chapter. Successful end-results are the best indication of the many sources of innovativeness, rather than the common but partial approaches that focus on support for manufacturing technology. Such measurement establishes no more than a base camp for exploring the many interdependent factors supporting innovation, of which only one is commitment to new technology. On this basis the contribution to any successful technological innovation of organizational innovation or inter-organizational cooperation might be given due weight, including better service support in training or marketing. Such methods remove the inherent bias of input-directed measures against service innovation. The significant and growing influence of consultancy might then be placed in its proper context.

Policy

Finally, the context of innovation has also been affected by the changing role of state support, which has become less direct but perhaps more pervasive. The 1996 European Commission Green Paper summarized this, indicating that public authorities 'have to teach, persuade, involve, stimulate and evaluate rather than order' (European Commission 1996: 58). The state needs to be a sympathetic regulator, provide stability through long-term investment and training, coordinate many agencies, and orientate policy towards collective needs. Direct state involvement, through the ownership of key sectors or spending on research and development, has diminished since the 1980s compared with that of private capital. Nevertheless, state-organized education and the regulatory environment of taxation, laws and supportive institutions are considered as important as ever. The Green Paper proposed establishing thirteen routes of public action, including improved technology monitoring; clearer direction of research; more effective training and mobility; the promotion of innovativeness; better finance and taxation regimes; more consistent arrangements for intellectual and industrial property rights; and simpler administrative and legal frameworks.

More directly, one of the Green Paper's 'routes of action' (No. 11: 56) emphasized the need to promote methods of economic intelligence, including private sector services and management training in this field, as well as consultation bodies at international, national and regional levels. Such measures should be especially directed towards SMEs, and towards managing information in relation to specific innovative projects. In most European countries, these activities include a strongly articulated regional component, encouraged by EU policies. Further action was thus proposed (see Route of action No. 12: 57) to organize external support for innovation especially for SMEs at local or regional levels. This requires both local collaboration and an international orientation, supported by networked and 'one-stop-shop' information services.

Adding to the impacts of new technologies, new forms of organization and broader conceptions of innovativeness, a more complex regulatory context has also increased the need for both access to information and the expertise required to interpret it. The state itself has a part to play, yet the policy debate, including that in the Green Paper, insufficiently recognizes major changes in the management and organization of commercial information and expertise. These are most clearly signalled by the growth of specialist KIS and the consultancy process.

The Green Paper emphasized the need to 'invest in the intangible' including, 'lifelong training, creativity, the exploitation of research results and the anticipation of technical and commercial trends', as well as improving 'the management of enterprises, their openness to external influences' (European Commission 1996: 5). Much of the growth of consultancy in recent years has been associated with supporting clients in precisely these processes. In the context of innovation management, the Green Paper acknowledges a role for KIS (ibid.: 20) but their wider influence is given little attention. A private sector transformation of intelligence gathering and management has occurred that is integral to the development of the 'learning economy'. This has linked interactive processes of innovation at the firm level to the impacts of global competition and new forms of regulation. It is marked most obviously by the rapid development of KIS and, more specifically, of consultancy. The implications of this transformation for innovation policy will be returned to in Chapter 12.

References

Amin, A. and Goddard, J.B. (1986) *Technological Change, Industrial Restructuring and Regional Development*, London: Allen and Unwin.

Amin, A. and Thrift, N. (1992) 'Neo-Marshallian nodes in global networks', *International Journal of Urban and Regional Research* 16: 571–87.

—— (1994) *Globalization, Institutions and Regional Development in Europe*, Oxford: Oxford University Press.

—— (1995) 'Institutional issues for the European regions: from markets and plans to socioeconomics and powers of association', *Economy and Society* 24: 41–66.

Aydalot, P. and Keeble, D.E. (1988) *High Technology Industry and Innovative Environment*, London: Routledge.

Benko, G. and Dunford, M. (1991) *Industrial Change and Regional Development: The Transformation of New Industrial Spaces*, London: Belhaven Press.

Bessant. J. (1994) 'Innovation and manufacturing strategy', in Dodgson, M. and Rothwell, R. (eds) *The Handbook of Industrial Innovation*, Cheltenham: Edward Elgar, 393–404.

Bessant, J. and Rush, H. (1995) 'Building bridges for innovation: the role of consultants in technology transfer', *Research Policy* 24: 97–114.

—— (1996) *Effective Innovation Policy: New Approaches*, London: Thomson Business Press.

Braczyk, H.J., Cooke, P. and Heidenreich, M. (eds) (1998) *Regional Innovation Systems*, London: UCL Press.

Brouwer, E. and Kleinknecht, A. (1996) 'Measuring the unmeasurable: a country's non-R&D expenditure on product and service innovation', *Research Policy* 25: 1235–42.

Carlsson, B. (1994) 'Technological systems and economic performance', in Dodgson, M. and Rothwell, R. (eds) *The Handbook of Industrial Innovation*, Cheltenham: Edward Elgar, 13–24.

Cooke, P. and Morgan, K. (1994) 'The creative milieu: a regional perspective on innovation', in Dodgson, M. and Rothwell, R. (eds) *The Handbook of Industrial Innovation*, Cheltenham: Edward Elgar, 25–32.

Coombes, R. (1994) 'Technology and business strategy', in Dodgson, M. and Rothwell, R. (eds) *The Handbook of Industrial Innovation*, Cheltenham: Edward Elgar, 384–92.

Coombes R. and Miles, I. (1998) 'Innovation, measurement and services: the new problematique', paper presented at the workshop on Conceptualizing and Measuring Service Innovation, Centre for Research on Innovation and Competition (CRIC), University of Manchester, 20 May 1998.

de Wit, G.R. (1990) 'The character of technological change and employment in banking: a case study of the Dutch automated clearing house', in Freeman, C. and Soete, L. (eds) *New Explorations in the Economics of Technological Change*, London: Pinter, 95–119.

Dodgson, M. (1992) 'The strategic management of R&D collaboration', *Technology Analysis and Strategic Management* 4: 227–44.

Dodgson, M. and Bessant, J. (1996) *Effective Innovation Policy: A New Approach*, London: Thomson.

Dodgson, M. and Rothwell, R. (eds) (1994) *The Handbook of Industrial Innovation*, Cheltenham: Edward Elgar.

European Commission (1994) *Growth, Competitiveness, Employment*, White Paper, Brussels: European Commission.

—— (1996) *Green Paper on Innovation* (*Bulletin of the European Union*, Supplement 5/95), Brussels: European Commission.

Foray, D. (1993) *Modernisation des entreprise, cooperation industrielle inter et intra-firmes et ressources humaines*, report for French Ministère de la Recherche et de la Technologie.

Freeman, C. (1994) 'Innovation and growth', in Dodgson, M. and Rothwell, R. (eds) *The Handbook of Industrial Innovation*, Cheltenham: Edward Elgar, 78–93.

Freeman, C. and Soete, L. (1997) *The Economics of Industrial Innovation*, 3rd edn, London: Pinter.

Gadrey, J., Gallouj, F. and Weinstein, O. (1995) 'New modes of innovation: how services benefit industry', *International Journal of Service Industry Management* 6(3): 4–16.

Gershuny, J. and Miles, I. (1983) *The New Service Economy*, London: Pinter.

Granstrand, O., Bohlin, E., Oskarsson, C. and Sjoberg, N. (1992) 'External technology acquisition in large multi-technology corporations', *R&R Management* 22(2): 111–33.

Hagedoorn, J. and Schakenraad, J. (1990) 'Inter-firm partnerships and cooperative strategies in core technologies', in Freeman, C. and Soete, L. (eds) *New Explorations in the Economics of Technological Change*, London: Pinter, 3–37.

Henderson, R.M. and Clark, K.B. (1990) 'Architectural innovation: the reconfiguration of existing product technologies and the failure of established firms', *Administrative Science Quarterly* 35: 9–30.

IMIT (Institute for Management Innovation and Technology) (1996) *International Transfer of Organisational Innovation*, European Innovation Monitoring System (EIMS) Publication No. 45, Brussels: European Commission, D-G XIII.

Kleinknecht, A. (1998) 'Measuring and analysing innovation in services and manufacturing: an assessment of the experience in The Netherlands', paper presented at the workshop on Conceptualising and Measuring Service Innovation, CRIC, University of Manchester, 20 May 1998.

Kleinknecht, A. and Reijnen, O.N. (1992) 'Why do firms cooperate on R&D? An empirical study', *Research Policy* 21: 347–60.

Kline, S. and Rosenberg, N. (1996) 'An overview of innovation', in Landau, R. and Rosenberg, N. (eds) *The Positive Sum Strategy*, Washington, DC: National Academy Press, 275–305.

Lundvall, B.A. (1988) 'Innovation as an interactive process: from user-producer interaction to the national system of innovation', in Dosi, C., Freeman, C., Nelson, R., Silverberg, G. and Soete, L. (eds) *Technical Change and Economic Theory*, London: Pinter, 349–69.

—— (1992) *National Systems of Innovation: Towards a Theory of Innovation and Interactive Learning*, London: Pinter.

Marceau, J. (1994) 'Clusters, chains and complexes: three approaches to innovation with public policy perspective', in Dodgson, M. and Rothwell, R. (eds) *The Handbook of Industrial Innovation*, Cheltenham: Edward Elgar, 3–12.

Marshall, J.N. and Wood, P.A. (1995) *Services and Space*, London: Longman.

Miles, I. (1993) 'Services in the new industrial economy', *Futures* (July–August): 653–72.

—— (1994) 'Innovation in services', in Dodgson, M. and Rothwell, R. (eds) *The Handbook of Industrial Innovation*, Cheltenham: Edward Elgar, 243–58.

Miller, R., Hobday, M., Leroux-Demers, T. and Olleros, X. (1995) 'Innovation in complex systems industries: the case of flight simulation', *Industrial and Corporate Change* 4: 363–400.

Nelson, R. (1993) *National Innovation Systems: A Comparative Analysis*, Oxford: Oxford University Press.

Porter, M. (1990) *The Competitive Advantage of Nations*, London: Macmillan.

Quinn, J.B. and Dorley, T.L. (1988) 'Key issues posed by services', *Technological Forecasting and Social Change* 34: 405–23.

Rothwell, R. (1992) 'Successful industrial innovation: critical factors for the 1990s', *Journal of General Management* 8(6): 5–25.

—— (1994) 'Industrial innovation: success, strategy, trends', in Dodgson, M. and Rothwell, R. (eds) *The Handbook of Industrial Innovation*, Cheltenham: Edward Elgar, 33–53.

Rothwell, R. and Zegveld, W. (1985) *Reindustrialization and Technology*, Harlow: Longman.

Teece, D.J. (1989) 'Inter-organisational requirements of the innovation process', *Managerial and Decision Economics* (Special Issue Spring) 10: 35–42.

Tulder, van R. and Junne, G. (1988) *European Multinationals in Core Technologies*, Chichester: Wiley.

Tushman, M. and Nadler, D. (1986) 'Organizing for innovation', *California Management Review* XXVIII(3): 74–93.

Wood, P.A. (1996) 'An "expert labor" approach to business service change', *Papers, Regional Science Association* 75: 325–49.

2 European consultancy growth
Nature, causes and consequences

Peter Wood

KIS supply patterns and trends

The rapid expansion of knowledge-intensive services (KIS) since 1980, especially computer and business consultancies, reflects two other major changes in the business environment. The first is a general move by firms and public agencies to seek expertise from outside organizations, as well as from their own employees. Then, in response to growing demand, KIS have themselves become increasingly active in transforming the business information and specialist expertise environment. An important component of this transformation has been the increasing 'tradability' of KIS expertise over longer distances, nationally, internationally, and even on a global, multinational scale. The innovative potential of consultancy has therefore been enhanced, while established patterns of localized knowledge exchange have been augmented and challenged.

KIS firms are very diverse in the types of expertise they offer, the size of their business, the scale and regularity of the projects they work on, and their relationships with clients. Some KIS firms are relatively large but most are small–medium in size, and they include many 'sole proprietors' and self-employed individuals. Table 2.1 is based on an EU classification of business services and, similar to the listing presented in the Introduction, groups activities that respond to client needs for expertise in management and administration (including personnel management), production, sales and information and communications. Measuring these services from official sources poses several problems. First, as already indicated in the Introduction, KIS are often inappropriately and inconsistently classified. Table 2.2 presents an attempt to adapt the European Union standard NACE classification (Nomenclature des Activités Communauté Européen) to provide a common basis for international comparison of the different German, UK and French classification systems up to the early 1990s (Gaebe *et al.* 1993). The categories are necessarily very broad, and often include consumer services (such as law, banking and finance firms and housing and estate agents). KIS are also treated differently by industrial classifications. Before 1993, most of them grouped much KIS activity into a residual group of 'other business services', usually including

Table 2.1 The main knowledge-intensive services

Business functions	Knowledge-intensive services

Management and administration	Organizational consultancy, including:
	    Strategy and change management
	    Quality management
	    General management
	Personnel, including:
	    Recruitment and staff selection
	    Training
	    Temporary staff services
	Legal services, including:
	    Legal professions
	    Legal consultancy
	Tax consultancy
	Accountancy and audit
	Public relations
	Administrative services (secretarial, language)

Production	Industrial engineering
	Engineering consultancy
	Technical services, including those related to:
	    Construction
	    Production
	    Environmental protection
	    Inspection and certification
	    Tests and trials

Sales	Advertising Market research Sales promotion Direct marketing Export services Trade fairs and exhibitions Commercial art and design

Information and communication	Computer services, including:
	    Hardware consultancy
	    Software consultancy and provision
	    Systems consultancy and development
	    Technical computer services
	Information services, including:
	    Data processing
	    Online database services
	    Specialist press and publishing
	Private communication services, including:
	    Courier and delivery services
	    Advanced telecommunications services

Source: Based on European Commission 1990.

Table 2.2 National and NACE classification of business services

Code (NACE)	Description
831	Activities auxiliary to banking and finance
832	Activities auxiliary to insurance
833	Dealers in real estate (excluding letting by owners)
834	Property (house and estate) agents
850	Letting of real estate by the owner
835	Legal services
836	Accountancy, tax collectors, auditors
837	Technical services
838	Advertising
839	Other business services

Source: Gaebe *et al.* 1993.

Notes
Activities identifiable from classification systems of France (NAE; NAP), Germany (WZ; WS; MZ), the United Kingdom SIC: (Standard Industrial Classification) and NACE: (Nomenclature des Activités Communauté Européen).
NAE: Nomenclature des Activités Economique; NAP: Nomenclature des Activités et des Produits; WZ: Wirtschaftszweigsystematik (Economic classification); WS: Systematik der Wirtschaftszweige für die Statistik der Bundesanstalt für Arbeit (Economic classification for Federal Labour Institute statistics); MZ: Systematik der Wirtschaftszweige für den Mikrozensus (Economic classification for Microcensus).

market research, management consultancy, data processing and employment agencies. Computer services were sometimes included, although they were more usually placed in technical services (or even electrical engineering). As a result, business and technical services often had to be aggregated for international comparison. Such well-known data problems were compounded at the regional level, where only data for all business services together was available, sometimes including financial services.

The difficulties of measuring the scale of KIS activities for the EU are well summarized by Rubalcaba-Bermejo (1999: 42–8). His estimates suggest that in the mid-1990s over 9 million personnel were employed by almost 900,000 business services firms in the twelve countries of the EU. About 2.5 million jobs were in operational services, such as cleaning and security, and a further 600,000 in temporary work agencies and equipment leasing. 'Core' KIS jobs, excluding activities with appreciable consumer markets, including staff training (957,000), engineering consultancy (940,000), computer services (680,000), marketing-related activities (584,000), inspection and control (222,000), export assistance (200,000), market research (133,000), fairs and exhibitions (70,000) and management consultancy (50,000), amounted to more than 3.8 million. There were also perhaps a further million jobs in business-related legal, accountancy and property management, giving a total figure of 4.8 million KIS jobs. Employment does not reflect comparative levels of output in such widely diverse activities, but these estimates at least indicated how extensive and diverse European KIS functions had become.

International comparisons generally demonstrate a strong relationship between gross national product (GNP) per capita and the share of business services in national employment and business turnover. Business services are best developed in richer countries, such as Denmark, France, Germany, Luxemburg, The Netherlands and, to an unusual degree, the UK. Typically, these have over 3.5 business service jobs per capita (from Rubalcaba 1998: Figure 1.1). In contrast, low-income countries, such as Greece, Portugal and Spain have fewer than two business service jobs per capita. Italy and Belgium fall between these, although their business service representation is lower than might be expected from their GNP per capita. The larger business service markets also show higher levels of productivity and value added per capita.

The European market for computer-related services amounted to over ECU 40 billion by 1994. Software production, worth about a further ECU 20 billion, was growing faster, following the shift to smaller computer systems, hardware standardization, distributed network services and the various associated advisory services. The main markets by 1992 were in the professional services (31 per cent), in applications software (18 per cent), systems software (16 per cent) and in processing services (12 per cent). By 1996, hardware maintenance had declined from 21 per cent to 16 per cent of the European market. Germany accounted for 28 per cent of the computer service market, followed by France (18 per cent), the UK (16 per cent) and Italy (10 per cent). In addition, electronic information services attracted ECU 4.2 billion of revenue in 1992, and employed 37,000 workers, mainly for financial and business online services, about half monitoring real-time developments.

The market for management consultancy was estimated at ECU 9 billion, following a 15 per cent annual growth rate over the previous five years (European Commission 1996: 24–34). About 40 per cent of the work was for manufacturing clients, 25 per cent for private services, 20 per cent for the public sector, and 15 per cent for other activities including the utilities. These proportions vary between countries and types of consultancy, but the dominant expertise offered was in information technology (44 per cent), human resources development (20 per cent) and corporate strategy (12 per cent). European engineering consultancy, which supports construction and manufacturing production, is longer established, remains larger than these other branches and has been engaged in significant global trade for many decades. The international activity of the larger firms, affiliated to the European Federation of Engineering Consultancy Associations, accounted for about 40 per cent of the total market. They employed 176,000 workers in 1996, including

Table 2.3a Great Britain: employment in 'management-related' services, 1981–94

SIC Activities	1981	1989	1991	1994	1981–89	1989–94
	× 1,000				% change	
8310/20: Ancillary to finance	87	148	142	153	70	3
8370: Professional and technical	169	274	230	212	62	−23
8380: Advertising	36	53	46	47	47	−11
8394: Computer services	55	138	148	157	236	14
8395: Business services not elsewhere specified	153	398	402	527	160	32
Total	500	1,011	968	1,096	102	8

Source: Wood (1996a), based on GB Census of Employment data.

Table 2.3b Great Britain: regional distribution of knowledge-intensive employment, 1981–89

	1989	1981–89
	% share	% change
London	34	79
Rest of South East	22	143
South West/East Anglia/East Midlands	13	140
West Midlands/North West/Yorkshire and Humberside/North	25	92
Wales/Scotland	6	97

Source: Wood (1996a), based on GB Census of Employment data.

51,000 in Germany, 45,000 in the UK and 24,000 in France. More generally, domestic markets accounted for about 75 per cent of turnover, with most of the rest exported outside Europe.

National developments in these activities have been widely reviewed in recent years (Ochel and Wegner 1987; Howells 1988; Elfring 1989; Illeris 1989; Daniels and Moulaert 1991; Daniels 1991; Moulaert and Tödtling 1995). In particular, Moulaert and Tödtling summarize the developing significance of transnational service corporations and present a series of country case studies. Tables 2.3–2.10 summarize information for various periods between 1978 and 1996 concerning the scale, growth and regional distribution of KIS/business services. The chapters of Part II of this book investigate these trends and subsequent development in much greater detail. As in all international comparisons, the data are based on different data sources, definitions and dates. They nevertheless demonstrate the universality of growth, especially in IT and information-based consultancies (France, Greece); computer services (the UK, Italy); advertising (France, Germany); management, business and organizational consultancies (France, Germany, Greece, Spain, the UK); technical certification (Italy); and accounting and fiscal consultancies (Greece, Spain).

One study employed Labour Force Survey data as a basis for comparing trends in France, Germany and the UK. This revealed both similarities and differences in patterns of concentration and change between the three countries (Gaebe *et al.* 1993; see also Marshall and Jaeger 1990; Moulaert and Gallouj 1995; Schamp 1995; Gaebe 1995; Wood 1996a; also Chapters 4, 5 and 7 of this book). For Great Britain (excluding Northern Ireland), 'management-related' services more than doubled employment, creating over half a million net jobs between 1981 and 1989 (see Table 2.3a). Growth rates significantly greater even than these were experienced by the computer and other business services. After 1989, however, the cyclical sensitivity of many of these activities was demonstrated by setbacks, especially in the professional and technical services (including architectural and engineering consultancies). Even computer services slowed down, although general business services, including much consultancy, continued to grow through the recession, stimulated at this time by demand from the privatization of state agencies. In 1991, private business service organizations employed 6.5 per cent of the British workforce, compared with 4 per cent a decade earlier. These figures exclude self-employed groups which often account for 20–25 per cent of employment in business service activities. Trends in the 1990s are considered in more detail in Chapter 7.

German social security payment data, used to count employees (again excluding the self-employed, as well as civil servants and part-time workers), showed a 75 per cent expansion between 1980 and 1991 in West Germany (Table 2.4), with a rise in the share of business services employees from 5.8 per cent to over 8 per cent. Over 426,000 jobs were created, comparable in scale to those in Great Britain. The fastest growing sectors were 'other business services', which more than doubled, advertising and accountancy-related activities. Technical business services are more important in Germany than in Great Britain, but they grew more slowly than other business services at this time (as they did in Great Britain). In France, the most spectacular growth, in terms of numbers of salaried employees between 1981 and 1991, also occurred in 'other business services' (more than doubling), and in IT and organizational consultancy (more than trebling) (Table 2.5a).

According to Eurostat data based on the Labour Force Survey for the three

Table 2.4 Germany: business service employment, 1980–95

	West Germany			West and East Germany	West Germany
	1980	1991	1994	1995	1980–91
	×1,000				% change
Technical services	205	330	378	497	61
Accountancy/tax collecting/auditors	148	269	329	397	81
Legal services	65	90	103	120	38
Advertising	37	66	72	81	95
Other business services	113	239	253	359	112
All business services	568	994	1,135	1,454	75

Sources: Bundesanstalt für Arbeit, *Statistik der Sozialversicherungspflichtig Beschäftigen* (unpublished data: calculations by Strambach), Nuremberg: Bundesanstalt für Arbeit (Federal Institute for Employment).

Table 2.5a France: business service employment, 1981–91

	1981	1991	1981–91
	×1,000		% change
Auxiliary to finance and insurance	55	71	29
Technical (engineering, architectural) consultancy	168	208	24
IT and organizational consultancy, data processing	57	180	216
Legal services	86	102	19
Accounting and financial audit	75	106	41
Real estate	128	136	6
Advertising	55	101	84
Economic, social, information, document studies	123	146	100
Temporary labour services	178	274	54
Other business and professional services	98	227	131
All business services	923	1,451	57

Source: Based on Table 4.1 (page 96).

countries in 1991 collected on a comparable basis, while the share of business services in all employment was 4.5 per cent, it was as high as 12 per cent in some regions. In Britain, London and the South East are predominant (Table 2.3b), but in France, while the Ile-de-France employed the largest numbers, two other regions had above average concentrations: Rhône-Alpes and Provence-Alpes-Côtes d'Azur (Tables 2.5b and 4.6). Germany has no single predominant centre, but the trading and port functions of Hamburg gave it the highest share of business services.

The patterns of growth in both Britain and France in the late 1980s suggested a spatial dispersion process at that time. The highest growth rates in Britain were in the industrial North West and West Midlands, 'catching up', as the booming services of the South East suffered from recession. In France, wide areas in the southern, central and north-west regions benefited, although often from low starting points. A similar dispersal trend was evident in The Netherlands (Table 2.6). In Germany, on the other hand, only Cologne showed the highest rates of growth, above 50 per cent. Regional patterns of business service development across Europe in the 1980s, at least, reflected the combination of spatial polarization favouring increasingly dominant core regions, and spatial dispersion in and around these regions (Moulaert and Tödtling 1995).

The Italian evidence in Tables 2.7a and 2.7b also shows high growth rates, especially of computer services and, in absolute numbers (130,000 in ten years), the professional consultancy services. Although the highest concentrations were in the industrial north-west and expanding north-east regions, the southern Mezzogiorno region appeared to show the fastest growth. In Greece, growth in the 1980s was again concentrated in computer-related and business consultancies, as well as accounting and fiscal advice (Table 2.8). In Portugal, a high proportion of activity is concentrated in the capital and adjacent regions (Table 2.9). The Spanish data suggest a strong move to the outsourcing of routine functions, including industrial cleaning and security services, as well as an over 80 per cent growth in the eight years before 1996 in knowledge-intensive legal, financial and management consultancy (Table 2.10).

It is widely recognized that the economic roles of consultancies vary significantly between countries. These differences reflect levels, patterns and rates of economic development, and national regulatory environments. This is illustrated by perhaps the best available evidence, for management consultancies in EC and European Free Trade Association (EFTA) countries, from a 1989 questionnaire survey by the Bundesverband Deutscher Unternehemensberater (BDU 1991; Keeble and Schwalbach 1995). The enquiry was conducted through professional associations, focusing on larger firms, and national response rates were far from uniform (5 per cent in Spain, 79 per cent in The Netherlands). The data are thus by no means a complete census, and exclude much of the small consultancy sector. For example, VAT data presented by Keeble, Bryson and Wood (1991) suggest that there were over 11,000

Table 2.5b France: regional distribution of business service employment, 1981–91

	1991		1981–91
	×1,000	% share	% change
Ile-de-France	567	39	52
Midi-Pyrénées, Rhône-Alpes, Languedoc-Roussillon, Provence-Alpes-Côte d'Azur, Corse	348	24	59
Upper Normandy, Picardy Nord-Pas-de-Calais	142	10	61
Lower Normandy, Pays-de-la-Loire, Brittany	120	8	69
Champagne, Ardennes, Bourgogne, Centre	89	6	59
Lorraine, Alsace, Franche-Comté	91	6	62
Limousin, Auvergne	25	2	54
Aquitaine, Poitou Charentes	70	5	62
Total	1,452	100	57

Source: ASSEDIC: Information supplied by Farcy, Moulaert and Gallouj (see Chapter 4).

Table 2.6 The Netherlands: business service employment, by province, 1989–95

Province (Location)	1995		1989–95
	×1,000	% share	% change
Groningen (North)	8.8	3.1	8.5
Friesland (North)	5.8	2.1	5.8
Drente (North East)	4.6	1.6	2.3
Overijssel (East)	12.6	4.5	7.5
Flevoland (Polders)	4.3	1.5	4.8
Gelderland (East)	28.8	10.3	2.6
Utrecht (Centre)	36.6	13.1	6.1
North-Holland (Amsterdam, Harlem, etc.)	52.9	19.0	4.1
South-Holland (Rotterdam-C.Randstad)	73.8	26.5	2.2
Zeeland (South)	3.1	1.1	5.9
North-Brabant (South)	32.5	11.6	4.7
Limburg (South East)	14.9	5.3	10.0
All business services	278.7	100	4.2

Source: CBS 1996.

Table 2.7a Italy: selected knowledge-intensive service employment, 1981–91

	1981	1991	1981–91
	×1,000		% change
Technical services and certification	124.0	196.4	58.3
Research and development	29.4	43.5	48.0
Computer services	51.9	181.0	248.8
Professional consultancy services	184.7	314.2	70.1
Advertising	22.2	40.8	83.8
Total	412.2	775.9	88.2

Source: Information supplied by Cavola and Martinelli (see Chapter 8), from Censimento Generale dell'Industria e dei Servizi, facsioli regionali (General Censuses of Industries and Services, regional tables).

Table 2.7b Italy: regional distribution of knowledge-intensive service employment, 1981–91

	1991		1981–91
	×1,000	% share	% change
North West Italy	289	37	80
North East–Central Italy	335	43	93
Mezzogiorno	152	20	97
Total	776	100	88

Source: Information supplied by Cavola and Martinelli (see Chapter 8), from Censimento Generale dell' Industria e dei Servizi, facsioli regionali (General Censuses of Industries and Services regional tables).

management consultancies in the UK in 1989, compared with only 1,529 recorded in the BDU study.

The evidence nevertheless indicates national variations in the activities of the larger management consultancies. In Italy, for example, in 1989 they appeared to rely much more on taxation business than elsewhere. Other national markets also show some specialization: on financial advice in Greece, software products in Belgium and advertising in The Netherlands (Table 2.11). These different national roles reflect different patterns of business fiscal regulation, and the behaviour of other support sectors, especially banking and accountancy. German consultancies, especially the larger ones, are particularly active in information technology and systems.

Table 2.8 Greece: employment, 1978–88

	Greece		Athens		
	1988	*1978–88*	*1988*	*1978–88*	*1988*
	×1,000	*% change*	*×1,000*	*% change*	*% share*
Banking and stock exchange	45.0	31	23.3	15	51.8
Insurance	13.2	56	9.4	42	71.2
Real estate	1.4	–24	0.7	–33	50.0
Accounting and fiscal consulting	5.7	175	2.0	88	35.1
Electronic data processing, etc.	1.5	1,366	1.2	1,140	70.0
Advertising	2.3	44	2.1	40	91.3
Business consulting	3.0	123	2.2	88	73.3
Miscellaneous	31.9	47	14.0	15	43.9
Total	104.0	46	54.9	25	52.9

Source: Based on Table 9.6b (page 259).

Table 2.9 Portugal: regional distribution of knowledge-intensive services, 1994

	% share
Lisbon metropolitan area	65
Porto	17
North coastal	5
Central coastal	4
Algarve	3
North/Centre	3
Alentejo	2
Madeira/Azores	2

Source: Data supplied by authors of Chapter 10, based on National Employment Census.

The main basis for management consultancy work in all European countries was the advice the consultancies give on corporate strategy and organizational development (Table 2.12), and this especially dominated business in Italy. Human resources expertise was more commonly provided than elsewhere in Greece and France, as well as the UK. German specialization in IT and systems

Table 2.10 Spain: employment in business services, 1988–96

	1988	1996	1988–96
	×1,000		% change
Legal, tax, management consultancy	106	192	81
Technical services (architectural and engineering) and testing	46	78	70
Advertising	26	32	23
Industrial cleaning	71	176	147
Security services	31	61	96
Miscellaneous services	28	57	104

Source: Data supplied by authors of Chapter 10, based on National Employment Census.

Table 2.11 National markets for large management consultancies, 1989

	Percentage share							
	Belg./ Lux.	*France*	*Germ.*	*Greece*[a]	*Italy*[b]	*Neth.*	*Port.*[b]	*UK*
Management consulting	73	79	69	82	60	80	87	81
Software products	21	6	4	3	1	3	6	8
Advertising	0	2	6	0	0	10	0	2
Tax advice	2	2	0	1	24	1	1	1
Financial advice	1	2	0	14	3	2	4	1
Auditing	4	8	2	0	7	3	1	3
Others	0	0	19	0	5	0	0	5

Source: Keeble and Schwalbach, 1995: Tables 6 and 7, based on BDU 1991.

Notes
a >ECU 240,000 turnover.
b >ECU 480,000 turnover.
All other countries, >ECU 2.4 million turnover.

Table 2.12 Types of service offered by large management consultancies, by country, 1989

	Percentage share						
	Belg./ Lux.	*France*	*Germ.*	*Greece*[a]	*Italy*[b]	*Port.*[b]	*UK*
Corporate strategy/ organizational development	23	21	20	19	52	19	19
Financial and administrative systems	14	4	5	12	18	5	7
Human resources	8	29	17	36	14	11	22
Production and services management	17	10	12	2	3	15	8
Marketing and corporate communications	2	13	9	10	1	5	12
Information technology and systems	25	10	21	8	6	4	17
Project management	7	11	11	4	3	32	5
Economic and environmental studies	3	4	5	10	3	10	10

Source: Keeble and Schwalbach 1995, Tables 6 and 7, based on BDU 1991.

Notes
a >ECU 240,000 turnover.
b >ECU 480,000 turnover.
All other countries, >ECU 2.4 million turnover.

advice was mirrored in Belgium/Luxemburg, and to a lesser extent in the UK. In Portugal, management consultancies appeared to be most commonly employed for project management and production and services management.

These patterns for management consultancy are reinforced and complemented in various ways across Europe by the activities of many other types of consultancy, including computer, technical and engineering consultancies, accountancy firms and human resources specialists. They are also affected by the competitive development of international KIS activities, which we shall examine in more detail in Chapter 3. Table 2.13 summarizes the scale of US direct business service investment in various European countries in 1994 (OECD 1998). Europe is the main target for US global investment, taking 56 per cent of world sales by US affiliates. Computer and data-processing services account for over 60 per cent of investment in these activities, with the UK alone taking 29 per cent of US foreign direct investment in computer-related services. The UK is also the main target of US investment in advertising in

Table 2.13 Direct investment in Europe by United States business service companies, 1994

US$ million, by sector, historical cost basis

	World	Belg.	France	Germ.	Italy	Neths.	Spain	Switz.	UK
Computer and data processing	5,628	398	329	200	137	241	75	427	1,629
R&D and technical testing	606	46	−5	13	n.d.	173	n.d.	3	100
Advertising	2,321	101	94	102	36	118	45	3	321
Management and public relations	2,038	181	−19	n.d.	−25	n.d.	−2	556	161
Personnel supply services	300	n.d.	n.d.	n.d.	0	n.d.	2	n.d.	41
Education services	85	n.d.	<	−6	0	n.d.	0	n.d.	<

Source: OECD 1998: Table 3.

Notes
n.d.: data not available.
<: less than US$ 1 million.

Europe. The Netherlands is the main base for US R&D and technical testing firms, and the European headquarters of US management and public relations consultancies are most concentrated in Switzerland. The apparently small direct US presence in several important countries, including France, Italy and Spain, may reflect the delayed impact there of US-type consultancy culture, reflecting language and regulatory and business attitudes. The figures probably also underestimate US influence, which may be exerted through headquarters based elsewhere and in collaboration with other firms in these countries, rather than direct investment.

Supply patterns of the main types of consultancy are evidently not uniform across Europe, at either national or regional levels. International trends may be making them more uniform, but they must be adapted to distinctive national and sectoral conditions. How these patterns relate to economic competitiveness and innovation more generally, and how they may evolve in the future, of course remain questions for later consideration.

Demand developments: why have consultancies grown?

The growth of consultancy is often explained in terms of the development of the 'information economy', supporting a society in which the control and exchange of information has become the predominant source of value added (Castells 1989). This, in turn, seems to reflect modern technological developments in the storage, processing and transmission of information. Specialist KIS businesses profit by offering expertise that assists clients to acquire, interpret and respond to appropriate information. Communications innovation has also enabled their expertise to be increasingly exchanged across widening geographical markets.

The growth of IT innovation and information exchange thus only partially explains the growth of consultancy. Organizations have other options than reliance on outside support for information-related functions. The rise of consultancies also marks a significant change in the division of expert labour between outside and inside expertise employed by their clients. This is driven by the interpretative needs of the modern information economy, and marks a similar organizational trend towards greater reliance on inter-firm relationships and complementary labour skills to many other business functions, including innovation (see Chapter 1).

Influences moulding the growing demand for consultancy may be summarized as follows.

1 *Organizational transformations*

Consultancy use is an element of general organizational change affecting all sectors in the 1980s and 1990s. This arises from:

1   the growing pace of capital restructuring, reflected in merger and take-over activity, especially at an international level;
2   broad processes of technological change, especially towards capital-intensive and 'flexible' production methods, including those based on development in information and computer technology (ICT);
3   the internationalization of production;
4   the growth of world trade and associated intensification of competition;
5   increasing consumer and producer market diversity and choice;
6   changing methods and levels of state economic regulation, at national and European levels;
7   changing skill requirements, in both production and management, which create new relationships between different occupations.

The stability of established commercial and organizational practice is being constantly challenged. It is thus increasingly difficult for staff recruitment and training procedures to provide all the skills needed to cope with contemporary change.

2   *The growing need for specialist expertise*

The information explosion is more a symptom than the primary cause of economic restructuring. Information exchange alone does not ensure successful technical or market adaptation. In fact, information 'overload' may make adaptation more difficult, creating confusion rather than enlightenment. Adaptability depends on the quality of technical and managerial expertise brought to bear on appropriately selected information. Staff skills have customarily been developed, especially by large organizations, through their recruitment and training policies. These remain critical to any organization's information-processing capacities, but rates of technical, competitive and regulatory change have overtaken and transformed them in recent years. Traditional forms of in-house expertise development have had to adapt to new conditions by becoming more open to outside influence.

3   *Widening sources of expertise*

The adaptability required by modern uncertainty explains the widespread shift by many large organizations from relatively stable, hierarchical, career-based organizational forms, to more fluid, 'organic', devolved, project-based modes of working. In general, the scope and scale of directly employed expertise have been reduced. This is marked by declining corporate head office employment and the closure of many specialist technical and commercial support divisions. Overhead costs have been cut by directing staff to focus on core expertise, including key innovative functions. New information and computer technologies have encouraged this, supplanting some jobs, but these changes also mark a recognition that single organizations can no longer afford to employ directly all the expertise and experience they require. The rapid growth of KIS thus reflects in part the 'externalization' of former in-house functions at various levels of the traditional corporate hierarchy. In addition, many new forms of specialist expertise are also now required which even large organizations cannot, or do not wish to, develop.

4 Functional and sectoral diversity

Patterns of KIS use vary between different types of expertise. Generally, routine procedures are now often devolved, or 'outsourced'. In many cases, as well as saving operational costs, this is accompanied by significant change in the quality and adaptability of these functions. Many specialized or occasionally required activities have long been undertaken mainly by KIS firms, including market research and advertising, building design, legal advice and, often as a statutory requirement, accountancy. Similar practices are now emerging for computer hardware and software development, overseas sales planning, property maintenance, financial processing and salary administration, and even staff recruitment and training. More radically, consultancies are also involved at the highest technical and management levels of organizations, contributing to strategic planning and comprehensive organizational change, technical development, computer systems innovation, research evaluation, market development, and human resource management. Such penetration of consultancies at high levels nevertheless varies widely between countries and sectors.

In the UK, research evidence suggests that sectors which face the most radical and urgent technical and competitive changes are the most frequent employers of consultancies (Wood 1996a, 1996b). Such changes include major cost-reduction programmes, the introduction of new production technology or ICT, post-merger or takeover adjustments, shifts towards 'market-orientated' corporate strategies, and the application of 'total quality management' and 'business re-engineering' methodologies, designed to change and integrate organizational cultures. Consultancy inputs were particularly influential in the manufacturing sectors most subject to change, and in the privatization of utilities in the early 1990s.

In more stable manufacturing and service sectors, consultancies appear to be employed less frequently, or for more routine functions. However, the general pressures for change, such as those arising from the introduction of information and computer technologies in banking, retailing or the travel sectors, have boosted consultancy business. Other expanding markets for consultancy include the restructuring of public organizations and support for client international expansion or multinational rationalization. Once again, such patterns vary between countries, reflecting both different pressures towards business change, and dominant corporate and regulatory regimes.

5   *Regional diversity of demand*

The environment of consultancy growth is also regionally highly varied. Strategic consultancy activity is stimulated in regions of high urban or sector-based demand, especially from large organizations. These regions also foster the development of skills which, through entrepreneurial spin-off or recruitment, offer a seedbed for training consultants. There is strong evidence that regional-level competition for corporate client business encourages service specialization and quality, thus improving KIS competitiveness in inter-regional and even in international consultancy trade (O'Farrell *et al.* 1992; O'Farrell *et al.* 1996).

Some regions are also the bases for technological or market-orientated patterns of specialist production. Within these, consultancies may engage with clients in processes of mutual support for innovation and change, following and leading change at different stages. Consultancies may also support regional economic systems in adapting to technological and global market changes. Such specialist consultancies may be associated with concentrations of high-technology industry, high-quality consumer production or financial and business services (Scott 1988). Demand may also be strong in other specialist industrial regions, around ports, airports or other communication nodes, in regions of primary extraction (such as North Sea oil), tourist-based regions, major educational or health service nodes, and possibly even in agricultural regions.

The emerging structure of the KIS sector

The basis of consultancy competitiveness

KIS activities have thus grown at an unprecedented rate in recent decades in a great variety of national, regional and sectoral circumstances. This phenomenon is based on a common need to adapt the availability and management of expertise to permanently changing market and technological conditions. KIS have also had to adapt competitively to change through the quantity, cost and quality of their expertise, and the variety of demands they respond to. Some sectors, such as management consultancy and advertising, are dominated by large consultancies, often operating on an international level. These thrive because of the economies of scale associated with access to high-cost automated forms of information. Economies of scope also arise when diverse skills and experience can be combined to support particular projects, especially where international standards of practice are sought. Much consultancy provision, however, is by small–medium enterprises (SMEs), which offer specialist and personalized expertise in close collaboration with client management. Both size segments of consultancy have been expanding rapidly in recent years.

KIS quality depends on the training of consultant experts, supplementing their educational background with technical and commercial experience. Such training may be provided by consultancies themselves, especially the larger firms, but consultancies may also acquire expertise through the recruitment of staff with experience, sometimes gained in other sectors. Such experience is also a vital foundation for the KIS small-firm sector. Much consultancy therefore depends on the inter-company mobility of staff. As noted in Chapter 1, this process may itself be an innovative influence on management and technology compared with more introverted corporate processes of training and career development. Working with many clients also supports the development of consultancy expertise, enabling consultancies to offer new 'products' in a competitive market. New approaches to business change and innovation were promoted in the 1990s by the consultancy sector, incorporating such practices as matrix strategy – the coordination of multiple functions; business process re-engineering – the comprehensive, often IT-influenced, redesign of business processes especially to improve market position; total quality management – continuous production quality improvement; team-based working; supply-chain partnerships; and 'just-in-time' production.

In the 1990s the competitive adaptation of the larger consultancies led these firms to expand their market and geographical reach, but also to emphasize the importance of tailoring projects to the needs of individual clients. More intensive forms of client collaboration developed, as was already common among smaller, specialist consultancies. The variety of expertise has also grown, with consultancies expanding their core business in areas such as software, hardware, human resources, technical, design or project management. More diverse or comprehensive consultancy services were being developed, often grouped around the impacts of information technology. By 2000, the overlapping and merging of large consultancies functions was a major trend in the UK (see Chapter 7).

In spite of commonly expressed reservations about the cost, quality and appropriateness of consultancy work, more organizations than ever now employ consultancies on a routine basis. Reputation and repeat business, based on familiarity and mutual trust, are among the main consultancy assets. In this context, consultancies may promote improved levels of technical and commercial practice among client groups on a continuous basis, although this depends on the receptiveness of clients. It appears that, especially at higher technical and managerial levels, the client–consultancy relationship has become increasingly symbiotic. If this trend is beneficial, businesses which have poor access to good-quality consultancies, because of their small size, sector or location, may increasingly be disadvantaged.

In summary, the competitiveness of any service has been seen as depending on:

1   the knowledge and skills of service workers and managers;
2   the effective organization of inputs to serve customer needs;
3   the costs of labour;

4   the availability of data-processing and communications infrastructure;
5   the institutional environment;
6   proximity/accessibility to markets;
7   for KIS especially, the cumulative benefits arising from past exchanges (based on Feketekuty 1988).

Among these, consultancy competitiveness universally depends on the quality of labour (1) and the organization of the consultancy process to serve client needs (2). The other influences are more selective. Cost of labour (3) is less important for high-level consultancy skills, but significant for more routine outsourced functions. The control of specialist data-processing infrastructure (4) is particularly significant for global and specialist information consultancies, but less critical for many smaller consultancies. The national or regional institutional environment (5) may affect the quality of client interaction, and this may be more difficult to sustain in some commercial, legal and cultural, including language, contexts than others (Moulaert and Martinelli 1992; O'Farrell and Moffat 1991). The problems of proximity/accessibility (6) may also be significant when consultancy work extends internationally, or across regions.

The final competitive factor for consultancy (7) acknowledges, 'the level and pattern of prior development' as a general advantage (Gibbs and Hayashi 1989: 7). For consultancy this is gained by sustaining the quality of staff and management, enhanced by experience of working in well-developed, competitive markets. It also derives from adaptability to client needs over time through organizational flexibility, internally and through new forms of collaboration. Consultancies also offer essentially intangible products, so that competitive benefits also arise from reputation, accumulated especially through repeat work for influential clients.

Types of consultancy

The functional segmentation of consultancy, serving many types of client requirement, has resulted in a diverse pattern of supply, especially among the rapidly growing management, computer and marketing consultancies (see Table 2.1). Among management consultancies, for example, the following types were identified by the UK Management Consultancy Association in the mid-1990s:

1   hardware manufacturers (IBM, Unisys, AT&T);
2   systems houses (Cap Gemini, CMG, CSC/Index);
3   accountancy-based firms (Andersen Consulting, Price Waterhouse, KPMG, Ernst & Young);
4   traditional firms (PA, PE-I, AT Kearney);
5   strategy houses (McKinsey, AD Little, Booz Allen & Hamilton, Boston Consulting);

6 niche players (Hay, Bossard, Towers Perrin);
7 in-house consultancies (British Telecom, British Airways, BAA);
8 academic organizations (LBS, Cranfield, Manchester, Warwick);
9 small to medium-sized consultancies;
10 sole practitioners (about 13,000 of 25,000 total in the UK).

The major players have expanded their core skills, for example in accountancy and computing, into more general, and often more profitable, consultancy work. Specialization is also an important basis for consultancy, both in traditional sectors (such as market research) and in new markets (such as human resources assessment), forming the basis for many small to medium-sized consultancies, including sole practitioners. Consultancy profitability has also encouraged firms in sectors such as telecommunications, travel and energy to allow in-house specialist units to work for outside clients.

Much management consultancy today builds on computer hardware, software and systems expertise. An overlapping range of specifically computing-orientated consultancies was identified by Gadrey *et al.* (1992) as follows:

a systems/engineering companies (systems/hardware management consultancies);
b large accountancy companies (Andersen, KPMG, Ernst & Young);
c technical engineering consultancies and consultants in industrial organizations (niche consultancies);
d computer and telecommunications hardware consultancies (IBM; Unisys, ICL, AT&T);
e applications consultancies;
f organizational/computing departments of major industrial/service firms.

The first two of these correspond to the management consultancy groups (1), (2) and (3), while the computer and telecommunications hardware consultancies (d) are similar in origin to many of the 'in-house' management consultancies (7). Throughout the 1990s, the universal impact of information and computer technology has meant that the work of management and computer consultancies has progressively overlapped. This has added to the complexities of both classification and measurement, as we shall see in the evidence from various countries in Chapters 4–11.

KIS tradability: from local to global provision

Modes of exchange and measurement problems

Among the most important transformations of the consultancy environment in recent years has been the growing 'tradability' of expertise. This has expanded KIS markets across different regions within countries, between countries (especially within Europe) and even globally. The consultancy

Table 2.14 European cooperative links of German-based management consultancies, 1989

	Foreign offices	Cooperation networks	Total
France	6	24	30
United Kingdom	3	24	27
Spain	8	13	21
Sweden	14	7	21
Italy	5	13	18
Austria	7	11	18
Netherlands	1	12	13
Belgium	2	6	8
Hungary	1	5	6
Switzerland	1	4	5
Denmark	2	2	4
Czechoslovakia	1	2	3
Ireland	1	2	3
Total	52	125	177

Source: BDU 1990.

process nevertheless still requires close interaction with clients, delivered mainly at local or regional levels, usually within countries. Successful consultancy therefore draws increasingly on a nationally and internationally expanding range of expertise while retaining 'local' delivery. Large consultancies achieve this through branch office networks, while smaller firms support tradability by specialization and extended collaboration. In principle, such developments should enhance the quality and value to clients of consultancy inputs and their potential for supporting business innovation. In practice, the range and quality of such inputs varies widely between countries and regions.

The ability of consultancies to deliver advice over longer distances depends on the ease of expertise exchange using various modes, singly or in combination, including:

1   improved verbal/written/digital exchange through communication technology;
2   greater mobility of skilled labour, either through short-term business travel or longer-term secondment;
3   growing collaboration between consultancies based in different regions or nations;

4   increased investment in branch office networks by both global consultancies and formerly national or regionally based firms.

'Tradability' is based on direct exchange between consultancies and clients in distant locations. It is fostered by improved communications, greater ease of travel, semi-permanent or project-based collaboration, and international investment, especially by larger consultancy firms.

The extent of collaboration between consultancies, through networking or joint ventures, is illustrated in Table 2.14, taken from a Bundesverband Deutscher Unternehmensberater investigation (BDU 1990). Such cooperative networks enable these consultancies to overcome language and regulatory barriers, linking them to French, British and Spanish markets. A survey of a large sample of small to medium-sized British consultancies has also shown that about 20 per cent had established foreign branches (O'Farrell *et al.* 1996). Sixty per cent had also entered into formal collaborative arrangements with other consultancies, over half in other countries (O'Farrell and Wood 1999).

The implications for innovativeness of the developing relationship between the global market for expertise and its local delivery will be explored more fully in Chapter 3. Attempts to measure the scale or economic impacts of the growing international tradability of expertise, however, face the well-documented problems of measuring international service exchange more generally. Service trade is measured as the net returns or profits from exports of identified service agencies, such as consultancies. Even if these can be reliably and consistently identified, it has been argued that this measure would significantly undervalue service compared with goods trade by 'at least an order of magnitude' (Quinn and Dorley 1988: 419). To this technical deficiency must be added inadequate measurement of the economic value of international expert labour mobility, and the exchanges underlying both foreign collaboration and branch investment activities (Martinelli 1991).

Perhaps more fundamentally, the worth to any economy of consultancy expertise can be properly valued only in terms of the benefits conveyed to clients. Since consultancy essentially requires co-production with clients, the benefits depend on the client response to the exchange. The value of consultancy exports to recipient economies, however conveyed, is therefore likely to be only indirectly and inconsistently related to the fee incomes received by exporters. The basis of these is notoriously unclear, but generally based on cost-plus accounting and reputation, rather than any close assessment of benefits. Consultancy inputs are also often embodied in other types of transaction, for example as part of a construction project or sales budget, incorporated into the trade earnings of other sectors. Many international consultancy relations are thus not separately identified or market priced. These problems are compounded for transfers between branches of multinational companies or international agencies that benefit from consultancy work. Such problems in evaluating expertise exchange are inherent in any approach based on

Table 2.15 Top ten global service suppliers, by sector, 1995, revenue, US$ billions

Information services[a]		Software		Advertising		Management consultancies	
IBM (US)	20.1	IBM (US)	3.4	WPP Group (UK)	12.9	Andersen Cons.	3.1
EDS (US)	12.4	Microsoft (US)	3.0	Omnicon Grp (US)	9.4	McKinsey	2.1
Digital Equipment (US)	6.5	Hitachi (Japan)	2.8	Interpublic Grp (US)	5.5	Ernst & Young	2.1
Hewlett-Packard (US)	6.3	Fujitsu (Japan)	1.9	Dentsu (Japan)	4.8	Coopers & Lybrand	1.9
CSC (US)	3.9	Computer Associates	2.1	Young & Rubicam (US)	3.2	Arthur Andersen	1.4
Andersen Consulting (US)	3.8	NEC (Japan)	1.2	Codiant (UK)	2.3	KPMG Peat Marwick	1.4
Fujitsu (Japan)	3.8	Oracle (US)	1.0	Grey Advertising (US)	2.3	Deloitte & Touche	1.3
Cap Gemini Sogeti (France)	3.6	SAP (Germany)	1.0	Havas (France)	1.7	Mercer Grp	1.2
Unisys (US)	3.5	Novell (US)	0.9	Hakuhodo (Japan)	1.2	Towers Perrin	0.9
ADP (US)	3.2	Digital Equipment (US)	0.9	True North Comm. (US)	1.2	AT Kearney	0.9

Source: OECD 1998: Table 1.

Note
a Professional computer services (58%), data processing, network and electronic information services.

Table 2.16 EC and non-EC markets of European business services, 1989

	Location of clients, %		
	Domestic	Other EC	Non-EC
Engineering	55	4	41
Consultancy	90–95	2–5	2–5
Commercial communications	50	40	10
Computer services	90	5	5
Operational services	98	1	1

Source: European Commission 1990.

traditional measurement methods applied to the goods trade. The evaluation of expertise and goods exchange requires inherently different approaches. The problems are even greater on smaller, inter-regional levels. In effect, there are no reliable ways of measuring the value of expertise exchange on any level. The growth of unmeasured Internet exchange and the integration of the European Market are in different ways each likely to make such monitoring even more difficult.

The international exchange of consultancy and other expertise is growing rapidly, yet there is no reliable evidence for its value to recipient economies, let alone its impacts. The relationships that create wealth in interdependent, knowledge-based economies are simply too complex and intangible. All that can be inferred from current evidence is that world consultancy exchange is growing, and is dominated by the US, Europe and Japan. Table 2.15 lists the major international consultancies in the main sectors in 1995. Within Europe, all that the limited available trade data suggest is that there are marked inequalities between exporting countries, such as Germany, France, The Netherlands and, more recently, the UK, and importing countries such as Spain, Italy, Greece Portugal and Ireland (OECD 1998).

The widening reach of locally based consultancies

An important aim of European Union market integration is to foster international exchange in services to increase choice and levels of competition for clients. Consultancy markets are seen to be among the least integrated in the EU because of the predominance of national and local exchange. Survey evidence suggests that well over 90 per cent of clients for most types of consultancy are found within their home countries (see Table 2.16). Communications and engineering services are more export orientated, although the latter mainly to outside Europe (European Commission 1990). Table 2.17 shows

Table 2.17 Origin of business services, by proximity, 1989, percentage of projects

	All services	Personnel	Technical	Computer	Marketing	Management
Place of origin						
Same region	71	95	81	73	70	59
Another region in same country	57	34	57	56	60	57
Another EC country	14	2	17	3	17	14
Non-EC country	10	2	11	6	6	6
Importance of proximity of supplier						
Very important	28	37	36	29	32	13
Important	33	36	32	46	29	28
Unimportant	38	26	31	24	38	58

Source: European Commission 1990.

Notes
Personnel = Temporary staff agencies, training and recruitment.
Technical = Engineering, maintenance and quality control.
Computer = Computer software, consultancy, etc.
Marketing = Advertising, market surveys, sales promotion.
Management = Organization and management consultancy.

the reliance of clients for different types of consultancy not just on their home country, but also their home region. According to these data, over two-thirds of clients used consultancies with offices in their local region, and well over half used other national consultancies. Only 14 per cent used consultancies based in another EC country. Regional sources were most common for personnel and technical support, and national sources for marketing advice. Proximity was a very important or important choice factor for over 60 per cent of clients, especially for personnel and computer consultancies, but was less influential in the choice of management consultancies.

This evidence may not reflect the origins of consultancy expertise, however, especially for higher-level technical and managerial skills. For example, 'local' sources may be branches of either international or SME consultancies, but both may explicitly seek to link clients to non-local knowledge and experience. From the supplier perspective, global consultancies and nationally based firms have actively pursued strategies of international market development, especially within the EU (O'Farrell *et al.* 1996). Local client–consultancy exchange is also important for multinational development. Global consultancies, for example, need to acquire experience of a huge variety of local circumstances, often doing so by acquiring or merging with national and regional firms. For accountancy, advertising, legal and consultancy work, this has been, 'by far the most significant trend which is structuring the international market in these services' (UNCTAD 1989).

Many small to medium-sized consultancies offer both specific technical or managerial expertise and 'local knowledge' of different markets and production conditions. They therefore compete with larger consultancies both through market or technological specialization, and through close relations with local clients. Outside consultancies may in fact subcontract work to them on this basis. Although most remain nationally based, successful small to medium-sized technical consultancies, for example in engineering and construction, are quite often international in their reach and organization. More generally, specialist technical, software applications, financial, market or human resources development expertise has supported foreign market development by many KIS firms, operating abroad but also serving foreign clients in the home market. This trend has become particularly active within the European Union.

Small to medium-sized consultancies are nevertheless commonly orientated towards specific regional markets, often in metropolitan core regions. Gallouj (1996), in a review of a wide variety of European research evidence, identifies three categories of regional pattern for different types of business services.

1   Those which normally have significant markets outside their local region, including engineering consultancies and architectural services, management consultancy, computer services, R&D, marketing, market research and economic and social consultancies.

2   Those usually orientated mainly towards local markets, including accountancy and audit, legal services and employment agencies.
3   A mixed group including advertising, communications and property management, and possibly personnel consultancies.

Gallouj also observed that specialization into niche services appears to be an important key to consultancy 'tradability' among small to medium-sized, usually single-office firms. The head and regional offices of multi-office firms, controlling other branches, are also likely to be export orientated. Their branch offices, however, normally operate only within their target regions.

There is also evidence that international expansion by small to medium-sized consultancies may be associated with the prior development of markets in other regions of the same country, where experience is first gained of operating with distant clients. This has especially been noted in continental European countries, where inter-regional exchange may relatively easily extend into international exchange. A similar pattern has been observed for Scotland, a developed peripheral region in the UK. Here local market limitations restrict the scope for consultancy specialization (O'Farrell *et al.* 1992; O'Farrell *et al.* 1996), so that entry into other UK regions enables such specialization to develop and provide a basis for subsequent international competitiveness. In contrast, many consultancies in the large and developed market of south-east England export abroad directly without having worked in other UK regional markets. The same appears to be true for the development of specialist consultancies in the metropolitan cores of Greece and Portugal. In these cases, they simply lack developed regional markets, and must export if they are to specialize and grow (see Chapters 9 and 10).

UK evidence suggests that national or regionally based consultancies may expand their home markets through diversification, collaborating with different types of company (such as other consultancies, software or hardware companies, and training companies). The market is thus enhanced by offering clients variety of expertise. In contrast, in foreign markets they tend to cooperate with the same type of consultancy to increase the potential market for their core skills. O'Farrell *et al.* (1996) show that small to medium-sized UK consultancies may first enter foreign markets by working there for multinational clients already served in the home market. Because of these established working relations, such clients can be served without a permanent presence abroad. The development of these markets beyond the initial entry requires a further commitment to specialist staff training, branch office investment or collaboration with foreign consultancies. Success in foreign markets also depends on the innovativeness of the consultancies' approach to foreign clients, adapting their expertise to new requirements. Again, collaborative links to local consultancies may offer a means of achieving such adaptation.

Barriers to consultancy internationalization

In spite of the trend towards tradability, there remain barriers to the opening-up of European consultancy markets, as with international service trade at large. The European Union has attempted to reduce these, by promoting the mutual recognition of professional qualifications, the liberalization and standardization of information and communication systems, and the harmonization of company law and the rules of competition and capital exchange. For the financial services, a 'home country'-based regulatory regime is being developed on these principles, to include banking, insurance and accountancy. The aim is to encourage the competitive development of international markets under relatively liberal forms of regulation (Nicolaides 1989; Feketekuty 1988).

Such liberalization would probably be less significant for consultancy activities. They have only relatively recently developed on a large scale to serve sophisticated, diverse, competitive and therefore generally unregulated business markets, rather than vulnerable consumer clients. Nevertheless the regulatory/institutional practices still constrain the exchange of expertise, and especially the activities in various markets of multinational consultancy firms (Vogler-Ludwig *et al.* 1993). For example:

1 Barriers to labour mobility. These include significant inequalities of education and training, and differences in social security provision remain within the EU. In spite of moves towards harmonization, there are variations in legal and accountancy controls and procedures, affecting the acceptability of qualifications and ease of market entry.
2 Different procedures for business establishment and operation. These affect such diverse issues as levels and procedures of company taxation, regulations governing surveys, and variable or weak intellectual property protection rights (Nicolaides 1989).
3 Different value added taxation regimes, for example in relation to rates and deduction rules.
4 Differences in national technical standards, for example affecting engineering, computer services, telecommunications and quality control procedures.
5 National occupation-based regulations confining certain tasks to specialist enterprises. These include auditing, some financial services, computer services, legal services and surveying. They require such functions to be operated under separate legal form or to be subcontracted, preventing their integration into more general consultancy activities.
6 The influence of public sector competition and policies, including supply monopolies and forms of contracting favouring nationally based firms.

Such regulatory differences may slow integration, limiting the access of clients in some countries to international expertise. This affects not just the activities of international consultancies, but also the quality of national and regionally based consultancy expertise. In Chapter 3, we shall argue that international, national and regional agencies are linked through competitive processes, and barriers to international expertise exchange thus affect the quality of the whole consultancy sector.

Even if such formal barriers were removed, the inherent localization of client–consultancy exchange will continue to reflect more pervasive differences of culture, language and business practice. These are likely to sustain locally orientated consultancies in the foreseeable future. To compete with such local specialists, for example, international consultancy would also need to overcome further barriers, related to:

7   Advanced linguistic skills.
8   Familiarity with the general 'host country' corporate, financial and regulatory environment, including prevailing attitudes and contractual practices in the use of consultancies.
9   Specific consultancy expertise attuned to host country needs and practice for different sectors and types of client in relation, for example, to management training and deployment, production regulation, labour practices, ICT standards, market research conventions or the needs of SMEs.
10   Awareness of consumer and business attitudes towards different styles of business, including foreign influence.

Differences of culture and custom are thus likely to mean that advice which applies well in one country may be difficult to adapt elsewhere, whether in the implementation of new technologies and production methods, new computer systems, novel approaches to training, quality assurance procedures or even styles of documentation.

The impacts of global consultancies

Localism is inherent in consultancy exchange, but the influence of international and even global consultancy is also growing, especially in promoting more radical technical and organizational changes. This influence may be exerted indirectly through the practices of small to medium-sized consultancies. The major consultancies have nevertheless extended their direct presence throughout Europe in the past twenty years. This is not a new phenomenon. Expertise in engineering, architecture and trade has been internationally exchanged for many decades. Even management consultancy first developed international markets as early as the 1920s when 'scientific management' methods were first promoted. The growth of global consultancies, controlling branches or subsidiaries in many countries also follows the wider globalization of trade, production and capital exchange, especially by multinational companies (MNCs). Many clients of global consultancy are nevertheless nationally based, including public agencies, but they increasingly need to know about the implications for them of international developments.

Most of the global consultancies originated in management, technical or financial consultancy but now they generally offer comprehensive strategic and management systems advice affecting many aspects of corporate change. Table 2.18 describes some innovation-related activities of major management and IT consultancies in various European countries, based on published company sources and interviews carried out during the KISINN project. They are anonymous and generalized, intended simply to convey a sense of the type of work undertaken. These companies generally support corporate restructuring, including mergers; cost control and efficiency benchmarking; international market development; technological changes, including improved communications and data-processing capacities; global investment and financial management; and the control of extended production facilities including staff training and deployment.

Table 2.18 Consultancy and innovation: some cases

---

1   Consultancy innovation is based on putting pieces together in different ways: exploring product differentiation, quality of delivery, design and service support, especially in computers innovation.

2   A major example of the widespread influence of KIS on process innovation is the application of SAP business management software and other information technology to the needs of many client through licensed consultancies.

3   The main source of innovation by major consultancies is in the transfer of ideas and procedures between different markets (e.g. offshore structures experience to London Underground) and project management skills. For example, KIS-dependent innovations are transforming many financial services by drawing on ICT and management methods developed in retailing.

4   Pharmaceutical developments in France and Switzerland, for example, are supported by the methodology of a major consultancy for assessing market risks of investment in various drugs, influencing the direction of investment, and shortening the period of development.

5   A major consultancy supports partnership between French aerospace company and multimedia specialists to extend application of its R&D to other markets.

6   Interactive multimedia system for rapid customer access to the services of a major building society implemented by a leading consultancy firm. Nominated for international scientific award. Involved assembling unique collection of specialist skills for project. Involved redesigning business processes and retraining staff, during an extended process of change.

7   Employment of consultancy services in Greece is growing, to support widespread mainstream business innovation, including business process re-engineering, improved consumer response methods, supply-chain collaboration and electronic data interchange. Main barriers are associated with uncertainty of the market, including speed of institutional and legislative change.

(continued on next page)

Table 2.18 (cont.)

SMEs

8   In France, medium-sized firms (200–300 employees) employ consultancies for marketing and computer services, rather than engineering services which in-house staff believe they can handle.

9   In Germany, when it occurs, the innovative contribution of consultancy service for SMEs is high; for large clients, consultancies are involved more often but in more routine changes.

10   Engineering consultancies may assist SMEs involved in subcontracting to adapt to the needs of their clients, for example through introduction of electronic data interchange (examples in food and vehicle subcontracting).

11   Small specialist Dutch biotechnology firms accelerate development of new products by engaging consultant engineers with knowledge of specialist plant design, including computer simulation software and specialist staff.

Internationalization of consultancy

12   International consultancies are developing online databases of projects to build on experience around four research/training centres in the US.

13   Large consultancies in Germany develop new products through international comparisons of experience and knowledge, integration of knowledge from different sources and links between consultancy products. Emphasis on organizational innovation, for example in relation to international experience. Also employ global networks to support client internationalization.

Global consultancy growth was especially associated with the introduction during the 1980s of new ICT. As well as providing technical advice, consultancies in the ICT sector have increasingly addressed the often fundamental management systems changes associated with ICT. In general, they thus promote methodologies for rationalizing, integrating and planning complex organizations, especially in relation to the development of new technologies and intensified global competition.

As well as promoting specific methods and techniques, major consultancies also offer qualities less often spelled out. Their value may lie partly in providing an independent assessment of strategies or developments already planned by client managers. Clients also wish to share responsibility and risk, especially over major technical or organizational changes. 'Corporate re-engineering' programmes, even when necessary, are seldom completely successful. Some even prove to be disastrous. Examples abound of problems associated with the introduction of new computer systems in both the private and the public sectors. Consultancies may bring essential knowledge and experience to such projects, but they can also take the blame for difficulties and failures, and may be required to bear some of the costs.

In spite of these hazards, in the 1990s global consultancies expanded to all sectors, including service sector clients and public agencies; and into new markets such as southern and eastern Europe, South-east Asia and Latin America. They are also in some cases adapting their approaches to support growing medium-sized firms. Meanwhile specialist consultancies continue to spin off, for example from the major computer manufacturing and software or logistics consultancy firms. In some markets these have broadened their expertise even to challenge the global consultancies, at least in regional markets such as Europe. Within the EU, the increasing emphasis on Europe-wide marketing and production provides many opportunities for the global consultancies (Moulaert and Martinelli 1992; Gadrey *et al.* 1992). EU development programmes, including those in southern and eastern Europe, have also increased demand for their work. In spite of the remaining barriers, European commercial integration has encouraged the development of branch networks by the global consultancies.

The leading management consultancies in Europe are global in scope and often nominally US-based, dominated by the top global companies (including Andersen, PricewaterhouseCoopers, KPMG, Ernst & Young, Deloitte, Cap Gemini, McKinsey). The US market accounts for over half of world demand, but Europe provides a further one-third of global consultancy income. Most operate Europe-wide staff recruitment and training procedures, computer and information systems, and networks of branch or subsidiary outlets. They often work with MNC clients whose requirements justify the high costs of international business. They are constrained by the labour-intensive nature of most consultancy services, the consequent need to control and motivate a range of specialist staff, and to sustain the quality of client–consultancy interaction. Such conditions require flexible organizational forms. The key organizational dilemma for consultancy is how to deliver a consistent quality of advice in close consultation with different clients, across inherently dispersed projects. Failures may have an adverse effect on company reputation. Even the largest firms have come to recognize that a degree of specialization is necessary, and that established clients offer the best basis for success.

Consultancies have also become key sources of technical know-how and management methods for nationally based companies, privatized utilities and public sector agencies. MNCs and other large clients also employ many smaller, specialist consultancies. Smaller clients, however, less commonly employ the larger consultancies because of their cost and predominant experience with large organizations. A nexus of global corporate and consultancy interaction thus increasingly dominates key sectors and core regions of the EU, with the availability of a growing volume and range of international consultancy expertise promoting its economic influence.

Overview: competition and the emerging environment of consultancy provision

The development of the consultancy sector, especially internationally, has undoubtedly enhanced the expertise available to European business and public agencies to support technical and managerial change. Although the value of international exchange cannot be measured accurately, it has probably grown rapidly. Even so, consultancy remains dominated by national, and even regionally based exchange. The evidence from both consultancies and clients nevertheless suggests that internationalization has exerted a disproportionate qualitative impact on nationally based patterns of consultancy provision. This has come about in three principal ways.

1   National and regional consultancy firms face increased competition. This may adversely affect some of these firms because of the greater resources and reputations commanded by the international (especially the global) consultancies. From the client perspective, however, these may offer higher-quality expertise, especially in relation to international technical and market conditions.

2   Within growing consultancy markets, imported expertise may stimulate demand for local consultancy activities. For example, while international IT and strategy consultancy is increasingly dominated by global consultancies, the changes induced create demands for specialized training, recruitment, professional or marketing skills. National or regional consultancies have thus adapted to the developing market by moving towards specialized niches that exploit their 'local knowledge'. This may sometimes be achieved through direct formal or semi-formal collaboration with foreign consultancy firms, including global consultancies.

3   International competition should encourage national or regionally based consultancy firms into competitive niche specializations, aligned to the needs of major clients. These may then themselves develop new, including international markets. This may occur through direct export, collaboration with foreign firms, or eventually through the establishment of overseas branch offices. These developments may be linked to work for major home-market clients, including multinationals. Their strategies or wider commercial contacts may draw local consultancy firms into foreign (or distant regional) markets.

The impacts of growing consultancy competition within each country and region therefore depend on how local consultancies respond to the opening-up of these markets. Specialization should enable them to exploit both local expertise and new international opportunities. The KIS markets, especially of the high-income economies of northern Europe, have evolved over the past twenty years. From a disparate, segmented and sectorally dependent group of support functions, a complex, integrated system of expertise exchange with clients has emerged, orientated primarily towards the needs of the corporate sector. Much day-to-day exchange still takes place at the regional level, reflecting the importance of close client–consultancy relations. The inherent nature of consultancy exchange also means that there are limits to the hegemony of large consultancies. Clients often prefer working with small to medium-sized consultancies. Nevertheless, the 'state of the art' established by the global consultancies, with their multinational clients, dominates the economic impact of the whole consultancy sector. This determines the competitive conditions and standards which, increasingly, smaller consultancies must satisfy. The degree of development of this system also clearly varies across Europe, both nationally and regionally. Before examining this in detail, it is necessary first to explore the specifically innovative implications of international consultancy developments in Chapter 3.

References

BDU (1990) *Annual Business Report*, Bonn: BDU (Bundesverband Deutscher Unternehmensberater).

—— (1991) *Der Markt für Unternehmensberatungsleistungen in Europa*, Bonn: BDU.

Castells, M. (1989) *The Informational City*, Oxford: Blackwell.

CBS (Centraal Bureau voor de Statistiek – Central Bureau of Statistics) (1996) *Employment Statistics*, Voorburg: CBS.

Daniels, P.W. (1991) *Services and Metropolitan Development: International Perspectives*, London: Routledge.

Daniels, P.W. and Moulaert, F. (eds) (1991) *The Changing Geography of Advanced Producer Services*, London: Belhaven Press.

Elfring, T. (1989) 'Evidence on the expansion of service employment in advanced economies', *Review of Income and Wealth* 35: 409–40.

European Commission (1990) *Business Services in the European Community: Situation and Role*, report to Director-General for Industry, July, III/89/2234/EN/Rev2, Brussels (internal report, available through EC libraries).

—— (1996) *Panorama of EU Industry, 1995–96*, Brussels: European Commission, D-G XIII-Eurostat.

Feketekuty, G. (1988) *International Trade in Services: An Overview and Blueprint for Negotiations*, Cambridge, MA: Ballinger.

Gadrey, J., Gallouj, F., Martinelli, F., Moulaert, F. and Tordoir, P. (1992) *Manager les conseil: strategies et relations des consultants et de leurs clients*, Paris: Ediscience Internationale.

Gaebe, W. (1995) 'The significance of advanced producer services in the New German Lander', *Progress in Planning* (April–June): 173–84.

Gaebe, W., Strambach, S., Wood, P.A. and Moulaert, F. (1993) *Employment in Business-Related Services: An Inter-Country Comparison of Germany, the United Kingdom and France*, report to D-G V, Brussels: European Commission.

Gallouj, F. (1996) 'Le commerce interregional des services aux entreprise: une revue de la litterature', *Revue d'Economie Regional et Urbaine* 3: 568–95.

Gibbs, M. and Hayashi, M. (1989) 'Sectoral issues and the multilateral framework for trade in services: an overview', in UNCTAD, *Trade in Services: Sectoral Issues*, New York: United Nations, 1–49.

Howells, J. (1988) *Economic, Technological and Locational Trends in European Services*, Aldershot: Gower.

Illeris, S. (1989) *Services and Regions in Europe*, Aldershot: Avebury.

Keeble, D. and Schwalbach, J. (1995) 'Management consultancy in Europe', Working Paper No. 1, ERSC Centre for Business Research, University of Cambridge.

Keeble, D., Bryson, J. and Wood, P.A. (1991) 'Small firms, business service growth and regional development in the United Kingdom', *Regional Studies* 25: 439–57.

Marshall, J.N. and Jaeger, C. (1990) 'Service activities and uneven spatial development in Britain and its European partners: deterministic fallacies and new options', *Environment and Planning A* 22: 1337–54.

Martinelli, F. (1991) 'A demand-oriented approach to understanding producer services', in Daniels, P.W. and Moulaert, F. (eds) *The Changing Geography of Advanced Producer Services*, London: Belhaven Press, 15–29.

Moulaert, F. and Gallouj, C. (1995) 'Advanced producer services in the French space economy: decentralisation at the highest level', *Progress in Planning* 43(2–3): 139–54.

Moulaert, F. and Martinelli, F. (1992) 'Le conseil en informatique: conseil en systèmes et systèmes de conseil', in Gadrey, J., Gallouj, F., Martinelli, F., Moulaert, F. and Tordoir, P. *Manager les conseil: strategies et relations des consultants et de leurs clients*, Paris: Ediscience Internationale, 79–103.

Moulaert, F. and Tödtling, F. (eds) (1995) 'The geography of advanced producer services in Europe', *Progress in Planning* 43 (April–June): 89–274.

Nicolaides, P. (1989) *Liberalizing Service Trade*, London: Royal Institute of International Affairs.

Ochel, W. and Wegner, M. (1987) *Service Economies in Europe: Opportunities for Growth*, London: Pinter/Westview.

OECD (1998) *Business Services in OECD Countries; Part II Country Monographs*, report produced by the Industry Committee, Directorate for Science, Technology and Industry, September 11, Paris: OECD.

O'Farrell, P.N. and Moffat, L. (1991) 'An interaction model of business service production and consumption', *British Journal of Management* 2: 205–21.

O'Farrell, P.N. and Wood, P.A. (1999) 'Formation of strategic alliances in business services: towards a new client-oriented conceptual framework', *Service Industries Journal* 19: 133–51.

O'Farrell, P.N., Hitchens, D.M. and Moffat, L.A.R. (1992) 'The competitiveness of business service firms: a matched comparison between Scotland and the SE of England', *Regional Studies* 26: 519–34.

O'Farrell, P.N., Wood, P.A. and Zheng, J. (1996) 'Internationalisation by business services: an inter-regional analysis', *Regional Studies* 30: 101–18.

Quinn, J.B. and Dorley, T.L. (1988) 'Key policy issues posed by services', *Technological Forecasting and Social Change* 34: 405–23.

Rubalcaba, L. (ed.) (1998) *Business Services in European Regions*, draft report to European Commission/Eurostat, December (available through EC libraries, Brussels).

Rubalcaba-Bermejo, L. (1999) *Business Services in European Industry: Growth, Employment and Competitiveness*, Brussels: European Commission, D-G III Industry.

Schamp, E.W. (1995) 'The geography of advanced producer services in a goods exporting economy: the case of West Germany', *Progress in Planning* (April–June): 155–72.

Scott, A.J. (1988) 'Flexible production systems and regional development: the rise of new industrial spaces in North America and Western Europe', *International Journal of Urban and Regional Research* 12: 172–85.

Vogler-Ludwig, K., Hofmann, H. and Varloou, P. (1993) 'Business services', *European Economy: Social Europe, Report and Studies* 3: 381–400.

UNCTAD (1989) *Trade in Services: Sectoral Issues*, New York: United Nations.

Wood, P.A. (1996a) 'Business services, the management of change and regional development in the UK: a corporate client perspective', *Transactions, Institute of British Geographers* NS 21: 649–65.

—— (1996b) 'An "expert labor" approach to business service change', *Papers, Regional Science Association* 75(3): 325–49.

3   How may consultancies be innovative?

Peter Wood

Specifying the innovative contribution of consultancy

What is the role of consultancies in promoting client change and in what circumstances might this be innovative? This chapter will review the qualities possessed by consultancies which are potentially innovative, based on the types of people they employ, how they are organized, and their technical skills, in recent years increasingly associated with implementing information and computer technology (ICT) systems. Consultancies codify and adapt knowledge to the specific needs of various clients. As we saw in the last chapter, some operate increasingly on an international and even global level. Consultancy is thus part of a system of expertise exchange which includes the international transfer of innovative ideas. As the influence of consultancies grows, this may have wider organizational and social impacts. For example, consultancy-supported innovation may differentially affect various types of enterprise (for example large v. small) and region (core v. periphery). This chapter concludes with an examination of the forces of polarization implicit in both consultancy supply and their influence on client behaviour.

First, it must be recognized that available evidence does not suggest that consultancies have become dominant forces in promoting change, still less innovation. It has already been argued that the outcomes of consultancy activity, like any service, depend on the response of clients, especially when radical change is involved. In any client–consultancy interaction it is seldom easy to unravel the part played by each actor, including inputs from various client staff and also from other organizations. Nevertheless, the manner in which consultancies work suggests that they can be significant catalysts of client change, for example influencing approaches to technical change and organizational adaptation. Only in-depth case studies can show the qualities required of such interaction, and any distinct contributions that consultancies make in comparison to client inputs.

Consultancy roles in supporting change

Consultancies may be agents of change and possibly of innovation by acting as:

1 *Facilitators of change*   This may involve no more than the routine outsourcing of specialist functions, such as legal, accountancy, patent or transport services. Sometimes, if such practices are extended, for example to include data processing or wage and salary payments, this may release resources for changes in core activities. Such an outcome depends on client management. Consultancy inputs may thus facilitate client innovation by offering complementary specialist support.

2 *Conveyors of change*   Consultancies may convey innovative ideas to clients from other firms, sectors or countries, although not developing these themselves. For example, they may prepare reports on alternative choices from which clients choose a course of action. Although the impact of their inputs would again depend on the client response, this might in turn be influenced by consultancy advice.

3 *Adapters of change*   Rather than acting simply as a 'messenger', consultancy increasingly requires adapting other experience to specific client needs. This implies specification of such needs by the client and an active analysis and selection of options by the consultancy. A consultancy process proper is thus in train. Any innovative impact once more depends on wider client strategy, but consultancy guidance may be influential. This adaptive expertise, common in management and especially in ICT consultancy, is probably the most general basis for an innovative contribution by consultancy.

4 *Initiators of change*   This may most clearly occur when specialist technical (including ICT) consultancies are directly engaged in new product or process development with clients. Clients, of course, must be receptive to such innovation. More generally, however, the modern universality and complexity of innovation (see Chapter 1) suggests that at least some type of consultancy support may be increasingly necessary for innovation to occur. Certainly, consultancies' ability to apply up-to-date management, organizational, information-processing, logistical or marketing experience to many different circumstances is nowadays often critical for much non-technical innovation. The success of much technical innovation also depends on managerial, logistical and marketing adaptation (based on Bessant and Rush 1995).

The innovative impacts of consultancy depend basically on the intensity of the client–consultancy interaction, as we have emphasized, on the *process* of consultancy taking place. 'Full' consultancy, in which its inputs are most likely to exert an innovative effect, must be fully integrated into the client's production strategy. In other cases, clients may simply support their own innovation strategies with management, technical, training or marketing advice or inputs from consultancies.

There may also be a difference between the consultancy role in relation to a client's 'core technology' and its support for process or organizational change; in effect, between client product and process innovation. Clients should be able to exert primacy over change in their core functions, and consultancies are likely to be less influential. If they are involved, however, their significance may be correspondingly greater. On the other hand, consultancy is now pervasive in promoting technical and non-technical process innovation. In practice, as we argued in Chapter 1, the distinction between product and process innovation is nowadays difficult to establish, not least because of the universal and overlapping influence on both of computer and information process technology. This is also generally the case for services, whose most innovative products are often based on the development of new processes of delivery. Even in continuously facilitating, conveying and adapting process innovations, therefore, consultancy may create the conditions that support client product innovation.

Processes of consultancy–client interaction

Strategic change

Although, of course, the quality of business and technical expertise offered by consultancies varies, in general it should be relatively high, since the technical, managerial, IT and computing skills they market are their most important competitive asset. The consultancies' contribution is nevertheless inherently difficult to isolate from the wider networks of inter- and intra-firm exchange within which they operate, including client personnel themselves. Perhaps because of these difficulties, innovation involving consultancies has been little researched in micro-level analysis (but see Marshall 1982; MacPherson 1988, 1991; Strambach 1994). Some guidance is available, however, from two UK studies of the use of consultancies in implementing strategic organizational change (sometimes including technical innovation), and in supporting technology transfer.

Some detail of the first study, based on over 120 large client companies, is presented in Chapter 7 (see Tables 7.6a–d). It showed that effective consultancy use in support of strategic change depends on the possession of broadly comparable expertise and experience by client and consultancy staff. When there are specific absences of in-house skills as is commonly the case, for example, in IT or market research, more specialist consultancies may be employed. Also valued, however, is the ability of a consultancy to bring an impartial perspective to management decisions, and its provision of extra capacity during periods of intensive strategic change. Consultancy outputs are essentially joint products with client staff, so that they must work closely and cooperatively together. The final impact of any consultancy recommendations depends on their utility to the client.

A client perspective also shows that the major consultancies are less dominant in change projects that much supply-orientated evidence might suggest. Only 60 per cent of the projects employed major consultancies, and most of these also involved smaller consultancies. For many projects, clients engaged combinations of smaller consultancies under their direct control. For firms undergoing major restructuring, the most important role of consultancies was often to support the emergence of a new set of business norms, orientated towards the more effective exploitation of innovation. This might involve moving from a 'production-based' to a more 'market-orientated' mode of working. In many other cases, however, corporations nowadays employ specialist consultancies almost routinely, as more or less permanent extensions of their own management capabilities.

Interdependent relationships between client and consultancy staff underlie these exchanges, and varying levels of client staff acceptance (Wood 1996). The balance depends on the relative levels of experience or technical skill between staff, and the urgency and degree of required change. The client organization's authority structure is also important, determining who initiates and controls change in relation to who is affected by it. The familiarity of client personnel with their own organization presents both barriers to, and the basis for, change. Senior managers frequently draw on the authority of consultancies to overcome internal resistance. Even where profound change is required, the ability of in-house staff to implement it, and still to deploy their own expertise effectively, is the principal determinant of the outcome.

Technology transfer

The importance of complementing and helping to exploit the skills of client staff applies *a fortiori* to the role of consultancies in technical innovation. Compared, for example, with broader organizational restructuring or market development strategies, this impinges most directly on client core skills, and often on established corporate norms. Innovation requires the development of client staff expertise, embedded in relations with suppliers, customers and other staff colleagues. It often also faces particularly strong internal resistance to change. The involvement of consultancies in technical innovation, as of any outside agency, may thus face special difficulties.

Technology transfer may generally be defined as the transfer and application of new process equipment, prototype products or codified knowledge. It is usually extended and complex, involving many actors at each stage (Bessant and Rush 1995; Dodgson and Bessant 1996). Of particular significance for its success are intermediary 'bridging institutions', including possible 'systems integrators', capable of coordinating diverse suppliers and agencies and adapting them to each company's needs. As well as the staff of the innovating firm, these institutions may include technology brokers, university experts, and national or regional technology or innovation agencies.

Consultancies also increasingly intervene in specific technical areas (see Appendix at end of chapter on page 85). Some may identify and help to select appropriate technologies to serve client needs, or seek out key sources of knowledge. Others may assist in financial appraisal and the search for funding, or help to devise a wider business strategy associated with innovation. Some may support the selection and training of staff. In many cases, consultancies may be involved in implementing change by supporting project management and organizational development. These various roles are often combined. In general, consultancies serve to disseminate and share experience gained with other clientele and may also advise in the selection of other consultants. They may act as sources of specialist knowledge, as diagnosers of client needs, or as advisers on strategic frameworks for change.

Two key characteristics of the innovation environment dominate the learning process in technology transfer. The first is uneven access to information, especially for smaller enterprises, and the second is the need to overcome this through continuous user–producer and other operational interactions. Opinion leaders and bandwagon effects may also direct the trajectory of technical and management change, for better or worse.

The importance of 'process consulting' is emphasized by Bessant and Rush (1995) as the basis for the adaptive or initiating role of consultancies. This requires long-term partnerships between clients and consultants, in which they improve access to information by encouraging and facilitating learning and change (Schein 1969). Consultants thus become familiar with clients' operational practices, and can assist in adapting wider technical opportunities to specific needs and capabilities. Such procedures generally pose difficulties of cost and resource commitment for the smaller firms, who may need the most specialist support. The UK evidence cited above also suggests that, even among large organizations, such close client–consultancy relations have only recently become a norm, at least in managing strategic change, and are still by no means universally accepted.

The appropriate choice of consultancies, individually or in combination, is also important, in relation to their strengths and limitations in supporting innovation. Experience from the introduction of advanced manufacturing technology (AMT) indicates that major management consultancies, while offering a large resource base, a good reputation and wide experience and contacts, are costly, and often lack a strong technical base (Bessant and Rush 1995: 107). The main engineering and technology consultancies possess the technical resource base and wide experience, but are also costly and may lack wider skills of organizational adaptation. Software, systems or hardware suppliers and human resources development consultancies have specialist reputations and technical expertise, but may lack resources, as well as manufacturing or other sector-specific experience. Industrial process contractors may be strong in practical experience and project management, but again are more limited in their resource base and lack adaptability between sectors. Finally, specialist research institutions are often thought to lack project management skills.

Technical innovation programmes are particularly unforgiving of perceived failure, while broader strategic change programmes may be accepted more equably, even if they are only partially successful. This is because of the 'core' qualities of many technical functions and the critical impacts of technical change. Projects extend over a long period, during which responsibilities are often diffuse. Consultancies may be involved at various stages including choosing and acquiring information and expertise; assessing risks and gauging market response; costing and acquiring finance; project management; and adaptation and training of personnel. The control and direction of change may thus be difficult and, if client managers do not maintain control, consultancies are easy to blame. Successful innovation requires both a combination of many actors and strong direction. It is within such networks of interaction that any consultancy contribution to innovation must be judged.

Four cases studies

The significance of the growing use of consultancies for management and technical innovation may be illustrated by reference to the four case studies in the Appendix, which explore the experience of major UK corporations, in different sectors and at different stages of strategic change in the early 1990s. Each corporation faced problems of innovation in different commercial situations. The cases include (1) a recently privatized utility, undergoing further competitive change; (2) a manufacturing company in need of radical transformation ('restructuring'); (3) a manufacturing company having already undergone such a change ('restructured'); and (4) a diversified service conglomerate adjusting through the progressive development of a market-orientated management culture. In cases (1) and (2), consultancies helped to initiate a virtually complete corporate restructuring; in the other two cases, (3) and (4), clients were undergoing a more continuous process of strategic change. New, semi-permanent relationships between these clients and consultancies had also been established. The accounts illustrate how integral consultancies have become to corporate change, and the complex processes of client–consultancy interaction that underlie it.

Sources of consultancy innovativeness

The case studies illustrate how consultancy support for client change, to improve a client's technological and market position, requires management, organizational and human resources, as well as technological innovation. More generally, in any client–consultancy interaction, the distinctive potential sources of consultancy innovativeness include the following.

1   Their ability to recruit specialist ICT, sector-specific and management personnel and employ them across a range of client applications. Their adaptive learning processes are augmented by a variety of project experiences.

2   Their particularly significant role in adapting ICT systems, including computer-based management systems, to many individual client circumstances. Global and specialist consultancies are also developing advanced systems to support their own activities, enabling them to link international experience across projects.

3   Consultancies are organized in innovative, flexible ways, cutting across the rigidities of formal organizations, employing project-based teams to sustain close working links with clients. They also devote resources to developing distinctive change methodologies and establishing specialist sources of information and intelligence. They largely depend for success on trust and reputation.

4   An important contribution of consultancies that enables organizations to respond to the learning demands of globalization is the codification of knowledge. This is a much wider process than that associated simply with new information technology. Consultancies primarily review the diversity of technical, managerial and marketing knowledge, through research and experience, and adapt and codify it for other clients. It is axiomatic that the innovative methods they promote now, if successful, will become standard practice in the future. This codification role is thus itself broadly innovative. The same principle applies to the application of new information and computer technologies, engineering and other technical consultancies, and to new management processes.

5   As already emphasized in Chapter 2, one of the most significantly innovative features of modern consultancy is the increasingly international level of its experience and intelligence gathering. International (including global) consultancy is thus becoming a distinctive source of new ideas and expertise for many clients, especially those operating at national or regional levels. These developments are exerting significant effects on the learning environment available to private and public agencies on all levels in a globalizing economy.

Consultancy internationalization and innovation

The expanding influence of global consultancies has transformed the potential support environment of national and regionally based firms wishing to grow, or at least sustain, competitiveness. In Chapter 2, we also saw that the majority of consultancies appear still to operate at national and regional/local levels. In innovative terms, however, the most influential of these are branches of global or international consultancies or their collaborators. National/regional consultancies must compete with or complement their influence. Within such a varied pattern of demand and supply, how can the system of consultancy provision be summarized, and where are the key sources of innovation located within it?

Figure 3.1 schematically summarizes the dominant levels of business demand for consultancy expertise, from global to national and regional clients, and the

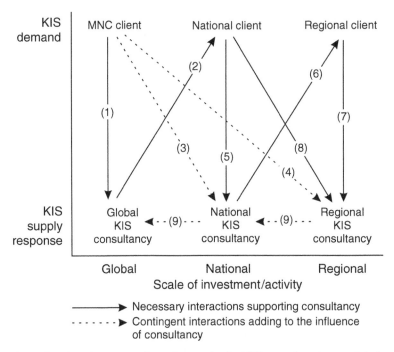

Figure 3.1 Dominant transfers of expertise by KIS consultancies at global, national and regional levels

patterns of consultancy supply response. It distinguishes 'necessary' interactions between demand and supply, upon which the development of consultancies at each level depends, from 'contingent' interactions, which may add to consultancies' influence on clients in specific circumstances. The direction of net dependence is suggested by arrows pointing from sources of innovation. This is not meant to imply that dependence is only one way. It has been emphasized that consultancy use must be validated through interaction with clients, and in relation to client innovative capacity. Sometimes, for example when global consultancies work with smaller companies, consultancies may lead clients' capacity to innovate. The nature of demand–supply interactions also varies widely between different levels of activity, and between different types of organizational, technical or marketing consultancy, as we shall see below. Different regional and national economic environments also exert their influence.

The following assumptions underlie the relationships described in Figure 3.1.

1 Consultancy is driven primarily by clients' need to supplement their own innovative methods. Consultancy exchange is a significant component of

the patterns of expertise exchange which remain dominated by corporate demand and supply.

2 Consultancies promote change through their technical and management expertise but also act as conduits, transmitting methods and experience developed with innovative clients to others.

3 Clients operating on larger (global) levels may generally test consultancy methods more rigorously than those operating on smaller (national/ regional) levels.

4 For smaller consultancy firms, successful specialist skills and/or local knowledge may originate from working with specialist national and regional clients. These may form the basis of wider markets for small consultancies.

The aggregate relations suggested by the diagram thus represent the general pattern of inter-company learning which has been facilitated in recent years by the expanding activities of global, national and regionally based consultancies. As summarized at the end of Chapter 2, this growing system of expertise exchange is a significant outcome of consultancy developments for the wider environment of innovation. It increasingly supports the transmission of expertise from global to national and local levels, through interactions with a wide range of clients.

The different forms of interaction indicate markedly asymmetrical relations of change and innovation. Consultancies may generally promote change, but are more likely to promote innovation when transmitting experience from larger to smaller clients, or from specialist to less specialist clients (such as within a particular industrial or regional innovation system). The numbered relationships in Figure 3.1 thus imply that:

1 Global consultancies respond primarily to the requirements of MNC clients and agencies. MNC requirements, which may include ICT developments, corporate restructuring, management systems or international market strategies, mould the primary contribution of global consultancies to change and innovation. Global consultancies nevertheless also assist MNC clients to change and innovate by disseminating current best management and technical practice, and supporting their expansion into new production areas and markets.

2 Global consultancies increasingly act as conduits of innovative ideas and methodologies between the global and national levels. They provide nationally based client firms and agencies with expertise, enabling them to respond to intensified international competition in the home market, or to expand into foreign markets or production. At the same time they develop familiarity with national market conditions which may be of value in their dealings with MNC clients.

3 Successful small to medium-sized, nationally based consultancies may develop internationalization strategies by serving MNC clients seeking

specialist expertise or familiarity with home-country conditions. This may also enhance the quality of national consultancy expertise in relation to national and regional clients (see (5) and (6) below).

4   Successful regionally based consultancies may also work for MNC clients operating in their regions on a basis similar to (3) above, although their growth more often depends on serving national clients (see (8) below).

5   Nationally based consultancies, serving national clients, both private and government, provide the predominant volume of consultancy exchanges across a wide variety of expertise. Such clients thus mould the contribution by national consultancies to change or innovation. In spite of growing competition from global consultancy (see (2) above), distinctive national characteristics, reflecting cultural conventions and economic conditions, are strongly reflected in such consultancy practice and expertise.

6   Within national systems of consultancy–client interaction, nationally based consultancies may offer regionally based clients key consultancy support. These may be active agents of regional innovation, as conduits of ideas derived from experience with national and international clients (see (3) and (5) above). Where regionally based clients possess specialist expertise, however (see (7) below), these exchanges may enhance the capabilities of national consultancies in dealing with other clients.

7   Regionally based consultancies originate largely to serve regional clients, and adapt to these needs on the basis of local exchange and innovativeness. In some regional systems, specializing for example in high-technology production, financial services, primary production or tourism, the amount of innovativeness may be considerable, forming the basis for wider consultancy markets.

8   Successful regionally based consultancies most often grow by serving national clients, as well as international clients ((4) above), once more on the basis of specialist skills or knowledge of local conditions.

9   Contingent links may exist between international, national and regionally based consultancies, either directly through subcontracting or networking relationships, or indirectly as a result of client tendering policy. Many clients generally combine the expertise of various types of consultancy in relation to their in-house expertise capacities, the types of skills required and the costs of acquiring them. When combined expertise is employed, international/national consultancy expertise is likely to be more innovative in relation to national and regional clients.

Multinational companies command many sources of innovative expertise, but national and regional systems also possess technical, organizational and cultural knowledge that is essential for economic success. This most obviously relates to personnel capabilities and market potential, but also include internationally competitive technical and production capacities. The exchange of expertise between these levels is a crucial function of the corporate and wider business system, and the rise of consultancy has provided new and more open channels for the diffusion of new practices.

Within this framework, many variations in the balance of innovativeness between clients and different types of consultancy might be anticipated. *Organizational* change has been the major area of influence, from global consultancies serving MNCs and national clients. The relationships represented in Figure 3.1 probably most closely reflect these activities, including a significant local or specialist role for small to medium-sized, nationally based consultancies, or even smaller consultancies operating in a metropolitan or specialist regional environment.

For *technical* innovation, ICT expertise is universally required. Consultancies have played a leading role in this type of innovation, except perhaps within the computer and IT industries themselves. Other forms of technical consultancy (manufacturing systems, logistics, construction engineering, R&D) are more sector based, often augmenting the core expertise of clients. Consultancy influence on technical innovation may therefore be secondary, especially for MNC clients. Specialist technical consultancies may nevertheless develop strong national/local supply specialisms, based on knowledge of oil exploration, for example; or the automobile industry; transport systems development; global financial systems; the defence–aerospace nexus; industrial design skills; agriculture; or even retailing or tourist services. Regional industrial complexes based on high-technology or design and sales skills also support the development of high-quality technical consultancies. By extending their activities more widely, these may also have an international impact on innovation. Many important engineering/technical consultancies have developed from such regional demand–supply relations and expanded through specialization and trade.

Market intelligence is widely regarded as crucial for successful innovation. It is also recognized, however, that in Europe corporate links between strategic market analysis and innovative functions are generally poor. The often low strategic status of in-house marketing expertise encourages this. One reason may be the fact that market research and marketing functions are traditionally externalized to consultancies and agencies, even by MNCs. Clients usually retain an in-house capacity simply to commission and interpret the work of consultancies in terms of short- to medium-term needs. Most such consultancies are nationally based, serving both MNC and national clients with an integrated range of expertise, from survey design to interpretation and strategy recommendations. Many have responded to competition, however, by specializing in particular markets, or research or design techniques. In recent years, the most innovative sources of strategic marketing ideas have come from the interaction of global consumer sector companies, in the clothing, food, footwear, electronics, and media and leisure industries, with international markets and advertising agencies. As the weight of innovation has shifted to the global level, the scope of national/regional marketing consultancies and agencies has increasingly become limited to more routine and medium- to short-term monitoring functions.

Consultancies and the learning economy

We have seen that consultancies exert influence on business change and innovation by conveying and adapting wider technical and managerial experience to individual client needs. They offer new sources of skilled personnel, promote the implementation of ICT systems, support new forms of business organization and generally act as codifiers of knowledge relevant to specific client situations. Internationalization has created a further dimension to their influence. More generally, in addition to these specific qualities, a 'knowledge-intensive service' perspective also promotes the wider processes of successful innovation, as summarized in Chapter 1. These emphasize the following important strategic elements.

1 Sources of expertise and interpretative skills as a driving force for innovation, rather than simply the processing of disembodied 'information'.

2 The importance of market awareness and delivery as keys to the success of new ventures. In other words, their 'service' qualities, rather than simply the production of new technologies.

3 The need for new organizational arrangements to support innovativeness, rather than relying on established hierarchies or market practices. Consultancies are both a significant example of new practice, and sources of expertise in such arrangements. These include a fluidity of organizational structures to support learning and change.

4 These arrangements also include novel, task-orientated approaches to the development of innovations, rather than organization-dependent, prodct-orientated approaches.

5 The active development and use of key human resources, not assuming that their development should follow prior capital commitment. This should include types of training, the project-based deployment of staff and the types of teamwork used and promoted by consultancies.

6 The management of inter-organizational relations in support of innovation, generally exemplified by successful consultancy, rather than a preoccupation with intra-organizational processes.

7 The importance of private sector agencies in transmitting commercially relevant innovation expertise, compared with the reliance on public intermediate institutions for much innovation policy formulation.

Consultancies may rarely be dominant forces in promoting innovation, but their growing influence and example have nevertheless made them significant catalysts of wider change. Perhaps more than any other factor, this explains their meteoric modern growth.

Conclusion: the frame of consultancy influence

It has been argued here that there is strong circumstantial evidence for the growing influence of consultancy on innovation. Consultancies augment the expertise base of clients at all levels, especially through the links they provide to international standards of technical and management practice. Successful consultancy, however, requires sustained interaction with the client and is highly contingent on these specific relationships.

The influence of consultancies on client innovation is thus unlikely to be measurable in isolation. The analogy of consultancies as catalysts, stimulating change across a wide variety of clients, suggests that their influence can be measured only by their effects. Consultancies seldom feature prominently in formal surveys of sources of innovativeness. Many studies simply do not ask about them, although this is changing (Cooke *et al.* 2000) and the nature of the innovations examined is often narrowly defined and technologically dominated.

One requirement in directly assessing consultancy influence is to draw the boundaries of innovation more widely to incorporate both organizational innovation and the organizational context of technical innovation, if these can be separated. This seems to be in line with current thinking in innovation studies where the significance of organizational innovation as a condition for firm growth and competitiveness now seems to be becoming generally accepted (IMIT 1996).

Consultancy appears to reinforce the environment of change and innovation especially among the more dynamic elements of the corporate sector. Many experienced clients now employ consultancies on an almost routine basis, engaging them selectively, with close liaison and control. There are also common situations of important one-off projects in which major consultancies support clients undergoing basic strategic change, with smaller specialists undertaking particular technical aspects of the work. Examples of this can be found in the UK in the early 1990s in the privatization and post-privatization of the utilities. But it was also important for many manufacturing firms undergoing radical restructuring, and in post-merger restructuring across all sectors. These same trends are now probably an important component of southern European developments.

The assumed influence of business consultancy clearly demonstrates the wider problems of evaluating various influences on innovation as its organizational and social systems contexts are increasingly acknowledged. While consultancy is extensive and evidently growing, corporate clients have also developed greater sophistication in the use and control of the consultancies' activities. Business consultancy, in other words, is being absorbed into the fabric of modern management, so that it becomes almost invisible, as yet another element contributing to the complexity and dynamism of modern economic change. Nevertheless, this chapter has suggested that a systematic and internationally orientated view of these developments demonstrates both the range of consultancy impacts so far, and the potential for their further impacts on the commercial expertise base of both developed and developing areas of the EU.

References

Bessant, J. and Rush, H. (1995) 'Building bridges for innovation: the role of consultants in technology transfer', *Research Policy* 24: 97–114.

Cooke, P., Boekholt, P. and Tödtling, F. (2000) *The Governance of Innovation in Europe*, London: Pinter.

Dodgson, M. and Bessant, J. (1996) *Effective Innovation Policy: A New Approach*, London: International Thompson Business Press.

IMIT (Institute for Management Innovation and Technology) (1996) *International Transfer of Organisational Innovation*, European Innovation Monitoring System (EIMS) Publication No. 45, Brussels: European Commission, D-G XIII.

MacPherson, A. (1988) 'New product development among small Toronto manufacturers: empirical evidence on the role of technical service linkages', *Economic Geography* 64: 62–75.

—— (1991) 'Interfirm information linkages in an economically disadvantaged region: and empirical perspective from metropolitan Buffalo', *Environment and Planning A* 23: 59–66.

Marshall, J.N. (1982) 'Linkages between manufacturing industry and business services', *Environment and Planning A* 14: 1523–40.

Schein, E. (1969) *Process Consultation: Its Role in Organisational Development*, Reading, NJ: Addison Wesley.

Strambach, S. (1994) 'Knowledge-intensive business services in the Rhine-Neckar Area', *Tijdschrift voor Economische en Sociaale Geografie* 84(4): 354–63.

Wood, P.A. (1996) 'Business services, the management of change and regional development in the UK: a corporate client perspective', *Transactions, Institute of British Geographers* NS 21: 649–65.

APPENDIX

FOUR CASE STUDIES OF CONSULTANCY INVOLVEMENT IN UK CORPORATE CHANGE
(based on a survey in Wood 1996)

1 A privatizing utility

A privatized regional utility in the UK needed to adapt its production and delivery systems to increasing competition. Its transformation in the 1980s from a state-owned monopoly into a private company required major changes in its strategic management culture, especially in relation to financial management and shareholder accountability. Because the company lacked such commercial expertise, and also needed to convince government that privatization was being undertaken effectively, these changes required intensive use of major management consultancy firms. By 1994, its markets, especially among large industrial and commercial customers, were also being threatened by deregulation, introducing intensified competition. Urgent action was needed to improve both cost efficiency, including rationalization of the number of sites from which the company's six divisions were managed, and the quality of customer service.

A series of strategic change projects was initiated, several of which employed outside consultancy expertise. A senior management review, with consultancy support, concluded that costs could be significantly reduced by administrative collaboration between the six divisional units. The consultancy was then commissioned to lead a staff team to design and test new arrangements. Client staff provided information, ideas and negotiating links to local managers, who were responsible for implementing the changes. Fifty posts were dispensed with, creating an annual saving of over £1 million, at a cost including the consultancy fees of £800,000. This accorded closely with the project brief, agreed in advance with the consultancy.

Another project concluded with the company concentrating its headquarters on one site. In this case, various alternative arrangements were reviewed by the human resources director. He was supported by a consultancy, which provided outside information and experience, a costing methodology for the alternatives, and a final report for the company board. Once the decision was taken, in-house managers implemented the change without consultancy support, although new staff training was sought from outside, in line with a brief developed during the review phase.

In spite of extensive use of consultancies, this company, now operating in an increasingly competitive market, retained management control over change. For some other significant projects, such as the introduction of a computer-based information system, in-house staff undertook the work without strategic consultancy support, although equipment/software suppliers were used for technical support. Consultancies thus selectively added particular expertise to the company's review and planning operations, as well as some business acumen not available in the formerly state-owned business. The experience of privatization had shown how some consultancies offer poor value. It was also recognized, however, that project failure could equally result from poor internal management. The speed of commercial change was the main driving force behind consultancy use, but the company also depended heavily on in-house management skills and drew on networking with other utility companies undergoing similar changes.

2   Restructuring manufacturing

Many manufacturing corporations have faced radical change since the 1980s, including extinction or takeover if they do not adapt. Corporate restructuring has gripped British business, requiring scrutiny of overhead costs, devolution of management responsibilities, flexibility in employment practices and sensitivity to customer needs. In many cases, a significant fall in employment has also resulted. The direction of corporate change is widely similar, even though the urgency and degree of change may vary. In a relatively typical case, a medium-sized chemicals company faced surplus capacity, cyclical demand, intense competition and growing customer quality expectations after being cast adrift by a rationalizing UK multinational. To survive, it required funda-mental restructuring of its operations, transforming the inherited 'big-company' culture to one based on low overheads and flexibility.

A major consultancy was employed to 'benchmark' the company's perfor-mance in relation to others, and to review its current operations. A 15 per cent cut in costs and 40 per cent redundancies were accepted as necessary over two or three years. A 'business process re-engineering' programme was devised, in collaboration with key in-house managers, to create a more responsive and competitive organization. Change was largely undertaken by company staff, building on their established understanding of the business, with the consul-tancy designing, supporting, monitoring and disciplining the process. A highly systematized procedure of task specification and implementation was laid out over a year, including management retraining and team development. Overhead functions were reduced and semi-autonomous product divisions created. Key to this was an internal project manager, who acted as a guide to the culture and politics of the organization, at the same time acquiring consul-tancy skills which helped to convince company management of the validity of the consultancy's diagnosis. A second component was training team leaders to define and implement particular tasks, involving other managers in the process, with consultants as 'counsellors'. The project was judged a qualified success, with a high proportion, but not all, of the target savings achieved. The longer-term implications of the organizational changes were viewed, more subjectively, as promising.

3 Restructured manufacturing

Where companies have achieved a high degree of rationalization, this itself creates dependence on outside consultancies. This international engineering group had reduced the number of its employees in the UK from 100,000 to 30,000 since the late 1970s, also reducing twenty-seven management subgroups to three operating sectors. The number of executives was halved, and each now had less administrative support. In a further change project, headquarters administrative staff were reduced by one-third in three years, cutting overheads by £10 million. A consultancy was used to support the deci-sion to change, and initiate a new training regime. For the management, it also legitimated an unpopular decision. The company then needed to expand into a third core business, having divested many peripheral activities in the 1980s. A major international consultancy was engaged to scrutinize the strategy and to identify suitable target activities. Although the takeover strategy proved difficult to implement, the company subsequently became the major shareholder in another UK business, providing it with its third core.

Thus, the use of consultancies had become routine, even at the most strategic level, and they were valued for their independence of view in relation to in-house analysis and decision making; the back-up they provided beyond the capacities of increasingly busy managers; and for their capacity in relation to the uneven demand for specialist management expertise. A culture is nevertheless required which incorporates careful preparation and targeting of consultancy activities, often favouring those already familiar with the client's operations. In effect, they may become semi-permanent, but dispensable extensions of client management.

4   Cultural change in a service conglomerate

In this case, events were directed towards the comprehensive corporate transformation of a diversified but loosely controlled service multinational, based on the systematic employment of consultancies. The company had a significant share in several markets, including freight forwarding, marine services and distribution. Its inherited organization was stagnant, preventing it from responding effectively to market and technological changes. The in-house planning director, with experience of service organizations and strategy consultancy, encouraged dialogue about change, choosing a succession of consultancies for their specific inputs.

The most striking characteristic of this process was the interaction of various forms of consultancy expertise with an evolving appreciation of needs by the client's management. First, a consultancy based at a major business school, already familiar with the company, reviewed the internal diagnosis of what was needed. This focused on reducing corporate overheads by devolving more responsibility to local offices. The consultancy opinion added credibility to the proposed process, as well as identifying specific needs. Another consultancy was chosen from several bidders to achieve specific targets for central cost reduction. Working with a team of managers from both corporate and business levels, this focused on the financial and personnel departments, acting as 'referee and facilitator' and achieving significant staff savings. It also became clear that the reduction of overheads was a necessary but not sufficient condition for improving the company's competitive position. As a dispersed service company, with few economies of scale, growth depended crucially on adding value to its products, and encouraging customers to pay a premium for them.

A highly selective search was undertaken for a consultancy that could advise on service marketing, and involve employees in changing the structure of relationships with customers. A new organizational framework was promoted which would enable local workers much greater discretion and initiative. A pilot project was undertaken, drawing on managers from various levels in the company and on a further range of consultancies that undertook client surveys and interviews, reviewing competition, and analysing internal processes and activities. A programme was then introduced to encourage local staff self-management, multiple task allocation within teams, and a flexible, project-based response to client orders.

This company was under no immediate threat of failure, but was able progressively to identify the need for innovative product change, based on quality of service. The tactics of change required a progressive approach and persuasion, with many consultancies employed on short contracts to provide specific inputs and insights. Their involvement was both a learning and a political process for the client management.

Part II

National and regional variations in knowledge-intensive service development

4 France

Knowledge-intensive services and territorial innovative dynamics

Hélène Farcy, Frank Moulaert and Camal Gallouj

Introduction

Knowledge, whether 'tacit' or 'codified', is one of the most critical assets in any innovative strategy. When presented through specialist markets, it is typically the product of business service providers and, in particular, of different types of consultancy firm. In reality, knowledge is also embodied in manufacturing products, especially those of high-technology or craft manufacturers, as well as in services offered by the service 'professions', in information packages provided by data-processing or information-networking firms. This chapter will focus on the knowledge-intensive services (KIS) provided by specialist business service firms, primarily management, organizational, technical engineering, IT and human resources consultancies. We will examine their role in the renewal of national and regional economies in France by evaluating their impact on the wider economic structure, and their influence as 'carriers of innovation' (Moulaert *et al.* 1990).

The evidence is presented in four sections. The first section sketches French business service employment trends, their patterns of demand, and the organizational structure of supply. The second section examines the same themes from a spatial viewpoint, asking three questions: What are the dominant regional and metropolitan trends and patterns of specialization? What is the relationship between financial concentration and spatial networking of service firms? And what are the main locational factors for KIS firms?

The third section examines regional and industrial policy for business service development. A distinction is made between policies that may directly affect KIS demand and supply, and those fostering a favourable business climate for KIS in general. In the final section, case studies of consultancy activities in two regions of France are employed to examine the innovative role of KIS arising from their interaction with clients. French experience also illustrates the influence of non-commercial service intermediation on innovation within the structure of inter-organizational networking. An important feature of French experience is the role given to local authorities and chambers of commerce.

National KIS developments

The main factors of KIS development

At the national level, three factors have played a significant role in the modern growth of the French KIS economy: a 'Cartesian' business culture; the impacts of sectoral restructuring; and the innovation policy of the French state.

First, there is the 'cybernetic' tradition of the French business world. Compared to other national economic cultures, the French business world favours the analysis and formalization of economic and business organization and planning (Moulaert *et al.* 1990). This has supported the spectacular rise of organizational and strategy consultancy, often hand in hand with information and communication systems development.

Second, the restructuring of traditional manufacturing sectors and the rise of new high-technology and service activities have transformed the French space economy from '*Paris et le désert français*' (Gravier 1949) to '*la métropole-réseau et le désert français*' (Veltz 1990). Although a number of regional metropolises have grown to complement Paris-dominated economic networks, polarization in relation to the rest of France remains very strong.

KIS have played a significant role in this transformation, while also displaying their own regional specializations (Moulaert and Gallouj 1995). For example, in Toulouse, high-technology services have grown spectacularly, driven by aircraft and space manufacturing. In Lyon, the control structures of rapidly growing industry require management consultancy and IT services of various kinds. In Lille-Roubaix-Tourcoing, the principal developments in advanced producer services are linked to the organization and computerization of trade, especially the mail-order business, printing and banking.

Third, the public sector has played a significant role in the 'informatization' and, more broadly, the modernization of French society and economy. Since the 1980s, public spending, educational and research policies have significantly encouraged the use and development of new technologies and new KIS, for example through the government *Rapport annuel sur l'informatisation de la France*. Moreover, public policy has reinforced the 'cybernetization' of the French business world and regional specialization in KIS development.

The development of KIS supply since 1970

The most reliable basis for analysing the evolution of the French KIS market is employment data. However, data sources have not been consistently compiled over a long period. Thus different sources, adopting different definitions and classifications of employment, must be combined. A significant change during the 1990s, for example, makes it difficult to bring the data in this chapter up to date.

The growth of service employment in France over the last fifty years has been concentrated in the health and social assistance sectors, consultancy and transfer of knowledge activities, and education (Gadrey 1996). From 1970 to 1994, the share of employment in trade, transport and communication remained practically unchanged, while that in health and cultural services, banking, insurance and *services aux entreprises* (producer services) grew significantly. Data in Table 4.1 for the period 1975–91, from ASSEDIC (Association pour l'Emploi dans l'Industrie et la Commerce, responsible for the national insurance scheme), show producer service trends, confirming that paid employment more than doubled during the period (the classification used is NAP: Nomenclature des Activités et des Produits). Among particular activities, only surveyors (NAP 7706) and 'various professional services' (NAP 7712) showed an absolute, although limited, decline in employment. Among the relative losers were technical engineering consultancy (NAP 7701), legal advice (NAP 7708), activities auxiliary to finance and insurance (NAP 78) and real estate services (NAP 79).

Very strong growth was experienced in information technology and organization consultancy (NAP 7703), which expanded more than tenfold, temporary employment agencies (NAP 7713) and 'other business services' (NAP 7714). The last two categories possess little internal coherence, and NAP 7713 includes workers actually employed in other sectors of the economy. More modest growth was experienced in information and documentation consultancy (NAP 7707), data processing (NAP 7704) and in services related to advertising and publicity (NAP 7710/7711). Taken as a whole, the business services rose from 5.08 per cent of total private salaried employment in 1975, to 7 per cent in 1981 and 10.57 per cent by 1991.

In 1993, there was a change in the French business classification, from NAP to NAF (Nomenclature des Activités de France), to align it with the European Union standard, NACE (Nomenclature des Activités Communauté Européen). Knowledge-intensive services are now regrouped under *conseil et assistance* (business consultancy and support). Table 4.2 shows the total paid employment in these sectors in 1996. A comparison of Tables 4.1 and 4.2 shows that the classifications are quite different. Given the limited availability of NAF data at the regional level, NAP data are used in the spatial analysis below, although this prevents examination of trends through the 1990s.

Comparison of the data in Table 4.2 for 1981, 1991 and 1996, using the NAF classification, confirms the shifts in sectoral composition of KIS supply. Between 1981 and 1991 KIS increased by 53 per cent, whereas from 1991 to 1996 they grew by only 4 per cent – a virtual stagnation. As well as this slowing down of KIS growth in the first half of the 1990s, declines occurred in data processing and in more traditional KIS such as architecture, engineering and technical studies.

The composition of KIS demand

These national patterns of KIS employment growth obviously reflect changing patterns of demand for KIS services, but it is difficult to acquire a

Table 4.1 France: paid employment in business services, end 1975, 1981 and 1991

		1975	%	1981	%	1991	%
NAP 7701	Technical engineering	107,175	16.0	125,210	13.6	165,529	11.4
NAP 7702	Economic and sociological studies	13,825	2.1	15,008	1.6	19,737	1.4
NAP 7703	IT and organization consultancy	13,220	2.0	30,758	3.3	139,681	9.6
NAP 7704	Data processing	13,063	2.0	26,443	2.9	38,758	2.7
NAP 7705	Architecture	26,114	3.9	28,318	3.1	31,050	2.1
NAP 7706	Offices of surveyors	14,799	2.2	14,477	1.6	13,517	0.9
NAP 7707	Information and documentation	5,299	0.8	8,492	0.9	25,567	1.8
NAP 7708	Legal advice	71,333	10.7	85,225	9.2	102,252	7.0
NAP 7709	Accounting and financial auditing	56,911	8.5	75,382	8.2	106,319	7.3
NAP 7710	Advertising creators and intermediaries	27,593	4.1	42,074	4.6	70,358	4.9
NAP 7711	Advertising	7,402	1.1	12,549	1.4	30,875	2.1
NAP 7712	Various professional services	20,517	3.1	17,496	1.9	18,493	1.3
NAP 7713	Temporary employment agencies	102,934	15.4	176,729	19.2	273,921	18.9
NAP 7714	Other business services	29,506	4.4	81,334	8.8	209,425	14.4
NAP 78	Activities auxiliary to finance and insurance	55,316	8.3	54,853	5.9	70,501	4.9
NAP 79	Real estate services	103,778	15.5	128,382	13.9	135,771	11.9
Total	All business services	668,785	100.0	922,730	100.0	1,451,754	100.0

Source: ASSEDIC: standard tables 1975, 1981, 1991.
Note
NAP: Nomenclature des Activités et des Produits.

Table 4.2 France: paid employment in knowledge-intensive services, 1981, 1991 and 1996

Code	Sector	1981	%	1991	%	1996	%
721Z	Computing systems consulting	5,283	1.2	33,903	5.2	52,116	7.6
722Z	Software development	6,197	1.5	45,585	7.0	55,837	8.2
723Z	Data processing	27,285	6.4	44,829	6.9	40,904	6.0
724Z	Data bank activities	1,472	0.3	4,875	0.7	4,985	0.7
725Z	Maintenance and repair of office equipment and computers	22,272	5.2	19,028	2.9	14,527	2.1
741A	Legal advice	84,356	19.7	101,228	15.5	100,107	14.7
741C	Accounting	75,382	17.6	106,837	16.3	106,100	15.5
741E	Marketing and market research	15,512	3.6	21,604	3.3	21,841	3.2
741G	Business and management consulting	20,118	4.7	68,033	10.4	84,062	12.3
742A	Architecture	26,880	6.3	30,299	4.6	24,357	3.6
742B	Offices of surveyors	14,477	3.4	13,637	2.1	11,821	1.7
742C	Engineering and technical studies	113,845	26.7	138,764	21.2	135,261	19.8
743	Technical analysis and control	13,802	3.2	25,110	3.8	30,588	4.5
	Total	426,881	100.0	653,732	100.0	682,506	100.0

Source : ASSEDIC: NAF classification.

systematic overview of these market trends. Recent regional surveys have nevertheless examined the sectoral composition of demand for business services quite well. The studies differ in their survey methods (postal surveys, interviews, input-output quotients); types of area (peripheral areas, urban regions, groups of regions); and definitions of business services included (all business services, consultancy, some types of consultancy). As shown in Table 4.3, however, in the late 1980s and early 1990s the share of demand by the primary sector was variously estimated at between 1.8 and 2.9 per cent. Other estimates included the share of manufacturing with construction, of between 27 and 56 per cent; the public sector, between 13.3 and 20 per cent; and private services, between 22 and 49.4 per cent. Exports varied between 5 and 18 per cent of the regional business services markets. These divergent figures are not necessarily inconsistent, but indicate a wide variety of regional and sectoral situations.

Small–medium enterprise (SME) demand for business services is hard to evaluate. Studies for the Nord-Pas-de-Calais suggest that it increased rapidly between 1978 and 1988 (Gallouj 1996). Service sector demand for KIS, which is spatially more widespread than that from manufacturing, has played a significant role in the spatial diffusion of KIS. Mayère and Vinot (1991) showed that, although the markets for engineering and technical services in Rhône-Alpes were dominated by primary, manufacturing and construction activities, KIS such as computer services, advertising, consultancy and accounting were mainly orientated towards other service firms, which accounted for well over half their turnover (Gallouj 1993, 1996).

The role of the public sector in business service growth is also hard to identify. For the Bordeaux agglomeration, Monnoyer's study showed that the public sector represented a 26–35 per cent share of turnover for consultancy, software services and engineering consultancies (Monnoyer 1995; see also Gallouj 1993, 1996). The 1982 Decentralization Law seems to have created a large public sector market for KIS activities. This Law left the regions with some autonomy with respect to their economic development. At the same time, decentralization intensified competition between cities and between regions in attracting firms, financial resources and employment. This competition also occurs at the international level, including in the quest for EU funding. Local authorities are being forced to innovate, to become more productive and to offer higher-quality services. They thus employ service firms, for socio-organizational and financial audits and consultancy, to increase their effectiveness without having to hire new staff. In the Nord-Pas-de-Calais a KIS supply structure directly orientated towards public sector *organisations non-entreprise* is taking shape. Gallouj (1996) has even shown that the demand stemming from local authorities is a condition of survival for many business service firms.

Table 4.3 France: share of KIS demand by sector, from various research sources

	Primary	Manufacturing and construction	Public services, administrations, local authorities	Services			Others (including KIS to households)
				KIS	Distrib.	FIRE[a]	
	%	%	%	%	%		%
Mayere and Vinot (1991)	—	56	20	—	22	—	2
Syntec (1989)	—	27	18	14 Services and misc.	17.0	6.0	18 Exports
Syntec (1993)	1.8	35.5	13.3 (local authority 4.0)	24.1	19.9	5.4	—
Syntec (1995)	2.9	37.4 (construction 1.6)	15.5 (local authority 4.0)	n.a.	17.2	6.6	Transport and telecom 17.2 Other market services 5.0 Other non-market services 1.5

Sources: Mayère and Vinot (1991). The survey was carried out by INSEE Rhône-Alpes in four Rhône-Alpes agglomerations (Lyon, St-Etienne, Grenoble and Annecy). The sample was stratified according to ty pe service, firm size and agglomeration. It consisted of 1,020 firms and 108 branch offices. Services included: NAP 7701, 7702, 7703, 7707 and 8202. 1989, 1993, 1995: National surveys among members of Syntec technical consultants.

Notes
a FIRE: Finance, insurance and real estate.

The organizational structure of KIS firms

The structure of KIS provision reflects this sectoral and regional diversity of demand, as well as the impacts of increasing inter-regional, and even international, exchange of expertise. KIS supply is dominated by small firms, operating predominantly at local and regional levels. According to INSEE (Institut Nationale de la Statistique et des Etudes Economique), in its *Enquête annuelle des entreprises de services*, there were 146,228 KIS (*conseils et assistance*) firms in France in 1994 (see Tables 4.4 and 4.5). Their average size was five employees, and 90 per cent of them had fewer than nine salaried workers. Total turnover was FF 512 billion, exports FF 38.5 billion, and investment FF 11.7 billion. Within these aggregates, however, there was also a great diversity of activities, within the four broad sectors shown in the tables.

Seventy per cent of the 21,000 computer-related service firms were in the most labour-intensive computer systems consultancy (*conseil en systèmes informatiques*) and software engineering (*réalisation de logiciels*). Data-processing activities, however, were on average larger and more capital intensive. Although dominated by small firms, the software sector showed a significant level of exports, and computer systems consultancy includes some of the major French companies in the field, such as Alcatel ISR, Dassault Systèmes, and Matra Datavision.

Most of the 60,000 professional service firms employed fewer than nine people, about 40 per cent being law firms. Accountancy had a higher proportion of medium-sized businesses. Once more, being in predominantly labour-intensive activities, the professions generally require a low level of capital investment, and only business and management consultancy showed significant exports. Among the 15,000 advertising and marketing firms, small size was associated with a relatively high financial turnover in the media management sector, and high exports from the marketing and opinion poll companies. More than half of the 46,000 technical consultancies were in architecture and surveying, with the great majority being single-person firms or firms employing fewer than five workers. Engineering and technical consultancies include more large firms, and show the highest levels of exports, on average twice the level measured for any other KIS activity.

Location trends in KIS

In view of the small scale and localized orientation of most KIS activities, their market relations in France, as elsewhere, are dominated by regional and metropolitan exchanges. Recent spatial trends in KIS activity have been as significant as their rapid national growth in judging their economic impacts. Although the French state had for long supported steady economic growth combined with measures to reinforce the redevelopment of the South (Braudel 1986), it was only after the Second World War, and especially in the 1960s, that French industrial and spatial planning policy attempted to counter

Table 4.4 France: firm size distribution of KIS activities, 1994

	Percentage of firms by size category					No. of firms
	0–9	*10–19*	*20–49*	*50–99*	*>100*	
Computer services						
Computer systems consultancy	89.3	5.2	3.5	1.3	0.7	8,518
Software development	84.1	8.5	5.3	1.2	0.9	6,437
Data processing	82.6	6.3	7.2	2.0	1.9	4,151
Data bank activity	51.5	21.0	14.4	6.6	6.9	763
Maintenance and repair of office equipment and computers	85.6	5.3	5.9	1.3	1.9	1,238
Professional services						
Legal advice	90.2	7.8	1.7	0.2	0.1	24,777
Accounting	82.6	11.0	5.2	0.8	0.4	14,527
Business and management consultancy	96.2	2.2	1.2	0.2	0.2	23,387
Advertising and marketing						
Marketing and polls	91.2	4.4	2.6	0.8	1.0	2,698
Managing of advertising media	89.1	5.6	3.2	0.9	1.2	3,391
Advertising	90.9	4.9	2.8	0.6	0.8	9,478
Architecture, engineering and control						
Architecture, surveyors[a]	96.5[a]	2.0[a]	— 0.5[a] —			24,310[a]
Engineering and technical consultancy	89.3	5.9	3.2	0.8	0.8	19,390
Control and technical analysis	93.6	2.4	2.2	0.7	1.1	3,163
Total (average of all KIS firms)	89	6.2	3.0	0.7	0.6	121,918

Source: INSEE EAE (Annual Business Enterprise database), 1994.

Note
a Data from different survey, not included in total.

Table 4.5 France: organizational structure of KIS firms, 1994

	Turnover		Salaried jobs	Exports		Investments	
	× FF 1bn[a]	× FF 1m[b]	× 1,000[a,b]	× FF 1bn[a]	× FF 1,000[b]	× FF 1bn[a]	× FF 1,000[b]
Computer services							
Computer systems consultancy	35.3	4.1	55.4	7	1.46	171	0.83	97
Software development	34.9	5.4	52.8	8	3.30	513	0.76	118
Data processing	36.1	8.7	51.4	12	0.94	226	1.69	407
Data bank activity	2.8	3.6	2.4	3	0.13	170	0.11	144
Maintenance and repair of office equipment and computers	8.4	6.8	12.3	10	0.12	97	0.20	162
Professional services							
Legal advice	63.1	2.5	98.6	4	2.42	98	0.95	38
Accounting	47.6	3.2	105.8	7	0.41	28	0.93	64
Business and management consulting	41.2	1.8	53.1	2	3.59	154	0.98	42
Advertising and marketing							
Marketing and polls	9.7	3.5	18.2	7	1.39	515	0.28	104
Management of advertising media	49.0	14.4	46.2	14	1.03	304	0.75	221
Advertising	48.1	5.1	71.7	8	2.28	241	0.75	79
Architecture, engineering and control							
Architecture, surveyors	25.0	1.0	38.2	2	0.37	15	0.50	21
Engineering and technical consultancy	98.8	5.1	118.4	6	20.46	1055	2.50	129
Control and technical analysis	12.1	3.8	28.4	9	0.63	199	0.48	152

Source: INSEE EAE (Business Enterprise database).

Notes
a   Totals.
b   Average per firm.

the economic attraction of the capital region and return a more significant part of economic activity to the provinces (Durand 1974; Labourie *et al.* 1985). This decentralization proved to be relatively successful for the location of industrial activities. High-technology industry, such as communication and telecommunication equipment, found a place in some of the southern regions, including Rhône-Alpes, and Provence-Alpes-Côte d'Azur (PACA) while some intermediate regions around the Ile-de-France, such as Normandy, Brittany, the Loire, the Centre, also benefited from decentralization policy.

This did not lead to the spatial dispersal of business services, however. Most KIS exhibited an employment concentration in Paris of over 70 per cent in 1987. In fact, national spatial planning reinforced the supremacy of Paris and second-tier national cities, leaving the provincial cities as largely routine service providers. More recently, some moderation in regional concentration has taken place. The second metropolis, Lyon, is playing a larger part and some regional capitals, including Lille, Strasbourg, Toulouse and Grenoble are beginning to occupy a more prominent position. This relative spread of KIS over the French city-system, with the rise of regional service centres, has been accompanied by intra-urban shifts in KIS location and employment, as demonstrated by case studies for Paris, Lyon and Lille (Moulaert and Bruyelle 1993; Moulaert and Gallouj 1995). The relative regional spread of KIS firms and jobs thus relates to the wider development of a multi-nodal KIS economy in France, based on the large cities. The emergence of network organizations at the regional, national and international levels, and the search for optimal accessibility to clients and suppliers in the urban milieu, explain the multi-nodal form of the hierarchy of dominant urban centres in France and Europe.

Regional KIS trends

If employment in KIS or the business service sector has always been highly concentrated in the capital region, regions outside Paris thus seem to have gained some influence in the last fifteen years. The 92 per cent growth of KIS jobs between 1975 and 1990 in the Paris region has been slower than in other regions. In the other leading KIS regions, Midi-Pyrénées, Rhône-Alpes, Languedoc-Roussillon and PACA, growth rates were 175, 105, 152, and 108 per cent respectively. Table 4.6 uses location quotients (the ratio of the regional to the national share of business service employment) to measure regional concentrations of KIS jobs. Quotients for the Paris region fell from 1.83 in 1975 (that is, 83 per cent above the national average), to 1.79 in 1982 and 1.74 in 1990. Only two other regions show location quotients above the national average: PACA and Rhône-Alpes. Their position remained stable from 1975 to 1990, with location quotients respectively of around 1.20 and 1.02 in 1990. The whole South of France appeared to be attracting KIS jobs over this period. Languedoc-Roussillon and Midi-Pyrénées have shown a

Table 4.6 France: location quotients of business services by region, 1975, 1982 and 1990

Regions	Location quotients		
	1975	*1982*	*1990*
Ile-de-France	1.83	1.79	1.74
Champagne-Ardennes	0.57	0.62	0.59
Picardy	0.77	0.68	0.72
Upper Normandy	0.87	0.82	0.83
Centre	0.73	0.71	0.69
Lower Normandy	0.57	0.63	0.68
Bourgogne	0.61	0.66	0.62
Nord-Pas-de-Calais	0.66	0.73	0.75
Lorraine	0.67	0.60	0.66
Alsace	0.95	0.76	0.75
Franche-Comté	0.56	0.54	0.55
Pays-de-la-Loire	0.65	0.71	0.66
Brittany	0.59	0.63	0.65
Poitou-Charentes	0.54	0.56	0.57
Aquitaine	0.75	0.78	0.79
Midi-Pyrénées	0.61	0.70	0.80
Limousin	0.48	0.51	0.46
Rhône-Alpes	1.04	1.03	1.02
Auvergne	0.52	0.57	0.53
Languedoc-Roussillon	0.77	0.88	0.86
Provence-Alpes-Côtes d'Azur (PACA)	1.22	1.29	1.20
France	1.00	1.00	1.00

Source: INSEE, Censuses of Population 1975, 1982, 1990 (place of work), authors' analysis.

sharp growth and this is reflected in the positive trends in KIS location quotients. A number of regions, with location quotients ranging from 0.46 to 0.59, show a very low presence of business services. The case of Limousin is particularly notable, related to the small scale of the urban economy and the absence of effective agglomerating factors.

Table 4.7 France: regional specializations in business services, 1990

Regions	Specializations (3 activities with the highest LQ >1)
Ile-de-France	7702, 7707, 7703
Champagne-Ardennes	7706, 7709
Picardy	7706, 7704, 7714
Upper Normandy	7713, 7714
Centre	7704, 7706, 7711
Lower Normandy	7706, 7701
Bourgogne	7706
Nord-Pas-de-Calais	7711, 7713
Lorraine	7713
Alsace	7705, 7713
Franche-Comté	7706
Pays-de-la-Loire	7706
Brittany	7706
Poitou-Charentes	7706, 78
Aquitaine	7706, 7709
Midi-Pyrénées	7705, 7704, 7709
Limousin	None
Rhône-Alpes	7713, 7701, 7706
Auvergne	7706
Languedoc-Roussillon	7705, 7708, 79
Provence-Alpes-Côte d'Azur	79, 7705, 7701

Source: INSEE, Censuses of population 1990 (place of work) – NAP classification (authors' analysis).

Within national patterns, the activities most concentrated in the Ile-de-France are information and documentation consultancy (NAP 7707), economic and sociological studies (NAP 7702), IT and organizational consultancy (NAP 7703), and advertising (NAP 7710) (see Table 4.7). In contrast, accounting and financial auditing (NAP 7709), legal advice (NAP 7708), and architecture and surveying are more dispersed across the French regions. Regional KIS specializations are highly influenced by the economic base of each region. In the Ile-de-France, only activities auxiliary to finance

Table 4.8 France: KIS and metropolitan development, 1982–90

Metropolises	Average growth rate of all employment 1982–90, %	Average growth rate of KIS employment 1982–90, %	Number of KIS jobs in 1990	Share of KIS in total metropolitan employment in 1990, %
Toulouse	2.1	9.7	17,496	4.8
Nantes	0.8	6.9	11,261	4.0
Strasbourg	0.9	6.8	8,692	4.1
Lyon	0.8	6.7	33,344	5.2
Lille-Roubaix-Tourcoing	0.0	6.6	14,560	3.5
Grand-Marseille	0.9	5.9	22,208	3.9

Source: INSEE, Censuses of Population 1982 and 1990, table produced by IFRESI.

and insurance (NAP 78) increased their relative concentration between 1975 and 1990.

The two other regions with a relatively diverse KIS structure were Rhône-Alpes and PACA. The latter ranks second in eight KIS activities: technical engineering consultancy (NAP 7701), in which the region has became more specialized since 1975; architecture (NAP 7705); legal advice (NAP 7708); accounting and financial auditing (NAP 7709); various professional services (NAP 7712); other business services (NAP 7714); activities auxiliary to finance and insurance (NAP 78); and real estate services (NAP 79). For the last activity, PACA shows the highest concentration of all the regions, including even Paris. Rhône-Alpes has the highest concentrations after Paris in IT and organizational consultancy (NAP 7703), and information and documentation consultancy (NAP 7707).

Table 4.8 provides some key statistics for the most important KIS metropolises in France. Between 1982 and 1990, the years of the two most recent censuses, most regional metropolises reinforced their position with respect to KIS employment, both in absolute terms and as a proportion of regional employment in KIS activities. Toulouse dominated in its region, with almost two-thirds of the regional KIS employment. It is the sole city in the Midi-Pyrénées, and is also one of France's main high-technology centres, with leading firms in the space and aeronautic industries, and a high demand for technology-led services. It doubled its KIS employment in the 1980s, retaining its second position as a regional KIS metropolis. In absolute terms, however, Lyon remains the most important regional KIS centre, sustaining a higher than average annual employment growth rate. Management, technology and IT consultancy are the main growth sectors. Even though total employment in Lille-Roubaix-Tourcoing fell between 1982 and 1990, it benefited from an increase in KIS employment of 5,800 personnel, especially in professional training and consultancy. By 1990, 44 per cent of regional KIS employment was concentrated in Lille, compared with 40 per cent in 1982. In terms of its share of regional KIS employment the *agglomeration du Nord* was the sixth most dominant regional metropolis in 1990.

Regional networks of KIS provision

The regional availability and spatial organization of KIS firms combines both polarization and diffusion trends. The rise of multi-establishment advanced producer service firms also follows a multi-dimensional spatial logic, involving intra- and inter-urban, as well as intra-regional, considerations. Their locational behaviour and spatial organization reflect at least four types of influence.

1   The nature of the advanced producer service.
2   The national and regional origins of suppliers.
3   The sectoral and spatial dynamics of client markets.
4   For intra-urban location, the quantity, quality and price of office space, communications infrastructure, socio-cultural environment, and proximity to and the interface with complementary services (Moulaert and Gallouj 1995).

The first two of these, relating to consultancy *supply*, are structured in relation to several types of spatial network.

1   Loose partnerships in which each partner has their own market and clients, based on product and professional specialization, without specific geographical limits. Cooperation between partners is based on marketing, exchange of information and professional complementarity. This model of organization is typical of small consultancies with considerable ambition, who 'unite' to undertake a job, bid for a tender, or to increase the visibility of the consultancy profession (see below). The spatial forms adopted by such partnerships vary. Most typically, at the local or regional level, joint supply is organized by or in connection with the chambers of commerce. Sometimes supply is also organized by local authorities within the context of an innovation or service network. This type of organization seeks to obtain a better match between local supply and demand.
2   At wider (national or international) levels, 'consortia' often enable smaller suppliers to respond to demands from clients 'going international' by opening or acquiring agencies or factories in other countries. Client preferences for familiar consultancies may drive the latter to find partners abroad. This *de facto* partnership may then become an effective supply structure, serving new clients.

3   A larger 'networked firm' can be spatially organized according to differ-
     ent or complementary professional specializations, products and client
     sectors. Since the beginning of the 1980s, the presence of networked KIS
     firms in the French regions has grown (Moulaert and Bruyelle 1993). A
     global analysis of networking is difficult, so that case-study evidence must
     suffice (see for example Moulaert *et al.* 1990; Djellal 1993; Moulaert and
     Bruyelle 1993; Moulaert 1996).

The organization of networked firms, or networks of firms, may take place at
several spatial levels.

The international level

International networking has been studied only at a very general level, for
example by counting subsidiaries and offices in different countries (Daniels
1990) or through case studies involving several countries and interviews at headquarters and regional offices. There is some interesting evidence from the
computing and organization technology sector, including Andersen Consulting
and Cap Gemini. However, a reliable, comprehensive picture of the international
organization of consultancy firms is not yet available (Sauviat 1995).

The national or inter-regional level

Several nationally networked firms have a regional origin. These include firms
that started their activity in Paris, as well as those based in other regional
metropolises that have also expanded their activities elsewhere. For Cap
Gemini, the establishment of an inter-regional network offered the leverage
for internationalization through partnership or acquisition.

The regional or local level

This level has become quite important. Firms at this level include local
networks in which the public sector plays an active part, through chambers of
commerce, DRIRE (la Direction Régionale de l'Industrie, de la Recherche et
de l'Environnement), and ANVAR (l'Agence Nationale de la Valorisation de
la Recherche). Increasing numbers of consultancy and manufacturing firms
are also involved in local or regional producer service networks. These have
been particularly well documented for Toulouse. In 1996, 80–90 per cent of
service firms, including those providing cleaning and transport as well as business services, which had more than one location in Toulouse were involved in
local networks. Regional networks were rare except for accountancy, and in
consultancy and advertising (see Table 4.9).

The low level of regional, compared to national, networking for technical analysis, computer services, and engineering and technical studies in Midi-Pyrénées is
related to the global organization of the high-technology production system.

Table 4.9 Geographical ambit of multi-location service firms in Toulouse, 1996

	Local networks[a] %	Regional networks[b] %	National networks[c] %	International networks[d] %	Total %
Technical analysis and control	87	0	13	0	100
Accountancy	72	17	11	0	100
Consultancy	79	11	6	4	100
Computer services	77	2	14	7	100
Engineering and technical studies	73	5	16	6	100
Interim labour offices	67	0	33	0	100
Advertising	85	11	4	0	100
Secretarial services	94	0	6	0	100
Security, cleaning	75	4	21	0	100
Transportation	81	9	6	4	100

Source: Centre Interdisciplinaire d'Etudes Urbaines, Toulouse, and the Toulouse Chamber of Commerce and Industry, 1996.

Notes
a All offices in the Toulouse agglomeration.
b Not all offices in Toulouse agglomeration, but in the Midi-Pyrénées region.
c Not all offices in the Midi-Pyrénées region, but in France.
d Some offices abroad.

This is especially important for Toulouse, because of the national networking of computing and engineering services. Many important local software service firms, for example, have been absorbed by national or international firms (Zuliani 1992). Spatial networking in advanced services is dominated by the organizational dynamics of high-technology firms which seek cooperative and subcontracting relationships with high-technology services (Zuliani 1997). The high proportion of nationally networked firms even in quite standardized services, such as temporary labour, security and cleaning services, is also a consequence of the strength of national, and even internationally linked firms in local markets.

Regional and industrial policy for business service development

A historical synthesis

Traditionally, regional and industrial economic policy has been orientated towards manufacturing. In France until the beginning of the 1980s, as in most

European countries, services were largely ignored by public policy, especially business services. During the 1960s and 1970s, the first policies in European countries towards services attempted to limit the concentration of new office space in large urban centres. Later, in certain countries, although not in France, this policy was complemented or replaced by policies to decentralize public administration. Gradually more active service-orientated policies were established, to create or sustain employment in peripheral regions. By the end of the 1970s the significance of business services in regional development began to be accepted. Established policy measures for manufacturing were extended towards certain business services, which could satisfy the criteria of spatial mobility, exportability and job creation (Marshall and Bachtler 1987).

In France, these policies provided some direct financial support to business service firms, but more often supported business service demand. These subsidized the cost to clients of using either independent business services (in the form of Fonds Régional d'Aide au Conseil en France) or developing their own service capacity, for example through assistance in hiring managers. More recently the creation of *parcs tertiaires* and *pépinières* (seedbed accommodation) has contributed to local policies for new firm formation. By the second half of the 1980s, it was evident that simply supporting supply and demand for business services separately was not sufficient. Attention needed to be paid to the relationships between supply and demand. Thus a number of *intermédiaires* or mediating agencies were created to support the dissemination of information, technology and human resources. During the 1990s this *mise en relation* of supply and demand was further elaborated (Debandt and Gadrey 1994).

Regional policies promoting KIS development: supply-orientated policies

Today's regional service policy in France has three components: supply-orientated policy, demand-orientated policy and policies focused on the improvement of the 'external environment'. Supply policies are designed to support, create or attract service firms. Their main elements include the following.

Incentives for the location of services in certain regions

These were among the earliest measures, established in the 1970s, and are meant to encourage service provision, mainly in backward regions.

Service provision by public institutions and intermediary organizations

Public institutions often supply services to enterprises. This may be carried out by providing financial support to agencies offering such services at low cost; by supplying services at low cost; or by free supply through activities such as inter-professional service centres, or centres supporting enterprise creation or innovation.

Direct aid to business service firms is unusual in France, as in the rest of Europe. Such policies are limited to the inclusion of certain types of business service firm in general regional support schemes. In France, the Fonds Régional d'Aide au Conseil (FRAC) is now available to *tertiaire supérieur* (advanced business service firms). Service firms can now also benefit from financial support for the use of training consultancies, through the Ministère du Travail, de l'Emploi et de la Formation Professionnelle. There are also developments in the policy of COFACE (Compagnie Française d'Assurance pour le Commerce Extérieur) towards subsidizing exporting service firms under certain conditions. In the Nord-Pas-de-Calais, support for the recruitment of specialized personnel for export activities is also available to business services.

Other supply-orientated measures

Other policies include financial support for the improvement and the consolidation of services, such as the certification of service quality criteria, to promote the export of services, and to support collaboration between suppliers. The latter will be returned to in the next section.

Demand-orientated policy

Financial support to firms using services

In France, the main instrument for this policy is the FRAC, established in 1982 and applied in the French regions since 1984. The FRAC encourages demand for consultancy by SMEs, by providing them with the necessary purchasing power. Another objective, less explicit at first, was to support the quality of local business services. The FRAC subsidy is directed towards manufacturing firms which are financially sound, employ fewer than 500 people and are not controlled by other firms, thus excluding subsidiaries of large firms. Service industries are eligible if they offer a high level of technical activity to support manufacturing users. Demand should be directed towards private sector consultancy, independent of professional associations. The subsidies for consultancy must also be directed towards the main business functions of clients, including technology, marketing, human resources and training, quality control and corporate strategy. Firms providing consultancy services lasting up to five days receive 80 per cent of costs, to a maximum of about FF 25,000. Longer periods are financed at 50 per cent, with a maximum of FF 200,000.

There is also a FRAC *qualité* scheme. Firms seeking quality accreditation can receive subsidized consultancy. This has now been broadened to include the organizational quality of firms. This change in focus was promoted by the ARACT (l'Agence Régionale pour l'Amélioration des Conditions du Travail), whose original job was to develop diagnostics on work conditions, and which has expanded its activities to include the organizational qualities of firms.

The development of internal skills

This type of policy, applied in numerous Organization for Economic Cooperation and Development (OECD) countries, supports managerial skills in firms, especially in SMEs, to increase their innovative and adaptative capacities. The best-known measure is subsidizing the recruitment of highly skilled personnel, who are also capable of interacting with high-quality service firms. The subsidy can also go towards the training of managers and the creation of service departments inside firms.

In France, support for the hiring of specialized managers by SMEs is mainly provided by a programme called Aide au Recrutement des Cadres (ARC). This was launched as an experiment in the Nord-Pas-de-Calais region in 1986, and applied more generally after 1987. The Nord-Pas-de-Calais was selected because of its lack of high-level managers compared to the national average. Financial support goes to manufacturing firms established for at least two years, with fewer than 250 employees, which are also not more than 25 per cent owned by a group with more than 250 employees. Assistance is for the creation of new management jobs in functions essential to the future of the firm. Recruitment should be under a contract of unlimited duration (CDI) for persons with at least 'Bac + 4' qualifications or ten years of professional experience. The Nord-Pas-de-Calais also benefits from similar national or regional programmes, including support for the recruitment of *créateurs* (ARCREA), for exports (ARE) and for innovation (ARI).

Structures d'accueil (hosting structures)

In most OECD countries since the late 1980s, a significant initiative has been the creation of new areas where specific activities or services can be fostered. In France, *pépinières*, seedbed areas or buildings, are designed for start-up enterprises. The Agence Nationale pour la Création des Entreprises (ANCE) identified more than 200 *pépinières* in 1992, mainly established and run by local authorities. Providing services to all types of new firms, *pépinières* are effectively a structural policy to improve the relationship between demand and supply in business service markets.

Other measures mainly orientated towards demand

These particularly encourage demand for business services by local institutions, authorities and service intermediaries, but also by outside firms, for example by encouraging branch offices to use local service providers.

The improvement of the 'external environment'

Two types of actions with recognized impacts on service development in France are:

1 infrastructure and urban development policies;
2 policies to improve information and transparency in service markets.

In general, the logic of such policies is to foster environments and networks which improve the interaction between service demand and supply. It implies improving the attractiveness of cities and regions and local business milieux, as well as a wider dissemination of information in regional service markets.

Infrastructure and urban development policies

These require a good understanding of the key factors influencing service location and mobility at both intra- and inter-urban levels. They are designed to improve communication and telecommunication networks, the built environment and the socio-cultural infrastructure, for example, related to health provision, culture, education and leisure.

Policies to improve information and market transparency

The development of services is often limited by a lack of information and transparency in their markets. To overcome these problems, local authorities may provide information, create data banks on service suppliers and products, and establish, or encourage others to establish, local arrangements to support the exchange of ideas and experience, as well as promoting synergies among various initiatives. In practice, and certainly in the case of the Nord-Pas-de-Calais, attempts to rationalize and classify business services involve local authorities promoting the certification of services, or the production and diffusion of information on local and regional supply, for example through directories or publicity for suppliers. This type of public action tends to widen its remit, to include advice on subsidy procedures.

KIS and innovation potential at the regional level

Two approaches to the presentation of KIS developments in France have been followed so far. The first was a statistically based review of KIS change, especially in management consultancy services, in the context of national and regional economic change. The second has summarized policies, mainly regional, to promote innovation in economic activities, including those involving producer services. The impact of KIS on innovation at the regional level requires a more detailed examination of the processes of interaction between consultancies and client innovation which, with the current state of data, can be undertaken only through qualitative methods. In this case, trajectories of innovation associated with management consultancy have been examined from different perspectives in two regions in the mid-1990s: the Nord-Pas-de-Calais and Midi-Pyrénées. Panels of experts, including economists, consultants and representatives of local chambers of commerce were consulted. The following dimensions of innovation through consultancy were examined by the panel.[1]

Management consultancies as carriers of innovation

Innovations associated with consultancy activities, such as organizational changes, new modes of communication, or research and development, can occur in the consultancy firm, in the client organization, or through the interface structures between clients and firms, possibly involving other agents. The innovative experience of management consultancy firms can promote innovation in client firms, although there are differences between types of consultancy firm. The following illustrate how changes in consultancy firms may influence innovation through clients.

Organizational changes affect functional relationships; modes of communication and coordination; spatial organization at regional and extra-regional levels; and through the development of partnerships, joint ventures, and other forms of collaboration. For Coopers & Lybrand in the early to mid-1990s, one of the main dimensions of innovation was globalization. This consultancy's organizational structures possessed a global dimension, and the company used its worldwide network to benefit both large and small clients, in seeking out new markets and suppliers.

Marketing can apply experience of extension to new markets; expand existing markets; develop new products; and make use of new marketing tools, such as the Internet. Many major consultancies possess elements of sector-directed organization. Client business is structured according to sector-related strategies. SME markets are also targeted, especially through accountancy networks. Nevertheless, SMEs generally still lag behind in the use of consultancy services, especially those supplied by the larger consultancy firms. Demonstration effects may also sometimes arise from a particular experience. Consultancy Internet sites, for example, may influence the communication behaviour of the client. Incremental innovations demonstrated by consultancies, like targeted advice, can be very successful with small clients.

Human resources and training also influence innovation. The recruitment strategy of many consultancy firms is based on a mixture of quantitative and qualitative criteria. In general the skill level of the consultancy staff has increased and their professional backgrounds and cultures have become more diverse. In this way, consultancy firms act as innovators by widening their interface with the business environment. The larger consultancy firms have developed their training departments to serve the needs of their markets, even establishing their own professional schools or 'university' (Moulaert 1996).

They combine formal training in the department with collective on-the-job training, or teamwork, informal individual training and *ad hoc* project work.

Within client firms, a distinction must be made between dynamic and non-dynamic firms. Dynamic firms will have a view of their innovation needs, whether technological, organizational or strategic, and will seek out the consultancy service needed to support such needs. Less dynamic firms may face more general problems of efficiency and effectiveness, and may not have the means to identify innovation requirements and translate them into real demand. For them, engaging a consultancy firm is not an automatic or natural process. For SMEs, there are also important differences in their relations with consultancies based on whether the interlocutor is a salaried manager or the entrepreneur or owner of the firm.

The impacts of consultancies on client innovative behaviour differ between various firms, in three ways.

1   The recognition of efficiency and efficacy problems, and the organizational, institutional, managerial or technological factors which underlie them.
2   Access to consultancies. This includes the role of 'third parties', facilitating networks or interfaces. Preference-revealing techniques may place clients in touch with their requirements, allowing them to be matched with appropriate consultancy firms.
3   The consultancy approach and the consultancy products offered.

Client demand also follows economic cycles. Up until the 1970s, 'daring' innovation was important and client firms used marketing consultancies to improve their image. Following the economic crisis of the 1970s, firms sought to reduce costs, so that high-risk innovation became less important. By the 1990s, innovation had gained renewed influence, but the human and organizational context was also increasingly recognized. This explains the spectacular rise of human resources and organizational consultancies, and growing collaboration between such consultancies, university social scientists and technological consultancy firms. It has also become established that the 'accountancy' logic which underpinned the growth of many large consultancies twenty years ago has had to be replaced by a multidisciplinary approach, seizing new technical and market opportunities in which organizational, human resources and technological expertise are integrated.

Consultancy firms can also help their clients to enlarge the application of their research and development (R&D). This is the case for Aérospatiale Bordeaux. To help Aérospatiale enhance the value of its research activity in other markets, a major consultancy offered a partnership with a specialized multimedia firm. Consultancies may also exceptionally intervene in the organization of R&D itself. For a pharmaceutical firm, for example, the same consultancy claimed to have shortened the development period between the appearance of an innovation and its marketing from twelve to six years.

Client firms and their needs for innovation

The convergence in innovation agendas between consultancies and clients can be promoted by encouraging clients generally to play a more significant role in the consulting process and its organization. Different views by clients about the role of consultancies in innovation processes need to be identified. Consultancies may be viewed as:

- sources of information;
- facilitators in innovation processes;
- analysts of restructuring problems and strategies;
- actual implementers of innovation strategies.

There are likely to be differences between regions, economic sectors and types of enterprise, including large firms or SMEs, and international, national or regional companies. Such differences might form the basis for organizing different types of innovation network. In general, SMEs tend to view consultancies as sources of information, or even as interfaces with other services such as computer services, innovation subsidies, export support or legal advice. Large corporations employ consultancies more as advisers and co-producers of innovation. Recently, the demand for technology-related services has increased spectacularly and can be regarded, for example, as the main source of consultancy growth in the Toulouse metropolis.

The role of clients in the innovation process

The contribution of consultancies must be placed in the context of client procedures to support innovation. How do clients become involved in the innovation process? What are their goals and their inputs? Are consultancy firms/client networks joint production environments or simply channels for the exchange of information? Can clients convey their norms to consultancies? Will they actively collaborate to reveal their needs? Can they work towards clear priorities for public policy measures? Some clients, especially those in high-technology or other highly competitive sectors, exert strong pressure on consultancy firms continuously to produce innovative methodologies and products. The same may be true of those working in some regions, such as Paris or Rhône-Alpes, or countries such as the US with an aggressive business culture (Gallouj and Gallouj 1996). This mutual participation of clients and consultancies in innovation is illustrated by the development of new software, for example when a consultancy firm creates new financial software for a bank. Here, user advice is vital to improving the product. Regular meetings of the principal users are required to detect problems and to update the software.

Management consultancy and innovation networks

Despite their growing presence in client organizational and managerial restructuring, the role of management consultancies in the client innovation process should not be exaggerated. Consultancies are not regarded as agents to be imitated or blindly followed. Instead they are supporters of change, who must adapt their service products, consultancy methodology and interaction with other innovation agents to match client needs. As far as the introduction of new technologies is concerned, consultancy firms play only a modest role. Hiring qualified personnel, subcontracting to other firms, and supporting R&D networks with peer firms and research institutions seem to be of prime importance here. Nevertheless, according to SESSI (the Ministry of Industry and Research), for most firms in France, technology consultancy plays a more important role than joint ventures or takeovers.

Consultants will continue to play a critical role in facilitating the learning process, revealing the needs of client organizations within their own specific sectoral and institutional contexts. Networks help them to achieve innovation goals. They can respond to different objectives, such as product development, information exchange, training, and co-servicing. They can be composed of users, of consultants and other service providers, of public and private agents.

The role of the public sector and of interface structures

Innovation network building

Clients must learn to plan innovation. Consultancies can foster a wider learning process among clients and adapt their products and methods to a diversity of specific situations. But what is the role of the public sector? Does this go beyond that of merely a provider of cash subsidy? Or is its position sufficiently neutral to enable public agents to facilitate innovation-orientated networking?

Public agents may have a variety of roles in innovation processes:

- as network facilitators, organizing meetings, communication channels or discussion panels, and encouraging interaction;
- providing a bridging function, for example between professional organizations, local enterprises, schools and local professionals;
- in an entrepreneurial role, organizing business services which are not available in the region but are essential for the innovation process;
- as industrial policy makers;
- as labour market policy makers.

The ultimate challenge for interface structures is to improve communication and collaboration between economic agents engaged in innovation. These can include potential clients, consultancies, industrial policy agencies and professional schools.

Several types of development in interface structures are possible:

- improving informal relations within network organizations;
- designing and implementing 'need identification' methods;
- setting up project groups for industrial organizations;
- promoting training programmes for consultants, client interface agents and public policy agents.

Chambers of commerce and industry (CCI) already play an important role in the diffusion of consultancy services to SMEs. For example, the Bordeaux CCI requests each applicant for financial help to design a development plan employing three consultancy firms. When the Toulouse CCI prepares a new directory, for example for multimedia services, promotional meetings are organized, which in this case resulted in a multimedia supplier network.

The ANVAR also plays a significant role in the promotion of consultancy services. Privileged links are maintained with consultancy firms to evaluate projects bidding for ANVAR financial support. Moreover, ANVAR subsidy policy seems increasingly orientated towards global projects, for example linking large-scale technical investments with service innovation. Technical or professional schools can also help establish relations between consultants and clients. L'Ecole Supérieure de Commerce de Toulouse, for example, helps to promote relations between SMEs and computer consultancy firms, by organizing round-table meetings. In the Nord-Pas-de-Calais region, the DRIRE and the regional council have sought a clearer overview of consultancy provision in the region. This could help improve subsidy mechanisms, by better adapting them to needs. The response has nevertheless been disorganized. Some firms, such as Exaconseil, are attempting to coordinate structures of regional firms by adopting national norms, such as those of SYNTEC, the national professional syndicate for technical engineering.

The role of the chambers of commerce and industry

In recent years, French CCI have developed various actions to support service developments. Some fall into the policy categories presented earlier, directed towards the supply and demand of services, and the external environment in which they may develop. CCI policies to support *supply* may be directed towards sectors and particular professions, or may adopt a greater 'transference' role. They offer three main forms of support: through the creation of networks and collectives of suppliers and professions; the actions taken by these networks and collectives; and the provision of common means and knowledge. The actions orientated towards *demand* mainly attempt to improve information on service provision. The challenge is to create a culture and practice in the intelligent use of services, sometimes by means of financial incentives.

Policies aimed at the *external environment* include initiatives to improve the relationship between supply and demand. The challenge is to establish

Table 4.10 Actions by French chambers of commerce in support of business service development

Actions mainly orientated towards supply	Promoting personal contacts leading to networks and collectives of suppliers (professional clubs, communities)
	Support projects with common interest, valorisation and investments
	Aid to creation of service enterprises
	Sectoral studies: markets, their problems, their perception by current and potential users
	Information on specific suppliers (or groups/sectors). Service observatory
	Specific professional training for suppliers
	Actions to extend initial education and training (e.g. Ecole de Commerce) to service domain
	Actions with suppliers to improve quality; information on standards, ISO norms and certification
	Role in the establishment of urban infrastructure, offices, service pépinières
	CRITS (Support for technology transfer to service enterprises)
Actions mainly orientated towards demand	Advice on use of services, consultancy about consultancy, information and advice about public support measures
	Promoting recurrent use of services by supplying of certain services directly at lower prices, or by public aid initiatives (such as the FRAC)
Matching supply and demand	Fairs, forums and other initiatives grouping suppliers and users
	User clubs
	Catalogues and other databases meant for clients
	Personal mediation
Actions to influence quality of the 'external environment' of business service markets	Extension of relations to other public agents (higher education, research, local authorities, DRIRE) or private agents (banks, insurance corporations, various prescribers)
	External communication towards the general public (service image) and public authorities
	Interventions with public authorities to promote the position of services in relation to manufacturing industry

Source: Gadrey and Gallouj 1994.

exchange networks with common understanding and language. Also, such policies may concern the general environment *sensu stricto*, through the enlargement of networks to include other public and private agents, or the production and diffusion of knowledge about services, to develop a wider service culture (see Table 4.10).

Conclusion

KIS may provide some answers to the problems of selecting and implementing innovations in client firms. Not least because, thanks to their actions, the conceptualization of innovation in relation to organizational and human resources management no longer focuses only on innovation in isolation, but has become one of the mainstays of entrepreneurial behaviour more generally (McKee 1990).

The lessons that the state in general, and the French public sector to a considerable extent, are drawing from this changing view of innovation is that its role in innovation policy is less to stimulate the development of technological hardware, but more to act as an animator and a consultant in the creation of networks which are themselves breeding grounds for organizational learning. The experiences of several French chambers of commerce and industry also show that many such networks perform better when organized on a territorial (regional, metropolitan) basis, with linkages to other similar networks in other regions and metropolises.

Note

1 KIS Strategy and Policy Panel – the Toulouse panel was composed of: three researchers: Marie-Christine Monnoyer, member of Réseau Européen Services et Espace (RESER – Services and Space European Network) and senior lecturer at the University of Bordeaux-Montesquieu; Michel Grossetti, Centre National de la Research Scientifique (CNRS), Centre Interdisciplinaire d'Etudes Urbaines, University of Toulouse-Le Mirail; and Jean-Marc Zuliani, Centre Interdisciplinaire d'Etudes Urbaines, Toulouse; two representatives of KIS firms: Alain Addes from Coopers & Lybrand Development; and Philippe Valdiguié of Groupe Taillandier, a firm specializing in recruitment; three staff members of the Service Working Party at the Toulouse Chamber of Commerce, Cécilia Joint, Chantal Valette and Didier Latapie.

In Lille, the panel included a CNRS researcher working on regional innovation, Christian Mahieu, Institut Federatif de Research sur les Economies et les Sociétés Industrielles (IFRESI), Lille; a senior staff member of the Lille-Roubaix-Tourcoing Chamber of Commerce, Pascal Duyck. The information obtained from these panels was related to previous studies on network organization for innovation. Most panel members also provided background material in the form of research reports.

References

Beckouche, P. and Damette, F. (1992) 'Le systeme productif en région parisienne: le renversement fonctionnel', *Espace et Société*, Paris: L'Harmattan, 66–7.

Benko, G.B. (ed.) (1990) *La dynamique spatiale de l'économie contemporaine*, La Garenne-Colombes: Editions de l'Espace Européen, Collection Géographie en Liberté.

Braudel, F. (1986) *L'identité de la France: les hommes et les choses*, Paris: Champs/Flammarion.

Cohen, J. (1988) 'Mutations de l'emploi en région parisienne', *Espace, Population, Sociétés* 3.

Daniels, P.W. (1990) 'Services and the international economy: tradeability and trade', Working Paper, Portsmouth Polytechnic, Service Industries Research Centre.

Daniels, P.W. and Moulaert, F. (eds) (1991) *The Changing Geography of Advanced Producer Services*, London: Belhaven Press.

Debandt, J. and Gadrey, J. (1994) *Relations de services et marchés des services*, Paris: Les Presses du CNRS.

Djellal, F. (1993) 'Les firmes de conseil en technologie comme agents d'un paradigme socio-technique: analyse de leur organisation fonctionnelle et spatiale', doctoral thesis, University of Lille I.

Dumoulin, C. (1991) 'Développement des entreprises de services et contrôle du réseau', in Dumoulin, C. and Flipo, J.P. (eds) *Entreprises de services: 7 clés de succès*, Paris: Les Editions de l'Organisation.

Durand, P. (1974) *Industries et régions*, Paris: La Documentation Française.

Fabre, J. (1991) *Le développement des villes françaises de dimension européenne et les réseaux de ville*, report for the Conseil Èconomique et Social, 13 February.

Fontaine, F. (1988) 'Métropoles régionales, Lille-Roubaix-Tourcoing et les autres', in INSEE, *Les dossiers de profils*, 25, 58.

—— (1990) 'Les métropoles régionales à la recherche de leurs points forts', *Économie et statistique* 230 (March).

Gadrey, J. (1996) *L'économie des services*, 2nd edn, Paris: Collection Repères, la Découverte.

Gadrey, J. and Gallouj, C. (1994) *Les services aux entreprises et les politiques de développement régional*, report for the Paris Chamber of Commerce and Industry (October).

Gadrey, J., Gallouj, F., Martinelli, F., Moulaert, F. and Tordoir, P. (1992) *Manager les conseil: strategies et relations des consultants et de leurs clients*, Paris: Ediscience Internationale.

Gallouj, C. (1993) 'Commerce extra régional de services aux entreprises et développement régional dans le Nord-Pas-de-Calais', *Cahier Lillois d'économie et de sociologie* 21(1).

—— (1996) 'Services aux entreprises et développement régional', doctoral thesis, University of Lille.

Gallouj, C. and Gallouj, F. (1996) *L'innovation dans les services*, Paris: Economica.

Gallouj, F. (1992) 'Economie de l'innovation dans les services: au-delà des approches industrialistes', doctoral thesis, University of Lille.

—— (1994) *Economie de l'innovation dans les services*, Paris: L'Harmattan.

Gravier, J.F. (1949) *Mise en valeur de la France*, Paris: Le Portulan.

Illeris, S. (1989) *Services and Regions in Europe*, Aldershot: Avebury.

Labourie, J.P., Langumier, J.F. and de Roo, P. (1985) *La politique française d'aménagement du territoire de 1950 à 1985*, Paris: La Documentation Française.

Lakota, A.M. and Milleli, C. (1989) *Ile de France: un nouveau territoire reclus*, Paris: La Documentation Française.

Lhéritier, J.L. (1987) 'Les activités tertiaires se développent rapidement aux portes des grandes villes du Nord-Pas-de-Calais', *Profils du Nord-Pas-de-Calais*, no. 2.

—— (1988) 'Les activités tertiaires dans le Nord-Pas-de-Calais', *Dossier de profil* 21, Lille: INSEE Nord-Pas-de-Calais.

McKee, D.L. (1990) 'On Schumpeter, services and economic change: his evolutionary economics could not have foreseen the burgeoning of issue-specific consulting', *American Journal of Economics and Sociology* 49(3) (July): 297–306.

Maléziuex, J. (1990) 'Services d'immobilier d'entreprise et mode de production flexible: le cas de l'agglomération parisienne', in Benko, G.B. (ed.) *La dynamique spatiale de l'économie contemporaine*, La Garenne-Colombe: Editions de L'Espace Européen.

Marshall, J. and Bachtler, J. (1987) 'Services and regional policy', *Regional Studies* 25: 471–5.

Mayère, A. and Vinot, F. (1991) 'Offre de services et dynamiques urbaines en Rhône-Alpes', *Les dossiers de l'INSEE Rhône-Alpes* 42 (December).

Monnoyer, M.C. (1995) 'Attractivité de l'agglomération bordelaise: enquête sur l'offre et la demande de services', *RESER*, 451–85.

Moulaert, F. (1996) 'Arthur Andersen: from national accounting to international management consulting firm', in Nilsson, J.E., Dicken, P. and Peck, J. (eds) *Transnational Corporations in Europe: Functional and Spatial Divisions of Labour*, London: Chapman.

Moulaert, F. and Bruyelle, P. (1993) *L'évolution des centres tertiaires des années 60: mobilité des entreprises et réorganisations urbaines*, report for DATAR and Plan Urbain, April.

Moulaert, F. and Gallouj, C. (1993) 'Agglomerations in networks: locational strategies of KIS firms', *Service Industries Journal* 13(2).

—— (1995) 'Advanced producer services in the French space economy: decentralisation at the highest level', *Progress in Planning* 43(2–3): 139–54.

Moulaert, F. and Martinelli, F. (1992) 'Le conseil en informatique: conseil en systèmes et systèmes de conseil', in Gadrey, J. *et al. Manager le conseil*, Paris: Ediscience Internationale.

Moulaert, F., Djellal, F. and Chikhaoui, Y. (1990) 'La localisation des firmes françaises de conseil en technologie de l'information', in Benko, G.B. (ed.) *La Dynamique spatiale de l'économie contemporaine*, La Garenne-Colombes: Editions de l'Espace Européen: 255–91.

Moulaert, F., Martinelli, F. and Djellal, F. (1990) *The Role of Information Technology Consultancy in the Transfer of Information Technology to Production and Service Organizations*, The Hague: NOTA (Netherlands Office of Technology Assessment).

Moulaert, F., Djellal, F., Mahieu, C. and Tramcourt, E. (1993) 'Les agents d'innovation dans les systèmes d'information en France et en Grande Bretagne: une approche socio-organisationnelle', Lille: IFRESI, report for PIRTTEM/CNRS.

OECD (1989) *L'amélioration des services aux entreprises comme outil de politique régionale*, Paris: OECD.

—— (1995) *Dynamiser les entreprises, les services de conseil*, Paris: OECD.

Phillipe, J. (1991) 'Mercapoles, un enjeu mercapolitain pour les services aux entreprises', *Economies et Sociétés* F(3é): 149–62.

Sauviat, C. (1995) *Les réseaux internationaux de services: le cas du conseil et de l'audit*, Paris: Institut de Recherche Economique et Sociale.

Srevens, J.F. (1989) *Lille Eurocité*, Lille: SGAR, Préfecture de la région Nord-Pas-de-Calais.

Thouard, A. (1993) *Le marché de l'immobilier d'entreprise en 1992: le marché des métropoles régionales*, report.

Veltz, P. (1990) 'Nouveaux modèles d'organisation de la production et tendances de l'économie territoriale', in Benko, G.B. (ed.) *La dynamique spatiale de l'économie contemporaine*, La Garenne-Colombes: Editions de l'Espace Européen: 53–69.

Vogler-Ludwig, K., Hofmann, H. and Vorlou, P. (1993) 'Business services', *European Economy: Social Europe, Report and Studies* 3: 381–400.

Zuliani, J.M. (1992) *Une combinatoire du développement des services de haut niveau: l'exemple de Toulouse*, Toulouse: Centre Interdisciplinaire d'Etudes Urbaines, Working Paper.

—— (1997) *Les services aux entreprises dans une métropole à forte composante de recherche-développement: Toulouse*, Toulouse: Centre Interdisciplinaire d'Etudes Urbaines, Working Paper.

5 Germany

Knowledge-intensive services in a core industrial economy

Simone Strambach

Introduction

There has been a rapid growth of knowledge-intensive services (KIS) since 1980 within the context of German economic globalization. These services contribute to the informal 'knowledge transfer structure', making them an important element in the national innovation system. The contribution of KIS to innovation arises from the interactions between suppliers and their clients and the learning processes associated with these interactions. Such processes are strongly influenced by institutional contexts. This chapter examines KIS development trends within the particular institutional framework of Germany and the contribution of such services to the German innovation system.

The first section deals with the demand side, the main patterns and development trends of the German economy in recent years. The second section looks at KIS supply, their development in Germany at national and regional levels and the organizational structure of supply. The third section focuses on the contribution of KIS to innovation, analysing the interaction and the complex cumulative learning processes that take place between KIS suppliers and clients. The role of public and semi-public institutions, and their interplay with private suppliers in the regional innovation system, are also examined using the important example of Baden-Württemberg. The fourth section considers technology and innovation policies and their influence on the development of KIS. Conclusions are then drawn about the implications and strategic challenges for political actors wishing to support KIS in the innovation system.

Patterns and recent trends in the German economy

The structure of demand: the distinctiveness of German KIS markets

As in the past, the manufacturing sector constitutes the main basis of the German economy. At its core are four key sectors: mechanical engineering, electrical engineering, automobile manufacturing and chemicals. These account for more than half of industrial employment, and Germany's international competitiveness has been based on them since the 1970s. In 1997 exports of basic materials (such as chemicals) and manufactured products accounted for about 75 per cent of German foreign trade (OECD 1998). Over 50 per cent of exports were engineering and automobile-related capital goods. The capital goods industries also dominate the value added industrial sector. While these technology-intensive export industries are an expression of competitiveness, they also make Germany very dependent on foreign markets.

A predominance of small–medium enterprises (SMEs) is also a structural characteristic of German industry. In 1997, 72 per cent of firms had fewer than 100 employees and only 4.5 per cent employed more than 500. These larger firms, however, provided 44 per cent of manufacturing jobs. The high proportion of SMEs partly explains why German economic policy since the 1970s has promoted them through the so-called *Mittelstandspolitik*. The aim of this intervention has been to offset some of the disadvantages of small firm size. The industrial core sectors remain dominated by well-known domestic companies operating in world markets, including Daimler-Benz (now DaimlerChrysler), Volkswagen, BMW and Opel in the automobile industry; BASF, VEBA, Hoescht, Bayer and Henkel in chemicals; Siemens and Bosch in electrical engineering; and Krupp Thyssen, Mannesmann and MAN in mechanical engineering. The branch structure in the locations where these firms operate dominates local and regional supply networks. Strong linkages between large firms and SMEs are widely regarded as important assets to Germany's competitive strength in world markets.

At the national level, German institutional relationships have also been identified as one of the factors supporting the success of the German 'economic model'. By international standards the degree of cooperation between the institutions responsible for labour and industrial relations is very high. The major strategic questions relating to labour, social and industrial policies are normally dealt with through cooperation between unions and management, as well as at the level of industry associations (Jürgens and Naschold 1994; Katzenstein 1989; Naschold 1996). In contrast to more deregulated systems such as the UK and the US, based on the liberal market economy, the social market economy has been the leading principle in Germany's economic development. Another essential feature is the occupational training and further education component of a predominantly dual education system. Overlapping qualification structures of skilled workers, technicians and engineers have developed which ensure the transfer of technology-orientated knowledge and technical/technological skills to the production system (Naschold *et al.* 1997).

In Germany, as in other highly industrialized countries, the proportion of employees in secondary industry has been declining steadily since the 1970s, with the proportion in the tertiary sector rising. About 1.1 million jobs were lost in the west German secondary sector between 1980 and 1994. The

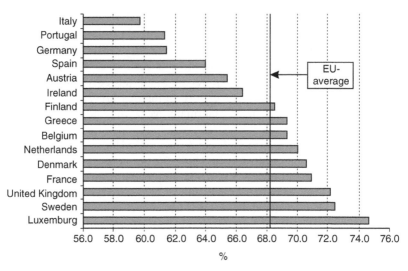

Figure 5.1 European countries: share of service employment in total employment, 1996

Source: Author's calculation based on Labour Force Survey data (Eurostat unpublished data).

employment figures for the service sector and its share in gross domestic product, at 65.9 per cent for Germany in 1998, indicate its increasing importance. However, international comparisons based on Labour Force Survey figures show that sectoral structural change has not proceeded as far in Germany as in other highly developed industrial countries. In 1996, Germany still had by far the highest share of industrial sector employment, with nearly 37 per cent. Tertiary employment in Germany in 1996 was well below the EU average and third from last in the share of service employment, with only Portugal and Italy having smaller shares (see Figure 5.1).

However, the growth of service-type employment is much higher when the occupational structure of employment within each sector is taken into account (Gruhler 1990; Hass 1995; Jagoda 1997). The service activities carried out within large firms through the 'tertiarization' of the secondary sector, especially in manufacturing, are not satisfactorily captured in the official statistics. Considered from a functional perspective, the relative position of services in Germany looks very different. Comparative analysis suggests that the share of service work done by all employees in 1993 was much the same in west Germany as in the US, at around 70 per cent (DIW 1996). International differences in the share of independent service activities thus appear to reflect the institutional organization of service functions, rather than a general deficiency in service support in Germany. The apparent 'institutional weakness' in services is a consequence of the German 'economic model',

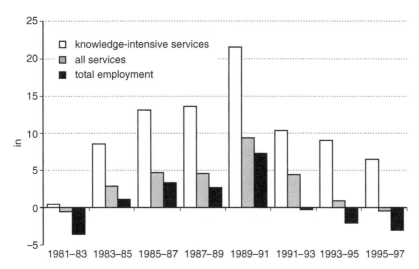

Figure 5.2 Germany: changes in employment, 1981–97

Source: Author's calculation based on Bundesantalt für Arbeit social security data.

through which success in world markets is based on the export-orientated core industries, whose products are both technology and service intensive (Strambach 1997).

Germany's traditional industrial strengths began to be regarded as potential weaknesses in the early 1990s when the world recession hit Germany. Saturation trends and the emergence of more intense international competition led to the loss of traditional markets, particularly in the dominant capital goods industries, which faced competition from the export-orientated newly industrialized countries in South-east Asia.

Germany's key industries suffered a severe structural and cyclical slump, which precipitated a large fall in employment in the highly export-orientated branches of manufacturing industry. Figure 5.2 shows the decline in total employment which began in the early 1990s. Structural unemployment in Germany was higher than in the earlier recessions. The crisis was exacerbated by the concurrent transformation of the east German economy, linked with a major process of de-industrialization. At the beginning of 1998 the unemployment figure reached its highest level at over 4.5 million.

The importance of regional KIS markets

Although the framework conditions are the same throughout Germany, there are nevertheless large regional differences in economic development. Germany's federal structure reflects the establishment of several internationally significant agglomeration centres. These have distinctly different economic structures and over time have become highly specialized. There are thus considerable inter-regional variations in KIS markets within Germany. The country's federal structure also allows state governments a relatively wide scope to carry out their own technology, innovation and regional policies. Some regional KIS developments have been the intended and/or unintended results of political intervention. Since reunification, the west–east disparity in Germany has dominated regional structure change. Other regional inequalities have been pushed to the background, including those of the traditional problem regions, the 'old industrialized areas' such as the Ruhrgebiet in North Rhine-Westphalia, undergoing industrial restructuring, with job losses and rising unemployment, since the mid-1970s.

The transformation process in east Germany since reunification is unique in terms of the speed and the radical nature of its structural change. Formerly, the east German economy was characterized by a subsidized agriculture sector, with a high share of employment, secondary industry as the leading sector, and a small and insignificant tertiary sector. The economy was dominated by giant enterprises (combines). SMEs, including the *Mittelstand* had been systematically destroyed. The transition from the planned to the market economy has been associated with a process of massive de-industrialization after 1991 which continued after 1994, although at a slower pace. Between these years around 1.4 million industrial jobs were lost, reducing employment in industry to one-third of the 1991 level. Industrial employment is now much lower in east than in west Germany. Even the service-dominated US economy has almost twice the proportion of the population working in industry (Nolte and Ziegler 1994). The former industrial regions, such as Thuringia, Saxony and Saxon-Anhalt, with the industrial agglomerations Bitterfeld and Halle/Leipzig, were particularly affected. Such regions have now become targets for structural and regional policy intervention in addition to the old west German problem regions.

The share of east German employment in the tertiary sector in 1997 was already higher (61.4 per cent) than in west Germany (58.8 per cent). There are, however, considerable functional differences in service structure between them. The degree of occupational tertiarization in east Germany is 10 percentage points lower than that in west Germany. This is due mainly to the low share of tertiary occupations in the primary and secondary sectors (Klodt *et al.* 1997).

In macro-economic terms, the economic changes that have occurred since the two Germanies became one may be positive. When, however, the high unemployment rates and the increasing numbers of bankruptcies which have followed the reductions in government subsidies are taken into account, the structural changes seem far less positive. In 1997 the unemployment rate in the new *Länder* was 17.6 per cent. If hidden unemployment is also taken into account, it rose to almost 30 per cent (OECD 1998). Many surviving east German industrial firms still have problems in finding a foothold for differentiated innovative products in west German markets. This can be attributed at least partly to a lack of productive and innovative services. It is quite obvious from this that the adjustment process has a very long way to go yet. The new *Länder* are currently the largest markets for consultancy firms in Germany. However, the institutional and economic frameworks in the two parts of Germany remain very different, and this influences the interaction of supply and demand between KIS firms and client firms.

The development of KIS supply in Germany

Key KIS attributes

There are two common attributes of KIS firms. The first is that their product is 'knowledge'. The use of the term 'knowledge intensive', by analogy with the terms 'capital intensive' and 'labour intensive', emphasizes the fact that knowledge is the most important factor of production for these firms. However, while capital and labour can be expressed in measurable economic units, the knowledge factor is difficult to define and even more difficult to measure.

The second common attribute of KIS is the intensive interaction and communication that take place between suppliers and KIS users, and which are necessary for service production. The purchase of KIS is not the same as the purchase of a standardized product or service. The exchange of knowledge products is associated with uncertainties and information asymmetries stemming from the special features of knowledge. The activity of consultancy is a process of problem solving in which KIS firms adapt their expertise to the needs of the buyer. To different degrees, this makes up the content of the interaction between KIS and their customers.

National KIS structure and trends

The lack of official statistics relating to services severely limits the possibilities for analysing KIS developments. For Germany, unlike some other European countries, there are no official statistics providing detailed information about the numbers and size of firms, sales, employment or types of service offered. The only data available to describe and systematically analyse current developments are the statistics of 'social security employment', that is National Insurance data of social security contributions, which include approximately 80 per cent of the labour force, but do not include the self-employed or freelance professionals, both of which are particularly important in KIS, or civil servants. The following analysis is mainly based on these data, still employing the limited subdivisions of the Nomenclature des Activités Européen (NACE) 1970 classification, rather than the more recent 1993 NACE schema. For detailed comment on statistical inadequacies for the analysis of KIS, see Bade (1990), Reim (1988) and Strambach (1995, 1997).

KIS have been growing rapidly in west Germany since the start of the 1980s. The two-year growth rates of social security employees in KIS have

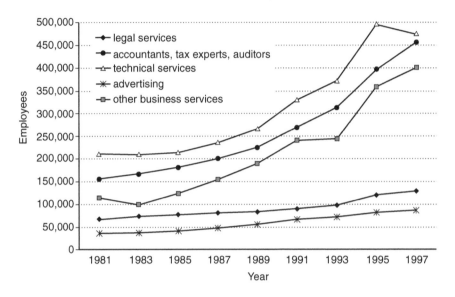

Figure 5.3 Germany: employment in business services, 1981–97

Source: Bundesanstalt für Arbeit social security data; includes east Germany from 1993.

been over 10 per cent since the mid-1980s. Growth continued even during the 1990s recession, although at a somewhat lower level. Between 1993 and 1995, KIS grew by 9.2 per cent, while service employment as a whole was stagnant (see Figure 5.2). Although KIS have grown continually, their direct employment effects should not be exaggerated. Skill requirements are high and the amount of economic activity is relatively small. Most of the firms are small and the segment's share of social security employment in 1997 in Germany, including east Germany, was 5.7 per cent. Nevertheless, between 1980 and 1997, 980,004 extra KIS jobs were created in Germany. Of these, 719,357 were in west Germany and the growth rate there was 127 per cent during this period.

At the national level, growth rates and shares varied between individual KIS sectors. In 1997, over 84 per cent of those employed in KIS were accounted for by three sectors – technical services; accountancy and related services; and 'other business services', including data processing and management consultancy, which also grew fastest after the early 1980s. In absolute terms, in both 1980 and 1997 the largest numbers of business service employees in Germany were in the technical services. Technical and technological consultancy, engineering, physical and biochemical laboratories, and architectural services are the strongest elements of KIS in Germany, being more important than in the United Kingdom or in France. More traditional

Table 5.1 Germany: share of KIS activities, 1981 and 1997

	1981	1997	1981–97
	%	%	% change
Legal services	11.5	8.3	–3.2
Accountants, tax consultants, auditors	26.5	29.5	3.0
Technical services	36.2	30.7	–5.5
Advertising	6.3	5.5	–0.8
Other business services	19.5	25.9	6.4

Source: Bundesanstalt für Arbeit social security data.

business services, such as legal services or advertising and marketing, grew relatively slowly over the fifteen-year period. However, technical services are becoming relatively less important, with a declining share of total KIS employment.

As Table 5.1 indicates, the share of technical KIS fell by 5.5 percentage points between 1981 and 1997. Between 1995 and 1997, a decrease in absolute numbers employed can be seen for the first time. Computer-related consultancies, market research, management and organizational consultancies (included in the 'other business services' group) and accountancy-related services had expanded faster, especially since the end of the 1980s. Over the years, industrial sector demand has supported the growth of technological KIS, and these structural shifts within KIS reflect changing demand requirements associated with the innovation processes of the 1990s.

The organizational structure of KIS supply: regulation, polarization and networks

Government regulations affect individual KIS sectors in quite different ways (Sahner *et al.* 1989; Strambach 1995). Formal regulations are characteristic of the established KIS sectors, including legal and tax consultancies, chartered accountancy and technical KIS such as engineers and architects. There are legally established barriers to entry, in the form of professional academic qualifications and membership of appropriate professional bodies. These professional associations monitor quality standards, the prescribed fee and charges structures, and restrictions on market conduct. The aim of these regulations is to prevent competitive behaviour detrimental to the members' interests. This is now changing with the deregulation of the European market for services. In future, factual advertising and the establishment of higher-level, cross-professional associations will be allowed.

These kinds of formal regulation and professional standards are completely absent in the newer segments of business services. There are no entry barriers in KIS branches such as management consultancy, human resources development and organization, marketing and software consultancy. There are also no legally prescribed professional associations to monitor the standards of performance. The occupational descriptions 'management consultant' and 'business consultant' are not protected in any way. Self-regulation through voluntary associations that represent the members' interests is the rule. The two best known are the Bund Deutscher Unternehmensberater (BDU – management consultants) and the Bundesverband der Wirtschafts-berater (BVW – economic consultants). Membership of these professional associations depends on meeting certain requirements in accordance with their aims of safeguarding quality. It is estimated that in Germany in 1998 there were 62,500 management consultants and 13,200 companies operating in the management consultancy area (BDU 1998; FEACO 1998). Of these, only 450 firms (around 3.4 per cent) were members of the BDU, and these were mainly the larger, German-based consultancies. Germany appears to be the biggest consultancy market in Europe with a turnover of ECU 9.5 billion in 1998 (FEACO 1998). Nevertheless, the degree of organization among the consultancies in Germany is considerably lower than in other European countries. In 1998 the market share of the consultancies belonging to the BDU was 24 per cent, one of the lowest in Europe. Rapid entry into the market is thus possible in many areas of KIS and competition is strong in some market sectors, particularly in management consultancy. This competition comes from both inside and outside each KIS sector. Industrial firms, banks and semi-public institutions increasingly also provide business consultancy services or operate in adjacent fields.

KIS are highly segmented, with a few large, mainly multinational, consultancies and many small and medium-sized firms. The 1987 census of firms is the only complete source of firm-level data in Germany. It showed that 78 per cent of KIS firms belonged to the smallest category, with fewer than five employees. KIS are thus quantitatively dominated by the small and smallest firms. Low entry barriers in many parts of the KIS market ensure strong growth in new firms. High birth rates are also connected with high death rates and, altogether, structural dynamism is at a high level. Most of the major firms, particularly in the management consultancy area, are international companies, based outside Germany. The only service sector in which German-based suppliers predominate is engineering services.

Only a very few German management consultancies have been able to retain a significant market position. In the 1990s, integration into international consultancy groups led increasingly to the disappearance of relatively large German consultancy firms. Internationalization and integration of the markets for services have greatly increased the competition faced by national KIS suppliers. Large German KIS firms serve large national companies and subsidiaries of international groups, the same clients as the multinational consultancies. Internationalization among the German firms is relatively small, so that the market position of foreign consultancy firms is stronger than that of German firms in international competition. In 1998, there was only one German member of the world's 45 largest management consultancy firms ranked by revenue, in 25th place (Roland Berger; cf. Unternehmensberater 1999). The pressure to enter international markets fell somewhat with the integration of east Germany. The opening-up of Eastern Europe induced the larger German-based suppliers to develop new markets there, and also to enter the markets in German-speaking Austria and Switzerland.

There is also an increase in formal and informal cooperation and networking. Suppliers of KIS often have to meet contradictory performance requirements. On the one hand, they have to offer highly specialized knowledge that is superior to the client's in-house know-how. On the other hand, the growing complexity of problems within the client firm requires interdisciplinary competencies. A further area of conflict emerges from the need to develop specific, high-quality ways of resolving problems for individual clients, while also standardizing services to serve efficiently as many clients as possible. This is more a problem for small KIS firms than for large multinational consultancies, whose internal organization is interdisciplinary and multi-functional. Although small firms cannot provide the same range of competencies internally, networks and cooperative relationships provide synergy potential and make it possible for consultancies to react quickly and flexibly to the frequently changing demands of the market.

A variety of evidence shows a high degree of cooperation in this sector. The low degree of formality, however, makes empirical measurement difficult. The range of services provided by firms significantly influences the tendency towards cooperation. Suppliers who primarily offer individual client services tend to cooperate more than suppliers who offer a more standardized range (Strambach 1995). Horizontal cooperative relationships, including cooperation with competitors, tend to be more important than in the manufacturing industry (see König *et al.* 1996, on innovative cooperation in the whole service sector). There is greater pressure to cooperate, not only because of the internationalization of clients but also because of the shorter shelf-life of knowledge and information. As suppliers specialize in core operations, a comprehensive solution for the client and the development of new products require cooperation.

The European Consultants Unit is an example of a formalized network covering different kinds of KIS. The network aims to support interdisciplinary cooperation between small KIS suppliers and to bring together professionals from the different sectors, either for a specific project or for longer-term collaboration, through interdisciplinary and/or multicultural teams in various organizational forms. In this way, small consultancies can act like large multinationals, assisting their clients to become internationalized. Overall, there is a high degree of self-organization between KIS sectors.

The spatial dimension of supply: region-specific profiles and development paths

There are large inter-regional differences in the distribution of German KIS. Two trends are apparent. First, KIS supply is concentrated in regions with large agglomeration centres. The highest concentrations are found in agglomeration areas where the economic range is international (Strambach 1995; O'Farrell and Moffat 1995). In more rural, peripheral regions KIS are not very important and this is unlikely to change in the foreseeable future. KIS are highly dependent on formal and informal information and contact networks in agglomeration centres. Because knowledge and information age rapidly, these networks also play an important role in interaction with clients, as will be shown below. Another factor is that the exchange of information, experience, skills and knowledge in cooperative and competitive network relationships stimulates the creation of specialized and innovative KIS products. In peripheral regions local demand does not reach the threshold level for the highly specialized innovative services needed to establish this self-sustaining process. The rapid development of KIS thus carries the danger of intensifying regional disparities.

Second, there are major regional differences in the spatial distribution of KIS, with individual regions having distinctive KIS profiles. In Hamburg, for instance, the concentration of marketing, advertising and media employment is three times higher than the national average. The share of technological services employees in the Stuttgart region is about 11 per cent higher than the national figure. These different profiles indicate the importance of regional economic and political institutional frameworks and their influence on the interaction between supply and demand. These specific regional level development patterns are hidden by macro-level national economic analyses.

This is particularly clear for east Germany. There were almost no KIS in the planned economy. Tax consultancy and management consultancy did not exist and other services were provided internally by the industrial combines and nationally owned enterprises. Unlike in market economies, where services are increasingly externalized, self-sufficiency for large firms was the rule. Official data now allow an insight into the recent development of KIS in east Germany. By 1997 the share of employment in KIS (5 per cent) had not reached the same level as in west Germany (5.8 per cent), even though the share of employment in all services was by then higher in east Germany. The structures of KIS branches also differed significantly in east and west Germany. While legal consultancy, economic services and advertising had not yet reached west German levels, the shares of technical services and 'other business services' were well above those in the west (cf. Figure 5.4). The predominance of technical services reflects the early boom in the building and construction industries, and demand arising from the restructuring of the industrial sector. 'Other business services' are mostly unregulated and have therefore provided significant self-employment opportunities. The different

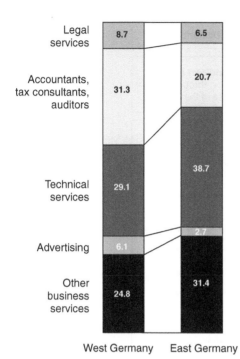

Figure 5.4 West and East Germany: share of different KIS activities, 1997

Source: Bundesanstalt für Arbeit social security data.

profiles of the KIS branches in the two parts of Germany, however, also reflect different institutional frameworks.

De-industrialization has meant that potential demand for KIS by the industrial sector in east Germany is very small. The economic position of potential buyers is often difficult, since they have lost their markets in Eastern Europe and are not well established in Western markets. The numbers of bankruptcies and firm deregistrations have grown since 1995 in sectors such as the building industry, previously the main growth sector, accompanied by a worsening of wage and productivity gaps (Bundesregierung 1996: 76). Subsidiaries and firms which are part of an outside group often have no opportunity to use regional KIS. Their decisions are made according to the strategic requirements of the head office. It is obvious that setting up KIS in the east German states is very different from doing so in west Germany and other EU countries. There is no well-functioning market to support supply-side growth. Suppliers face unpredictable, opaque demand and rapidly changing markets, even though the economic situation has become more stable than during the first years after unification. Overall, an approach which considers the specific regional economic and institutional context is required in analysing the KIS contribution to innovation in Germany.

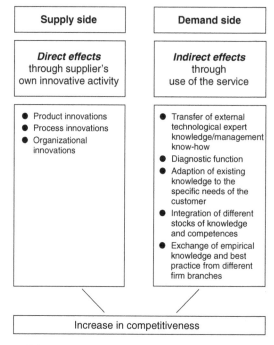

Figure 5.5 Germany: contribution of KIS firms in innovation systems

The contribution of KIS within innovation systems: direct and indirect effects

The contribution of KIS to innovation systems is twofold. First, there are the direct effects which emerge as a result of the suppliers' own innovative activities, including product, process or organizational innovations. Second, there are indirect effects which can emerge on the demand side when clients use KIS suppliers (Figure 5.5).

Direct effects: innovative activities of KIS firms and measurement problems

It is difficult to make quantitative statements about the productivity and innovativeness of KIS. Indicators and traditional tools used to evaluate productivity and innovativeness in the production system can only be used to a very limited extent for services, for a variety of reasons.

1   The traditional research and development (R&D) concept has been shaped by technological innovations in the manufacturing sector.

2   Internal innovation and knowledge organization is as a rule only weakly formalized in service firms.

3   In contrast to manufacturing firms, most KIS firms do not distinguish R&D activities in organizational terms. Quantitative indicators of innovation inputs such as investment and employment in R&D therefore cannot be used.

4   Patent applications are of limited use as output indicators for service firms because of the extremely short innovation cycle for KIS products (sometimes only six months for software products). In addition, the KIS firms' advances in know-how are difficult to protect by patents, as they are to a very large extent personnel and context bound.

5   It has not yet been possible to measure investment in intangible capital assets in available statistics (Brouwer and Kleinknecht 1997: 1236).

Given the limitations of innovation and service statistics, primary research is needed to obtain firm-level information about the complex contribution of KIS to innovation. Recent empirical results underline the fact that service innovations can be compared with those in the manufacturing sector only in a very limited sense (Licht *et al.* 1997; DIW 1998). There are differences both in inputs to innovation and in the innovation processes themselves. It is also evident that, because of the heterogeneity of services, standardized surveys do not do justice to the complexity and the specific characteristics of KIS innovation processes, and this also makes it more difficult to define and distinguish innovations by KIS. In particular, there are heavily overlapping areas of innovation. Most organizational innovations are also process innovations and there are strong interdependencies between technological and organizational innovations.

The contributions of the different KIS branches to innovation also need to be examined separately. For example, there are differences between more and less technically orientated KIS. For the latter, which include management consultancy, interdisciplinary collaboration and the combination of different competencies are of prime importance for product and process innovation. The informal character of the transfer of knowledge is particularly marked. For large consultancy firms, international comparison of empirical knowledge, the integration of different types of expertise and the adaptation of existing consultancy products may each be involved in the development of new products. Innovation by consultancy firms in the management and human resources area cannot be looked at independently of customer behaviour, as shall be discussed below.

Technology-intensive innovations are concentrated in the software and technical consultancy sectors. KIS play an important part in the adaptation and diffusion of such innovations. Computing is characterized by many product innovations, whose implementation in turn leads to process innovations elsewhere. A significant trend in the innovation activities of KIS is the dissolution of the boundaries between sectors and disciplines. The fields of engineering science and computing know-how overlap. The borders between management consultancy and software design are becoming increasingly blurred, and combining the two fields is especially important in the dominant re-engineering processes. The separation of sectors no longer appears to be appropriate for new developments.

Indirect effects: the interaction of supply and demand for innovative expertise

The contribution of KIS to the emergence and transfer of innovations can at present be determined only in qualitative terms. The role of KIS in the national and regional innovation systems is closely tied to the 'products' these services supply to the market, such as specialized expert knowledge, research and development ability, and problem-solving know-how. Given increasing differentiation and the acceleration of the growth of knowledge and information, significant indirect benefits can be expected when clients succeed in utilizing this external knowledge, such as the early recognition of problems and the more rapid adjustment to current economic and structural change.

The process of interaction between KIS firms and their clients is therefore crucial because it provides both the stimulus to innovate and the innovations themselves. The use of KIS cannot simply be equated with the purchase of external services. Demanding these services requires in-depth interaction between supplier and client, and is associated on both sides with cumulative learning processes. These learning processes must take place if the transfer of knowledge, or the solution to a problem, is to be successful. The results of the interaction process depend on the competence of both the KIS supplier and the client. The strategic significance of KIS in national and regional innovation systems consists primarily of indirect effects and positive feedback. These in the long run increase the capacity of clients to adjust and contribute to improving competitiveness.

These interaction processes are shaped within national and regional conditions, reflected in the specialization and development of these services. It is here that the recognition of systems of innovation is significant. Some typical features of the relationships involved in the interactions between the actors are discussed in the following section. Three groups of actors take part in the interaction at the firm level. A variety of public and semi-public KIS suppliers have emerged in recent decades, alongside the private suppliers in Germany. These provide intermediary consultancy services, particularly for the SMEs. The range of services supplied by the institutions differs from region to region and the institutions themselves differ considerably with regard to their market strategies (Figure 5.6).

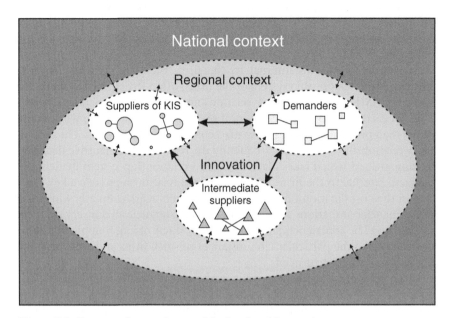

Figure 5.6 Germany: interaction model of regional innovation

The interaction process: different types of KIS–client relationships

The consultancy relationship emerges from a learning process whose features are:

- continual formulation and reformulation of expectations;
- recognition of the changes that occur during the consultancy process;
- adjustment of goals;
- a high degree of uncertainty because of information asymmetry.

The relationships between KIS and their clients are influenced by the nature, size and complexity of each project, and by the know-how available within the client firm. Different types of relationship are shaped more by client size than by sector. SMEs are most likely to benefit from external know-how, but their access to it is particularly difficult (Weitzel 1987). The special characteristics of consultancy services – their non-material nature and the interactive character of the service which often only emerges in the process of collaboration – make it extremely difficult for SMEs to anticipate the results of the consultancy process. It is even more difficult for them to set up formalized procedures within the firm which would help them to participate in the process. Large firms are in a better position to specify the external knowledge they need and to use the services of the KIS suppliers selectively. Uncertainty in the formulation of problems and the communication of requirements, and difficulties associated with evaluating the quality of the service in providing an adequate solution to the problem, are typically associated with the demand behaviour of SMEs.

As well as conditions internal to the firm, these differences also arise because of the relationship system of the firms, often structured on a regional basis. For SMEs, the consultancy relationship is proximity dependent. 'Proximity' is used here to indicate the local social and institutional environment, rather than proximity simply in a spatial sense (Strambach 1996). The consultancy relationship is itself very frequently the result of an exhaustive decision-making process which takes place through a relationship system made up of contacts external to the firm. Specific personal relationships have a key function in the coordination of activities. Small clients work primarily with small KIS. Financial limitations alone deter large consultancies from serving these small firms. The experiences, methods and products of large suppliers are also not tailored to the particular circumstances of small firms and are thus often not appropriate to their needs.

Large KIS firms work principally with large customer enterprises. The reasons for this are consultancy costs and the reputation of large consultancy firms, based on publications, the development of new consultancy products, and other clients' recommendations. These are often the important factors for large enterprises when choosing a particular consultancy firm. Large enterprises also employ small KIS firms. They are less costly, less bureaucratic, often more flexible than the large KIS firms, and specialized and innovative in particular knowledge fields. Large consultancy firms are often forced to standardize a great many of their activities, consultancy procedures and consultancy products, with a resulting loss of flexibility *vis-à-vis* customers' requirements. The willingness to learn during the interaction process with the client and to adapt to the client's specific character, perhaps even to undertake something quite new, tends to be greater in the case of small firms, giving them an advantage over the large suppliers. While small suppliers are increasingly pushing forward into the large-customer area, the large consultancy firms are also trying to open up markets in the greater area of medium-sized firms. Such clients have begun to acquire experience with consultancy services and, because of this, are no longer so cautious about buying external know-how. Moreover, they are also becoming increasingly dependent on external know-how to maintain their competitive position in the current situation of structural change.

The roles of large and small KIS firms in systems of innovation

These differences in the nature of KIS–client interactions indicate that the major multinational suppliers of KIS and the small KIS firms play different innovative roles. The importance of the large, mostly international, firms stems primarily from the fact that, on the one hand, they develop new consultancy products in the form of methods, instruments and models based on their own know-how and experience. Unlike smaller national or local suppliers, they often have formal internal R&D functions which further the creation of new expertise. On the other hand, the internationalization of their clients helps them diffuse these 'best practice' products. The transformation of consultancy innovations into standard products occurs more quickly when it is carried out within a specialized department of an organization. In this way international KIS firms hasten the standardization process in the areas of management and technology. Organizational innovations such as lean production, lean management and re-engineering, based on research such as that undertaken at the Massachusetts Institute of Technology, have been taken up, developed further, and transferred to other countries by large, internationally orientated KIS (see European Commission 1996a; Schneider 1995; Kipping and Bjarnar 1998; Rassam 1998 for a more historical perspective).

Innovations are generally associated with the communication and learning processes of firms. These are to a very large extent determined by social and cultural factors which differ not only nationally and locally but also at the firm level. These processes strongly influence the introduction of innovation. This culture- and context-linked knowledge, which is very difficult to codify, represents a major barrier to the internationalization of consultancy services. The significance of national or local KIS firms is that they can adapt knowledge of innovative methods through codification to the client-specific situation by using their local knowledge and experience. It is these firms which, in a learning and interaction process, can use the knowledge about models and instruments as an element in the problem-solving process (see Chapter 3).

Intermediary institutions at the regional level: the case of Baden-Württemberg

There is a long tradition of sponsoring innovativeness among the SMEs in Baden-Württemberg. The widespread decentralized knowledge and technology transfer structure for SMEs, established on the basis of public sector intermediary institutions, is widely recognized as a major factor in the state's successful growth (Pyke and Sengenberger 1992). The European Commission 1996 *Green Paper on Innovation* also viewed this as a model network (European Commission 1996b). The most important of these institutions are the chambers of handicrafts, the chambers of commerce and industry (IHKs), the German Rationalization Board (RKW), and the Steinbeis Foundation for Economic Development.

The RKW has been operating a consultancy service since 1995, focused mainly on economic management consultancy. The Steinbeis Foundation was established back in 1971. Initially, its aims were to provide technological consultancy services and to aid technological innovation. Over the years, the role taken by these federal- and state-funded KIS intermediaries in the regional innovation process has changed. Two distinct phases can be recognized. In the first, up to the end of the 1970s and the start of the 1980s, intermediary KIS mainly operated mediation and development functions with client SMEs. Their individual consultancy and knowledge needs were identified by the intermediary agency through short-term, free consultations, and they were then referred to external KIS specialists. Consultancy relationships were thus established between client and supplier who were then left to proceed on their own. In this phase there was a clear separation between the intermediaries and private suppliers of KIS. The intermediaries took a neutral position between supply and demand. Their function was to diagnose problems and help formulate requirements.

In the second phase after the early 1980s, the Steinbeis Foundation gave up its purely development and mediation functions and began dealing directly with the client's problems. It now no longer focuses on handing the client over to a competent partner. All stages in solving a client's problem are handled, from analysing the problem, to working out a solution and helping to implement it. Where the Steinbeis Foundation itself does not have the necessary know-how, it collaborates with private suppliers. The Steinbeis Foundation is now performance orientated, just like private suppliers. The Baden-Württemberg RKW has also followed this strategy since the mid-1980s. This change in strategy of the KIS intermediaries is connected with increasing competitive pressures from the rapid development and differentiation of the private supply of these services.

The shift also includes a widening of the range of expertise required. This affects the relationships between the intermediaries. The original subject boundaries in consultancy and knowledge transfer areas are breaking down, in particular between economic and technologically orientated consultancy. This can be seen in the expansion of the services offered by the intermediaries. Competition now takes place not only between the intermediaries which formerly cooperated with one another, but also between them and the private suppliers. In this second phase, the boundary between the intermediate KIS activity and the private KIS market has become blurred.

The successful role played by the intermediaries in the regional innovation system in Baden-Württemberg is related to:

1   Their process orientation, which allowed client trust to be built up during the interaction process. Trust is essential for the learning process. In most cases, long-term relationships between the clients and the suppliers have emerged, and, as a result, outside expertise is called upon again when new problems arise.
2   The comprehensive nature of the problem solving carried out by the KIS intermediaries links the various kinds of knowledge stocks (technological/economic) held by the different institutions and directs them towards the individual needs of the SMEs.

3   Process orientation means that the intermediaries take on responsibility for the results of the consultation. This not only affects the quality of the consultancy service but also helps the self-organization of demand for consultancy services. Every firm receiving successful consultancy services becomes a potential intermediary for other firms in its relationship system.
4   The Steinbeis Foundation, in particular, recognizes the coordinating function of informal networks of personal contacts, through explicit management of cooperation with both private and intermediary suppliers, and has expanded the scope of its external relationships with great success.

The procedures undertaken in Baden-Württemberg have resulted in overlapping supplies of services to SMEs and thus competition in the intermediate consultancy market. This supply surplus is an indication of the strength of the regional market (see Pyke 1994; Schmitz 1992; Grabher 1994 for the importance of redundancy in regional development). Strong competition has also forced the non-profit organizations to orientate themselves emphatically towards the customers and the market. They take active steps to acquire new customers by offering new kinds of needs-orientated products. KIS intermediaries have made a considerable contribution to the emergence and diffusion of high-quality incremental innovations in technical products and processes. However, the economic crisis of the 1990s, which hit the regional economy of Baden-Württemberg particularly hard, revealed that the intermediaries tended to reinforce the technically orientated development path of the region, with possible negative effects upon prompt adjustment to changing markets (see Braczyk and Schienstock 1996; Schienstock 1997; Reinhard and Schmalholz 1996 for an evaluation of technology transfer in Germany).

KIS development and the role of policy in the national innovation system

In summary, the following are characteristic of KIS structure and development in Germany.

1   There is an 'institutional weakness' in services in general, and KIS in particular, as a result of the different institutional organization of KIS functions compared to other countries.
2   Technical engineering KIS dominate the sector, although to a greater extent in the 1980s than by the end of the 1990s.
3   There has been a steadily declining share of technical services in KIS employment in recent years and a corresponding rise in integrated consultancy services such as management consultancy or services associated with computing.

4   On the regional level, there are major differences in the spatial distribution of KIS and there are distinctive regional KIS profiles.

KIS developments are thus clearly interdependent with other essential features of the German political and institutional context.

The role of policies in the development of the KIS market must also be seen against a background of vertical and horizontal fragmentation in the German innovation system. The vertical dimension is determined at the German federal, state and municipal levels and, since the 1980s, the European institutional level. So far, private KIS supply has not been the subject of direct government measures or economic development strategies. From a systemic point of view, however, the political and institutional framework created by public actors in research, technology and innovation policies has indirectly influenced the development and structure of KIS in Germany.

Economic, research, technology and innovation policy measures in Germany have been, and still are, heavily biased towards the key productive sectors and certain areas of technology that are held to be responsible for the country's competitiveness and growth. Previously, the productivity of services was generally considered to be very low, and the service sector was regarded as far less relevant for value added in the economy than a strong manufacturing sector. The basic characteristics of the German innovation system are the relatively small government share in the national R&D budget compared to that in other industrial countries, a high proportion of industrially financed R&D, and a highly decentralized and differentiated research system. The private sector – mainly the large firms in the key industrial sector, which provide two-thirds of the total funds for national R&D – is thus the most important actor in the German innovation system.

The federal government is the second largest provider of funds, and the state governments occupy third place (Keck 1993; Klodt 1995). The aims of national technology policy are reflected in the high value placed on basic research in all promotional programmes (OECD 1995), and by the support for selected key technologies and for technologies in as yet uncompetitive fields. Emphasis here is on cross-sectoral technologies such as information technology, production technology, biotechnology, material research and physical/chemical technologies. Government support is designed to create a broad scientific and technological base in these fields and thus provide an improved basis for production-orientated industrial R&D. The sectoral structure of government support has not changed for many years (BMBF 1994; Klodt 1995). By far the largest share of government funds goes to firms in the manufacturing sector.

The established research and innovation networks are mainly orientated towards the German core industries and special technology fields. Krull and Meyer-Krahmer (1996: 13) point out that, as far as the main technological flows are concerned, the most significant channel of diffusion in Germany predominantly reflects the features of the manufacturing processes. For many

years this focus on technical and physical science dominated political intervention. In the 1980s technical and technological scientific knowledge acquired
status as a factor of production. The introduction of technical innovations and
the transfer of technological knowledge, increasingly directed towards SMEs,
determined support strategies.

With respect to technology and innovation policy, since the 1980s the
federal level has been eclipsed by increased promotion activity by the states
and municipalities. For example, at the end of the 1970s Baden-Württemberg
was the first state to formulate a technology policy. Technology policy at
the national level does not have any explicit regional or spatial dimension.
In contrast to France, for example, there are no support programmes in
Germany specifically directed towards high-tech regions. Only very recently,
with the reunification of the two parts of Germany, has a spatial goal had
priority: the building-up of a research and technology structure in east
Germany. The regional distribution of funds, which are mainly given out for
particular projects, shows that more than half of the R&D expenditure of the
federal government is received jointly by Bavaria, North Rhine-Westphalia
and Baden-Württemberg (BMBF 1994). These indicators show that there are
indirect regional effects of government policy arising from the regional concentration of certain industries, as well as research institutions in these states.

As at the national level, private KIS are not part of regional technology
policy development strategies or the regional economic promotion programmes. These programmes, like the intervention of the federal government, are geared towards the demand side and thus only indirectly affect the
development of the KIS sector. The aim of regional policy agencies is to
increase the adaptability and competitiveness of the regional economy by
promoting links between the scientific sector, as the supplier of innovation-
relevant knowledge, and potential users of this knowledge. These links are
directed towards speeding up the transformation of innovative technical
knowledge into new marketable products and processes through a variety of
promotion measures. A major feature of the German innovation system is
now the blanket coverage of technology transfer and research infrastructure.

The current research and innovation infrastructure in Germany thus
reflects a technology-based competitive model which assumes that competitive advantage results mainly from the technical efficiency of products and
processes (see Braczyk and Schienstock 1996). The extensive structures and
networks that have been built between government, science and economy
have fostered patterns of innovation to support the long-term economic
success of technology-intensive and service-intensive industries. However,
this development path has also been an impediment to innovation in other
fields of knowledge and other sectors, reflected in the traditionally low profile
of KIS and the low importance of service exports.

The dominant position of technical KIS also reflects the political and institutional framework of the innovation system. This favoured their development compared with services based on economic or social science knowledge.

Technological and engineering science KIS correspond to the areas of operation of industrial firms in the key sectors. The orientation of the education and training system, and the overlapping qualification structures of the technicians and engineers aid the rapid conversion of technically orientated knowledge not only into production concepts but also technical service products.

The Federal Ministry of Education, Science, Research and Technology (BMBF), now the Federal Ministry of Education and Research, the most important actor in federal research and technology policy, has only recently begun to take more interest in economic questions (Meyer-Krahmer 1993). Since the early 1990s, when the structural decline of secondary industry became a problem, political actors have begun to realize the importance of services for future economic development. This is evident in the large BMBF initiative, 'Services 2000 plus', which in 1995 for the first time provided a definition of the fields of operation and research in the service area. Services are still mainly discussed in relation to manufacturing industry. KIS are not the research objects of this initiative and are completely neglected in the first phase. Only in the latest follow-up project was a team established to address producer-orientated services (Preissl 1998). The fact that the importance of KIS in the national innovation system is not perceived by political actors is a particular feature of the national innovation system and of the persistence of institutional structures.

The R&D policy of the EU has had little influence on the establishment of the dominant institutional structures of the national innovation systems, at least measured in terms of R&D expenditure. Thus EU funds were about 1.8 per cent of those made available by the federal government in the period from 1987 to 1991. In other European countries, for example Greece, where the amount of EU money is 54 per cent (for universities) and 85 per cent (for industry) of that made available to support innovation policy by the national government (see Krull and Meyer-Krahmer 1996: 24), the influence and the direct impacts of the EU policy are greater than in Germany. However, even in Germany, the indirect effects and cumulative impacts of EU R&D policy should not be underestimated. The European Union's support programmes are concentrated in individual subject areas and this increases the influence of EU funding quite significantly. An example is the field of information and communication technology where the EU is more active than the federal government. The expenditures by the EU amounted to between one-fifth and one-quarter of that of the federal government in absolute terms (Reger and Kuhlmann 1995). These priorities also stimulate developments and trigger innovative impulses at national and regional levels. Nevertheless, the contribution of socio-economic insights to innovation policy, which has been discussed at EU level since the early 1990s and reflected in the EU Framework IV Research Programme, was discussed at the national level in Germany much later, most notably in a BMBF conference in Bonn in July 1998.

Conclusion: the strategic challenge for policy

The strategic challenge consists of stimulating not only the established technology and industry development paths, which are certainly still important, but also new fields of activity, especially those marked by KIS growth. In the process of economic globalization KIS are developing increasingly into a non-institutional informal 'knowledge-transfer structure' and are thus an important element for the wider innovation system. As has been shown, these services link existing technological and management knowledge from separate disciplines, combine them in new ways and adapt them to the appropriate business context. This occurs not only through their innovative service products but also through the indirect and feedback effects which stem from these products. Because they are under continual pressure from intensive competition to build up new competencies in their core businesses, they spread both collective know-how, based on experience, and best practices across different sectors, and so produce further new knowledge. By doing this they encourage the transformation of knowledge into marketable products and contribute to the emergence, diffusion and adoption of technological, organizational and social innovations. Through these interdependent processes, KIS contribute to the acceleration of the innovation dynamic. The high degree of self-organization, which is one of their features, must therefore be seen as a strength with respect to innovative capacity. From this point of view, the integration of this service segment into European, national and regional innovation and development strategies appears to be urgently needed.

Delayed development of KIS supply, characteristic of the German system, may result in less rapid innovation dynamic and thus a reduction in adaptability. Taking KIS into consideration in innovation strategies, however, does not imply that there is an obligation to develop new sector strategies with reference to KIS. Any action to support the KIS function in innovation must promote interaction between the demand and supply sides. Such measures can be effective only if they take into account the specific national and regional socio-institutional context of the interaction. Political strategies for action thus require a decentralized, process-orientated perspective, which takes account of the high degree of self-organization in broad sections of the KIS market.

Without setting out the concepts too precisely, the fields for action include, above all:

- providing support for networks between suppliers and clients;
- promoting and stimulating cooperative relationships between small KIS suppliers;
- encouraging the build-up of transnational cooperation networks;
- making access to external knowledge easier, particularly for SMEs;
- developing quality standards. The question of the quality of KIS is one of

the key factors and is directly connected with the qualification and the competence of the workforce.

The strategies followed so far in technology and innovation policies in Germany originated as answers to the institutional operating conditions of the 1980s. The framework conditions and the pattern of innovations, however, changed in the 1990s as indicated by the increasing importance of organizational and social innovations, and discussions about the 'learning economy'. In future, the production of services, information and knowledge will be much more important for the competitiveness of national and regional economies in relation to the production of investment goods. The challenge for political actors is to maintain the strengths of the national innovation system while, at the same time, initiating processes of change in the institutional landscape. This should enable the opening-up of established research and innovation networks, and making extensive readjustments to the education and training systems.

Note

The author would like to acknowledge the support of the research by the EU Framework IV Programme of Research (Targeted Socio-Economic Research Area 1) Project KISINN SOE1–CT96–1017 coordinated by Professor Peter Wood.

References

Bade, F.J. (1990) 'Expansion und regionale Ausbreitung der Dienstleistungen: eine empirische Analyse des Tertiärisierungsprozesses mit besonderer Berücksichtigung der Städte in Nordrhein-Westfalen', *ILS-Schriften* No. 42 (special edition), Dortmund: ILS (Institut für Landes- und Stadtentwicklung des Landes Nordrhein-Westfalens).

BDU (1998) *Fact & Figures zum Beratungsmarkt*, Bonn: BDU.

BMBF (1994) *Jahresbericht*, Bonn: Bundespresseamt.

Braczyk, H.J. and Schienstock, G. (eds) (1996) *Kurswechsel in der Industrie: Lean Production in Baden-Württemberg*, Stuttgart: Kohlhammer Verlag.

Brouwer, E. and Kleinknecht, A. (1997) 'Measuring the unmeasurable: a country's non-R&D expenditure on product and service innovation', *Research Policy* 25: 1235–42.

Bundesanstalt für Arbeit (various dates) *Statistik der Sozialversicherungspflichtig Beschäftigten* (unpublished data, author's analysis), Nuremberg: Bundesanstalt für Arbeit (Federal Institute for Employment).

Bundesregierung (ed.) (1996) *Jahreswirtschaftsbericht 1996 der Bundesregierung*, Stuttgart: Metzler-Poeschel Verlag.

DIW (Deutsches Institut für Wirtschaftsforschung – German Institute for Economic Research) (1996) 'Keine Dienstleistungslücke in Deutschland: ein Vergleich mit den USA anhand von Haushaltsbefragungen', *DIW-Wochenbericht* (Berlin) 14: 221–6.

—— (1998) 'Innovationen im Dienstleistungssektor', *DIW-Wochenbericht* (Berlin) 29: 519–33.

European Commission (1996a) *International Transfers of Organisational Innovation*, EIMS (European Innovation Monitoring System) Publication No. 45, Luxemburg: European Commission.

—— (1996b) *Green Paper on Innovation* (*Bulletin of the European Union*, Supplement 5/95), Brussels: European Commission.

FEACO (Federation of European Management Consulting Associations) (1998) *Survey of the European Management Consultancy Market*, Brussels: FEACO.

Grabher, G. (1994) *Lob der Verschwendung: Redundanz in der Regionalentwicklung: ein sozioökonomisches Plädoyer*, Berlin: Edition Sigma.

Gruhler, W. (1990) 'Dienstleistungsbestimmter Strukturwandel in deutschen Industrie-unternehmen', *Materialien des Instituts der deutschen Wirtschaft* No. 6 (monograph), Cologne: Deutches Institut der Wirtschaft Verlag.

Hass, H.J. (1995) 'Industrienahe Dienstleistungen', in Institut der deutschen Wirtschaft Köln (ed.) *Beiträge zur Wirtschafts- und Sozialpolitik* (Cologne) 223(3): 11–40.

Jagoda, B. (1997) 'Neue Arbeitsplätze durch Dienstleistungen', in Mangold, K. (ed.) *Die Zukunft der Dienstleistungen: Fakten, Erfahrungen, Visionen*, Frankfurt: Gabler, 120–9.

Jürgens, U. and Naschold, F. (1994) 'Arbeits- und industriepolitische Entwicklungs-engpässe der deutschen Industrie in den neunziger Jahren', in Zapf, W. and Dierkes, M. (eds) *Institutionenvergleich und Institutionendynamik: Jahrbuch des Wissen-schaftszentrum Berlin für Sozialwissenschaft*, Berlin: Edition Sigma, 239–70.

Katzenstein, H. (1989) *Industry and Politics in West Germany: Towards the Third Republic*, Ithaca: Cornell University Press.

Keck, O. (1993) 'The national system for technical innovation in Germany', in Nelson, R. (ed.) *National Innovation Systems: A Comparative Analysis*, New York and Oxford: Oxford University Press, 115–57.

Kipping, M. and Bjarnar, O. (eds) (1998) *The Americanisation of European Business: The Marshall Plan and the Transfer of US Management Models*, London and New York: Routledge.

Klodt, H. (1995) 'Technologiepolitik aus ökonomischer Sicht: theoretische Anforder-ungen und politische Realität', *Aus Politik und Zeitgeschichte* (Bonn) 49: 11–18.

Klodt, H., Maurer, R. and Schimmelpfennig, A. (1997) 'Tertiärisierung in der deutschen Wirtschaft', in Siebert, H. (ed.) *Kieler Studien* No. 283, Tübingen: JCB Mohr.

König, H., Kukuk, M. and Licht, G. (1996) 'Kooperationsverhalten von Unternehmen des Dienstleistungssektors', in Helmstädter, E., Poser, G. and Ramser, H.J. (eds) *Beiträge zur angewandten Wirtschaftsforschung, Festschrift für Karl Heinrich Oppenländer*, Berlin: Duncker und Humblot, 217–43.

Krull, W. and Meyer-Krahmer, F. (eds) (1996) *Science and Technology in Germany*, London: Cartermill.

Licht, G., Hipp, C., Kukuk, M. and Münt, G. (1997) 'Innovationen im Dienst-leistungssektor: empirische Befunde und wirtschaftspolitische Konsequenzen', *Schriftenreihe des ZEW Zentrum für Europäische Wirtschaftsforschung* Band 24, Baden-Baden: Nomos.

Meyer-Krahmer, F. (ed.) (1993) *Innovationsökonomie und Technologiepolitik: Forschungsansätze und politische Konsequenzen*, Heidelberg: Physica-Verlag.

Naschold, F. (1996) 'Jenseits des Baden-Württembergischen Exceptionalism: Strukturprobleme der deutschen Industrie', in Braczyk, H.J. and Schienstock, G. (eds) *Kurswechsel in der Industrie: Lean Production in Baden-Württemberg*, Stuttgart: Kohlhammer Verlag, 184–212.

Naschold, F., Soskice, D., Hancké, B. and Jürgens, U. (eds) (1997) 'Ökonomische Leistungsfähigkeit und institutionelle Innovation: das deutsche Produktions- und Politikregime im globalen Wettbewerb', *Wissenschaftszentrum Berlin für Sozialforschung* (WZB-Jahrbuch), Berlin: Edition Sigma.

Nolte, D. and Ziegler, A. (1994) 'Neue Wege einer regional- und sektoralorientierten Strukturpolitik in den neuen Ländern: zur Diskussion um den "Erhalt industrieller Kerne"', *Information zur Raumentwicklung* (Bonn) 4: 255–2.

OECD (1995) *OECD Wirtschaftsberichte Deutschland 1995*, Paris: OECD.

—— (1998) *OECD Wirtschaftsberichte Deutschland 1998*, Paris: OECD.

O'Farrell, P.N. and Moffat, A.R. (1995) 'Business services and their impact upon client performance: an exploratory interregional analysis', *Regional Studies* 29(2): 111–24.

Preissl, B. (1998) *Knowledge-Intensive Business Services and Innovation in Germany*, DIW-SI4S Report No. 5, Berlin: DIW.

Pyke, F. (1994) 'Small firms, technical services and inter-firm cooperation', *Research Series* No. 99, Geneva: International Institute for Labour Studies.

Pyke, F. and Sengenberger, W. (eds) (1992) *Industrial Districts and Local Economic Regeneration*, Geneva: International Institute for Labour Studies.

Rassam, C. (1998) 'The management consultancy industry', in Sadler, P. (ed.) *Management Consultancy: A Handbook of Best Practice*, London: Kogan Page, 3–30.

Reger, G. and Kuhlmann, S. (1995) 'Europäische Technologiepolitik in Deutschland: Bedeutung für die deutsche Forschungslandschaft', *Schriftenreihe des Frauenhofer-Instituts für Systemtechnik und Innovationsforschung, Technik, Wirtschaft und Politik* No. 11, Heidelberg: Physica-Verlag.

Reim, U. (1988) 'Zum Ausbau statistischer Informationen über Dienstleistungen', *Wirtschaft und Statistik* 12: 842–8.

Reinhard, M. and Schmalholz, H. (1996) 'Technologietransfer in Deutschland: Stand und Reformbedarf', *Schriftenreihe des ifo Instituts für Wirtschaftsforschung* No. 140, Berlin and Munich: Duncker und Humblot.

Sahner, H., Herrmann, H., Rönnau, A. and Trautwein, H.M. (eds) (1989) 'Zur Lage der freien Berufe 1989, Teil I/II Empirischer Überblick; soziologische, wirtschaftswissenschaftliche und rechtswissenschaftliche Betrachtung', *Schriften des Forschungsinstitutes Freie Berufe* No. 1, Lüneburg: Buchheister KG.

Schienstock, G. (1997) 'The transformation of regional governance: institutional lock-ins and the development of lean production in Baden-Württemberg', in Whitley, R. and Kristensen, P.H. (eds) *Governance at Work: The Social Regulation of Economic Relations*, Oxford: Oxford University Press, 190–208.

Schmitz, H. (1992) 'Industrial districts: model and reality in Baden-Württemberg, Germany', in Pyke, F. and Sengenberger, W. (eds) *Industrial Districts and Local Economic Regeneration*, Geneva: International Institute for Labour Studies, 87–121.

Schneider, U. (1995) 'Experten zwischen verschiedenen Kulturen: Ist Beratung ein globales Produkt?', in Walger, G. (ed.) *Formen der Unternehmensberatung: Systemische Organisationsberatung, Organisationsentwicklung, Expertenberatung und gutachterliche Beratungstätigkeit in Theorie und Praxis*, Cologne: Verlag Dr Otto Schmidt, 139–58.

Strambach, S. (1995) 'Wissensintensive unternehmensorientierte Dienstleistungen: Netzwerke und Interaktion: Am Beispiel des Rhein-Neckar-Raumes', *Wirtschaftsgeographie* Band 6, Münster: Lit-Verlag.

—— (1996) 'Organisation versus Selbstorganisation des regionalen Wissens- und Informationstransfers: die Beratungsbeziehungen kleiner und mittlerer Unternehmen im regionalen Kontext von Baden-Württemberg und Rhône-Alpes', in Heinritz, G. and Wiesner, R. (eds) *Wettbewerbsfähigkeit und Raumentwicklung*, Band 3, Stuttgart: Franz Steiner Verlag, 162–71.

—— (1997) 'Wissensintensive unternehmensorientierte Dienstleistungen: ihre Bedeutung für die Innovations- und Wettbewerbsfähigkeit Deutschlands', *Deutschen Instituts für Wirtschaftsforschung (DIW), Vierteljahreszeitschrift* (Berlin) 66(2): 230–42.

Unternehmensberater (1999) *Wachstum und Ranking im Beratungsmarkt*, Vol. 4, Heidelberg: Dr Curt Haefner-Verlag, 4–7.

Weitzel, G. (1987) 'Kooperation zwischen Wissenschaft und mittelständischer Wirtschaft. Selbsthilfe der Unternehmen auf regionaler Basis: kritische Bewertung bestehender Modelle', *Ifo Studien zu Handels- und Dienstleistungsfragen* No. 31, Munich: Institut für Wirtschaftsforschung.

6 The Netherlands

Knowledge-intensive service markets in a small open economy

Walter Manshanden, Jan Lambooy and Chris van der Vegt

Introduction

The service sector can be studied from a macro-economic point of view (Fourastié 1949; Clark 1940; Fuchs 1964; Baumol 1967), or through a micro-economic approach focusing on characteristics of the firm. Much micro-economic work has also been carried out on research and development (R&D) intensity and innovation in The Netherlands (Brouwer and Kleinknecht 1994, 1995, 1996). The importance of business services, however, lies primarily in their relationship to, and impacts on, other sectors of the economy. Knowledge-intensive services (KIS) are generally assumed to employ highly qualified labour, to be located in urban areas, to attract other industries (including manufacturing) and to contribute to innovation in other firms. What is the evidence therefore for the impacts of KIS on other sectors in the Dutch economy? This chapter explores this and related aspects of KIS in The Netherlands, on the basis of various quantitative and qualitative sources.

The national context: features of the Dutch economy

The Netherlands: a service economy

This chapter will first examine some key features of the Dutch economy and review policies related to technological development. Data will also be presented to show the growth of business services, serving also as the basis for some general explanations of trends. In the 1970–86 period, the Dutch economy grew at a slower pace than its European counterparts. Until the late 1980s, real gross domestic product (GDP) per capita, which in the early 1970s had been nearly 15 per cent above the EU average, continued to fall behind other EU countries (OECD 1998: 29). There was, however, a strong economic recovery after 1987. The 1993 downturn was considerably milder in The Netherlands than in other EU countries. As a result, since the late 1980s Dutch GDP per capita has rebounded relative to the EU average (OECD 1998: 30).

Another feature of the Dutch economy is a relatively low labour-participation level. GDP per employee is one of the highest in the world, whereas GDP

Table 6.1 The Netherlands: value added, employment, labour productivity and annual growth rates, 1970–95

	Value added	Employment	Labour productivity growth	Number of jobs 1995 (×1,000)
Agriculture	3.0	−0.3	4.1	315
Manufacturing	2.3	−1.2	3.8	1,047
Building	0.9	−1.2	2.2	408
Wholesale and transport	3.7	1.0	2.9	835
Producer services	4.6	3.9	1.0	803
Consumer services	2.8	1.5	1.8	1,368
Non-profit services	2.2	2.2	0.6	1,841
Total economy	2.8	0.9	2.3	6,647

Source: SEO calculations, from Central Bureau of Statistics data, Van der Vegt and Manshanden 1996.

Note
Real estate not included.

per capita is only moderate. The Netherlands thus has a relatively small but highly productive labour force. However, to reduce budget deficits without undermining the welfare system, economic policy has targeted expansion of the labour force and structural reductions of the economically inactive population. The main instrument has been to restrain labour costs, a policy begun in the early 1980s which has had a significant positive impact on the labour market. Since 1987 employment has increased by 100,000 jobs per year. This recovery is recognized as the 'Dutch model', although, as the OECD notes, the large number of inactive persons, or broad unemployment, is still an economic weakness (OECD 1998: 6).

The service industries dominate Dutch employment, including the professional and knowledge-intensive services (Elfring 1988). During the 1960–84 period, Dutch employment growth rates in the business and professional services were comparable to those in Japan and the United States. Manufacturing industry comprised 16 per cent of total employment in 1995, whereas services contributed 74 per cent. The picture is somewhat different when measured by value added, owing to different productivity levels, but 67 per cent of value added still comes from services. Within services, 'wholesale and transport' is the largest sector, with 15 per cent of national value. This compares with 24 per cent from manufacturing. 'Producer services', including banking and insurance, are the third largest sector, with 13 per cent. In absolute terms, producer and financial services provide more than 800,000 jobs. They

Table 6.2 Use-table of the Dutch economy in 1990: use of commodities, total and KIS (Fl 1 million)

	Commodities			KIS sectors						
	Total use	KIS	%	Archit. and enging.	Computer service	Legal	Account and tax consult.	Econ. consult.	Market and advert.	Research
Agric/mining	24,780	580	2.3	10	155	31	328	9	20	27
Food	59,738	1,890	3.2	24	159	10	324	77	1,165	131
Chemical	33,956	1,723	5.1	19	341	3	332	186	479	363
Metal	50,639	2,371	4.7	35	433	9	713	278	457	446
Other manu.	50,639	2,465	4.9	21	396	20	826	208	651	343
Construction	42,778	688	1.6	27	124	7	306	67	132	25
Trade/repair	40,576	4,998	12.3	74	854	201	1,554	318	1,925	72
Trans and storage	20,216	1,272	6.3	0	413	1	374	14	450	20
Bank and insurance	12,436	2,054	16.5	0	435	159	1,016	2	417	25
Government	26,960	3,639	13.5	132	541	277	1,165	305	9	1,210
Other services	52,076	3,119	6.0	349	487	160	1,083	221	500	319
KIS	10,602	3,642	34.4	542	863	92	681	585	793	86

KIS sectors

Architects/engineers			664	147	214	7	102	187	0	7
System dev. & IT consult.			625	85	165	21	132	112	109	1
Legal advisers			218	3	46	26	88	51	2	2
Accountants/tax consult.			479	61	103	15	106	83	109	2
Economic advisers			856	179	292	9	193	146	37	0
Advertising agencies			750	47	33	10	54	1	531	74
Research institutes			50	20	10	4	6	5	5	0
Final expend.	666,594	12,878	2.0	6,553	1,242	1,715	972	987	229	1,180
Exports	234,133	6,612	2.9	1,315	1,242	1,189	586	987	229	1,064
Total	1,091,711	41,319	3.9	7,786	6,443	2,685	9,674	3,257	7,227	4,247

Source: Data from Centraal Bureau voor de Statistiek (Central Bureau of Statistics), Voorburg, 1993.

Note

Data adapted from national input–output make-tables and use-tables. Column totals affected by rounding figures.

were the fastest growing sector during the 1970–95 period (see Table 6.1), measured by both value added and employment. It should be stressed that the absolute levels of value added and labour productivity in manufacturing industries are still at a higher level.

The structure of demand

The market for services: the demand side

The demand for knowledge-intensive services is well demonstrated by the so-called 'use-tables' derived from national input–output data. Use-tables allow the quantitative study of the production and demand structure of sectors. They show which industries use which commodity. Together with 'make-tables', they constitute aggregate national input–output tables. The basic data are quite detailed, enabling aggregation of individual industries, and disaggregation of commodity classes. Thus, a precise profile can be derived of the consumption of the commodity defined as 'knowledge-intensive services', by various sectors.

Total annual production of KIS amounted to Fl 41 billion in 1990; 3.9 per cent of total production in The Netherlands. By examining demand for KIS by sector (Table 6.2), KIS intensity can be measured in terms of the use of KIS as a percentage of all the commodity inputs of each sector. From the use-table it is clear that KIS firms are themselves the most intensive users of KIS, drawing in 34.4 per cent of the value of their inputs from a wide array of other KIS. The second most knowledge-intensive sector is banking and insurance, in which 16.5 per cent of inputs constitute KIS, half from financially related accountancy and tax consultancy. The government sector ranks third and especially depends on accountancy, research and computer inputs. The trade and repair sector also purchases a high proportion of KIS, with the highest share from market and advertising services. The various manufacturing industries purchase relatively low levels of KIS, although accountancy and tax advice are generally important. The chemical industry is the most intensive user of KIS of the manufacturing industries shown, especially marketing and advertising services. Food, beverages and tobacco, a strong industrial sector in The Netherlands, appears to have a relatively low dependence on KIS, again mainly for marketing and advertising expertise.

From the point of view of the KIS sectors themselves, economic and organizational consultancy has strong markets in manufacturing, as well as trade and repair and government. The main outlet for marketing and advertising consultancy is in trade and repair, and in food, beverages and tobacco. Computer services are widely used to a similar degree by every sector. Legal advice, however, is strongly geared towards the government and other service industries. The most forward-looking KIS, research services, are most used by government and manufacturing. This reflects broad patterns of R&D expenditure, indicating how some KIS may be implicated in basic manufacturing innovation.

The use of KIS by other KIS sectors reveals an interesting pattern of mutual support. Economic consultancies act as a focal point in these cluster relationships. They are geared towards the manufacturing industries but are also large users of computer services, architectural and engineering agencies, and accountancy and tax consultancies. Legal advice is an exception, being employed only to a small extent by the other KIS. Marketing and advertising companies seem to be closely interdependent with other firms in the same sector. In no other KIS sector is input from the same sector so high. It is clear that the growth of the service sector has a positive effect on KIS. For example, engineering consultancies require information technology advice. Within manufacturing, it is also striking that the demand for commercial services is concentrated in a few industries: food processing, engineering and chemicals. These are also industries in which R&D intensity is quite high. Thus R&D, innovation and demand for KIS inputs appear to be positively related.

The environment of KIS development and KIS influence: traditions and recent changes

In The Netherlands, the growth of KIS has been influenced by several factors, including the national government budget cuts which started in the mid-1980s. Management consultancies and organizational consultancies especially benefited from these policies, as two developments were set in motion. The first was that ministries and public bodies sought support from consultancies to reorganize their structures. Second, many educated and skilled personnel left public bodies and established management consultancies with public agencies as their main clients. These moves were supported by a more positive general attitude towards entrepreneurship. Public agencies and their managers needed to work more flexibly, so that the number of project and interim managers grew very rapidly. It is estimated that the national government is spending Fl 100 million per year on outside organizational and management consultancy alone. This has raised concern among policy makers that the distinction between influential decision makers in public bureaucracies and private organizations is apparently becoming blurred (one organizational consultant declared, 'We are no Rasputins'! *De Volkskrant*, 19 April 1997).

Another factor favouring the growth of engineering consultancy is the large traditional demand for physical infrastructure expertise. As well as serving this stable source of demand, engineering firms have developed a strong position in environmental technology, such as in reducing soil water pollution. There is also a legacy of shipbuilding and aviation production, for which the technical universities provide an important cornerstone, especially in Delft. The failure of Fokker, the principal aerospace company in 1996, was a major setback. Until recently, The Netherlands possessed a unique spatial concentration of a major international airport, an aviation industry, an aerospace and aviation research institute and a high-grade technical university within 100 kilometres of each other.

A further important characteristic of The Netherlands is its small home market. Although this may favour some consumer sectors such as the music industry, which uses The Netherlands as a test market, more generally it is a disadvantage because of the lack of economies of scale. This has contributed to the failure of the Dutch motor and aviation industries. Success thus depends especially on developing export markets and it must be assumed that this applies to KIS.

There are general driving forces for the growth of KIS which are not specific to The Netherlands. Growing internationalization and globalization are the most important, as well as increasing regulation and legislation concerning the environment, product safety and the control of quality standards. This enhances the growth not only of firms specializing in quality control advice, for example in relation to ISO certification, but also that of legal services. Another driving force is the need for information in the production process, not simply its collection, but also its storage, analysis and application, each of which enhance the growth of KIS.

A final process supporting the growth of business services is externalization, the shift from internal towards external production ('make or buy'). Externalization itself does not explain the growth of business service demand. The result, however, is that small–medium enterprises (SMEs) too small to employ, for example, lawyers or software engineers themselves, can buy in such services. Externalization leads to a more evolved market in which SMEs potentially have better access to KIS.

Supply structure and KIS trends

KIS sector employment size, growth trends and geography

An important feature of KIS supply is its highly fragmented nature. Many firms are small or one-person businesses and the level of concentration is low. On the demand side, KIS requirements are limited by the educational and professional qualifications of the client (Tordoir 1993). Mutual communication in KIS transactions requires that the qualifications of the user are similar to those of the supplier (Tordoir 1993). Many SMEs do not meet this condition. They need skilled labour themselves to encourage the proper use of KIS.

KIS include various activities. This chapter examines several sectors in more detail, especially technical services (Nomenclature des Activités Communauté Européen (NACE) classification 742); computer services (NACE 72); law and economic consultancy services (NACE 741); and marketing and advertising services (NACE 744). Business services comprised 278,800 jobs in The Netherlands in 1995, 5 per cent of total employment (Table 6.3). Between 1989 and 1995, they grew by 4.2 per cent per annum, nearly three times overall employment growth at the time. Regionally, the largest absolute number of jobs in business services is concentrated in South-Holland, although this showed the slowest provincial growth rate (2.2 per cent annual

Table 6.3 The Netherlands: business service employment, 1989–95

	1989	1995	1989–95
	1,000		% growth
The Netherlands			
Employment	5,138.4	5,627.1	1.5
Business services	218.7	278.8	4.2
By province			
Groningen	5.5	8.8	8.5
Friesland	4.2	5.8	5.7
Drente	4.1	4.6	2.3
Overijssel	8.2	12.6	7.5
Flevoland	3.3	4.3	4.8
Gelderland	24.8	28.8	2.6
Utrecht	25.8	36.6	6.1
North-Holland	41.9	52.9	4.1
South-Holland	65.2	73.8	2.2
Zeeland	2.2	3.1	5.9
North-Brabant	24.7	32.5	4.7
Limburg	8.8	14.9	10.0

Source: CBS 1996.

growth). Growth, in absolute and relative terms, is concentrated in North-Holland, Utrecht, North-Brabant and Limburg. The so-called north wing of the Randstad, concentrated on Amsterdam, Utrecht and around Schiphol Airport, appears to be the most attractive area, followed by the south-east region.

Business services may be divided into the sub-categories shown in Table 6.4, where the absolute numbers are not comparable to those in Table 6.3 because of small definitional differences. Technical services include architects, engineering and other design and graphics consultancies, and testing and control firms. They included around 12,000 firms in 1994, employing 80,000 people. They had a turnover of more than Fl 11 billion in 1994, Fl 2.1 billion of which (18 per cent) was exported. Apart from accountancy and tax consultancy, engineering employs the largest numbers and, with design consultancies, is the largest exporter. Engineering services have a turnover of Fl 7.9 billion, of which Fl 1.9 billion is exported. Although employing fewer,

Table 6.4 The Netherlands: principal characteristics of business services, 1994

Category	Number of enterprises	Net turnover	Jobs	Export
	1,000	*Fl 1 bn*	*1,000*	*Fl 1 bn*
Technical services	11.9	11.5	80	2.1
Architects	5.6	2.9	23.4	0.1
Engineers	5.9	7.9	51.1	1.9
Computer services	9.7	8.9	54	—
System development/analysis	4.7	5.2	31.2	—
IT-consult	4.1	2.0	13.2	—
Professional services	28	15.1	122.4	—
Legal	3.4	3.9	28.4	—
Accounting/Tax cons.	12.7	7.9	68.0	—
Econ./Organ.	12.7	3.3	26.0	—
Marketing and advertising	11.0	10.1	36.6	—
Market research	2.4	1.0	9.1	—
Advertising	8.6	9.1	27.5	—

Source: CBS 1995.

advertising had a higher turnover, but close behind, and growing rapidly were the computer and information-technology firms. The much larger numbers of small professional service firms are orientated more towards home markets, including law, accountancy, book keeping, tax consultancy, and economic and managerial consultancies. They supported some 28,000 firms, with 122,400 jobs and a total turnover of Fl 15.1 billion.

Firm size

In general, firm size in business services is smaller than in manufacturing (Table 6.5). Business service firms include many single-person businesses, and even many with zero employees, so-called 'post-box firms'. These are characteristic of recent employment growth in The Netherlands but they do not all perform well. Smaller firms tend to have weaker skills and operate in domestic and regional markets, with lower export rates and learning capacities. The larger SMEs, the backbone of regional economies, tend to export more and thus appear to learn and innovate more. The business service sector has smaller proportions of such SMEs than the manufacturing industry.

Table 6.5 The Netherlands: firm size distribution of KIS activities, 1994

	Percentage of firms by size category							*Total 1,000*
	0	*1*	*2–4*	*5–9*	*10–49*	*50–99*	*>100*	
Manufacturing	13	30	24	11	16	3	3	52
Business services	21	49	20	5	4	—	—	85
Technical services	21	45	19	6	6	1	1	1
Computer services	22	50	18	4	4	1	—	12
Professional services	23	50	19	4	3	—	—	30
Market and advert.	16	53	22	5	4	—	—	12

Source: Kamers van Koophandel en Fabrieken. NV Databank 1995.

Regional distribution

It has already been noted that the growth of business services is comparatively slower in South-Holland which, with Rotterdam and The Hague, is the country's most urban province. During the last twenty years, regional economic development in The Netherlands has undergone spatial deconcentration. In 1970, 21.4 per cent of GDP was produced in the three largest cities, Amsterdam, Rotterdam and The Hague, but by 1995 this had decreased to 15.5 per cent (van der Vegt and Manshanden 1996).

During the strong economic recovery after 1984, the three largest cities had the lowest employment growth rates, and the south-east region the highest. Lower rates in the Randstad reflected losses in manufacturing, while the higher growth rates in the south and east regions were driven by trade, transport and communications. Even in these sectors, the Randstad showed lower growth. For business services, the pattern is somewhat different. They showed higher rates in the south-east area than other sectors, but also grew rapidly in suburban areas of the Randstad, including medium-sized cities and the 'Green Heart'.

Thus KIS are mainly located in the suburbs of the large Randstad cities, and in the radial zone extending to the East. Accessibility by car is the predominant reason for this spatial pattern (Manshanden 1996). Research shows that within the city regions, business services located on the urban periphery predominantly serve the national market, whereas firms in the urban core mainly serve local and regional markets (Hessels 1992). However, other spatial processes are also important. In the urban region of Amsterdam, many business services have left the inner city and settled in the so-called 'south axis', along the beltway with access to Schiphol Airport, Utrecht and The Hague. Nearly all of these firms serve national or international markets, and they have been forced to locate here through lack of space and poor accessibility in the inner city of Amsterdam. Such a drift away from the inner city has also affected manufacturing, wholesaling and transport activities, which have moved out of the city region altogether. Banking, insurance and KIS have largely remained within the Amsterdam city region. This suggests that these services are strongly tied to the urban environment, although, like all industries, they also need space and accessible locations.

The institutional basis for KIS development

Engineering firms and technical design consultancies

These services had a turnover of Fl 35 billion in 1994. Exports mainly relate to 'wet infrastructure', such as oil and gas exploration and environmental projects. Many Dutch engineering firms specialize in infrastructure technology, with a comparative advantage in the construction of harbours and canals. The relationship of many engineering firms to the Rijkswaterstaat (Department of Public Works) gives them strong institutional support and a consistent source of work. For example, an important market in The Netherlands is related to water level control. Engineering firms have nevertheless also developed systems to control traffic flows around large cities. Another important basis of engineering expertise is The Netherlands' position in trade and oil refining. This follows from the key position of the port of Rotterdam in the European oil and chemical markets, as well as the influence of Shell, a Dutch–British company. The gas industry and the former shipbuilding and aeroplane industries have also enhanced the development of engineering expertise in The Netherlands. Related service activities remain, as well as some high value added manufacturing (such as Fokker Special Products).

Computer and IT consultancies

No such institutional framework supports computer and information technology. In this market, however, as shown in Table 6.2, the strong development of the service economy is itself an advantage. An impetus for the development of computer and IT firms comes not only from public services and industrial firms, but also from investment by companies in banking, insurance, wholesaling and transport. These sustain high demand for hardware and software in computer and IT services. Initially, branch offices of Anglo-American companies dominated such services. Nowadays, specialist software producers in The Netherlands, such as Syllogic and Baan, are being purchased by American companies or are operating in the US. Internationalization is also widely developed in computer and IT services. As well as exporting services (see Table 6.2), consultancies in this sector have set up branch offices in other countries, or have merged with companies in other countries to serve foreign markets.

In The Netherlands, these activities can be divided into six segments: facilities management, database management, maintenance, software production, system development and education. Exports are dominated by software; the other activities are mainly produced and consumed locally, mostly by local employees. The international software market (estimated at US$40 billion in 1991 and projected to be about US$100 billion by 2000) is dominated by the United States, followed by the United Kingdom and Germany. The Netherlands performs at an average level (in terms of its share of the European market compared to the size of its economy), with France and Italy performing below this level and Belgium and Spain having an insignificant role. The small domestic market is a disadvantage, favouring the development of software products in 'niches' where The Netherlands is relatively strong, for example in software for transport and distribution. The quantity and quality of education and training in computer and software technology nevertheless needs to be sustained and improved.

The role of regulation and policy in KIS development

The central state: problems of national technological development

The moderate rise of wages in The Netherlands has resulted in strong growth in exports and slower growth in domestic consumption. However, in Dutch macro-economic debate it is recognized that labour productivity must rise to keep up with other countries. Kleinknecht (1994) argues that the low marginal labour productivity, which results from the rapid expansion of employment and moderate wage growth in The Netherlands, has a negative impact on technological development. Firms compete on price instead of on quality, technology or high value added goods and services, resulting in less innovative output. To boost innovation, wages should rise, encouraging both domestic consumption and firms aiming to be more productive, selective and innovative. However, Kleinknecht's (1994) recommendations have met objections because higher wages would undo the current progress in employment growth. Also, the success of a high-wage strategy would depend on other policies, for example on education and training.

The private sector contributes a little over half of the absolute expenditure on R&D, with the public sector and academic institutions accounting for the remainder. Dutch R&D spending is lower than in competing countries, especially in the private sector, largely because corporate R&D tends to be drawn abroad. Since 1987 total R&D expenditure as a percentage of GDP in The Netherlands has decreased compared with a stable situation for the OECD countries as a whole. Thus public sector R&D needs to focus more on supporting private sector competitiveness, through both R&D and an improved knowledge base.

Sectoral policies

The comparative advantage of The Netherlands is based on its central location in Europe. Wholesaling and transport are the most important parts of the service sector in The Netherlands and are regarded as cornerstones of the future Dutch economy. Reflecting this, Schiphol Airport and the port of Rotterdam are nominated as so-called 'mainports' in national economic policy, and the infrastructure development of these ports is supported by national and regional planning agencies. It is recognized, however, that physical infrastructure alone is not sufficient; the development of KIS requires a broader approach to what is known as the 'brainport'.

Slow technological development in The Netherlands was emphasized by the Ministry of Economic Affairs in its 1995 White Paper on innovation strategies, *Knowledge in Action* (Ministry of Economic Affairs 1995). The reduced R&D expenditure of the five largest multinationals in The Netherlands (the 'Big Five', including Shell and Philips) was the main cause for this pessimism. Whereas The Netherlands has benefited from policy aimed at employment growth, its economy is thought to be experiencing a decline in knowledge intensity. *Knowledge in Action* proposes an array of policy measures, including financial incentives, deregulation, more independence of public research institutes and the establishment of four so-called 'leading technological institutes'. One sector it recommends for further strengthening is the environmental industry. The White Paper mentions the service industries as a source of information to enhance industrial innovation, mainly through interaction with industrial firms. It suggests that developments in information technology might especially prompt developments in industrial know-how.

Such interaction has an impact both on services and on their client sectors, raising innovation potential in both. According to the White Paper, however, KIS do not play any strategic or direct role in these policy measures. Many measures focus on existing public research institutions and programmes. An exception is improvement in accessibility to the knowledge infrastructure by SMEs, especially at the regional level. The only way in which KIS may have some impact is through the 'electronic highway'. It is surmised that the information economy and globalization will grow rapidly. The main beneficiaries of the information services will be branches such as audiovision, publishing, software and telematics. In general, however, the conclusion of *Knowledge in Action* is that KIS can play a very modest role in national policies on innovation and R&D expenditure.

A more recent White Paper on the issues of environment and economy (Ministeries van VROM, EZ, LNV en V&W 1997) also proposed a set of policy measures. These mainly concern agriculture, transport and the physical infrastructure. Industry and services are mentioned, but the interaction between these two, and the necessary improvement of knowledge intensity to combine economic growth with a better environment, are not considered. One positive idea is the concept of 'brainport', which, although it does not especially apply to services, aims at a better knowledge infrastructure in The Netherlands, contrary to the existing mainport concept.

An additional problem is that sectors with knowledge-intensive production (such as semi-conductors, medical equipment, pharmaceuticals) are under-represented in The Netherlands. Dutch manufacturing exports are rising less rapidly than world trade. The main reason is not export volume, but the composition of export markets. Dutch industrial exports are over-represented in sectors with a slow rise of price levels and under-represented in the sectors with high-technology, quality or design products.

Local and regional government

Here, we shall mainly deal with the situation in North-Holland and Amsterdam as representative of policy attitudes in The Netherlands. The levels of innovation in firms in Amsterdam and North-Holland appear to be worse than the national average. Amsterdam shows a very modest innovation rate compared to the industrial south-east of The Netherlands (Budil-Nadvornikova and Kleinknecht 1993). Nevertheless, there are knowledge clusters in the Amsterdam economy that have a strong international position, such as the medical, information, financial and fundamental natural science sectors. However, supply and demand of this knowledge in the region do not match (Bartels 1996). The strongest economic sectors, such as tourism, multimedia and metal processing, do not have links to the academic and educational world. Conversely, the strong medical sector is not reflected in the regional economy or in regional institutions.

Common evaluations of innovation are not apt to recognize or measure innovation in the service sector. This is important because KIS firms mostly supply KIS and other services, such as banking, insurance, trade, transport and the non-profit sector. Many innovations, such as those in retail banking, are not measured. Thus in The Netherlands an unknown quantity of innovation goes unrecorded. In view of this, apparent inflation in the service sector should be interpreted with care: rising prices may not be inflationary if services are also improved (Centraal Plan Bureau 1997). It is therefore desirable that the common evaluation of what is innovative should be extended. A new or improved product or process could also be a service that is performed more quickly or specifically. This does not require a new definition of innovation, but the common definition needs to be interpreted with more imagination. For example, since services are produced and consumed at the same place and time, innovation in services must affect these characteristics. Services should be available at more places, such as through the Internet, and/or produced and consumed more quickly. Policies on innovation in services should thus focus primarily on space, time and the interrelationships between producer and consumer.

In the Amsterdam region the service sector is important. The rising number of call centres, for example, is itself innovative, with many associated opportunities for innovation. Amsterdam is an important European location for call centres because many languages are spoken in The Netherlands, especially in Amsterdam. The scarcest factor of production is labour: people who speak foreign languages. This sector is thus considered to be knowledge intensive. It is also believed that it cannot be outsourced to low-wage countries such as India, as has occurred with activities such as standard bookkeeping or other back-office activities.

Another important activity in Amsterdam is the medical sector, in which manufacturing and knowledge-intensive services interact. Because the sector is large, the Centre of Life Sciences and Technology was established in the 1980s. The main problem faced by this centre is the limited coordination and self-organizing ability of the sector. Can the city council, or any other public body, improve or control this? From the point of view of public policy the centre has argued that targets are too often defined in economic terms and that the role of public bodies is limited to promotion and initiation. Medical markets often contain potential needs which public agencies may trigger through coordinating action. So, public bodies on a regional level should develop a support basis for company cooperation to enhance market as well as technological information. The viability of these cooperative bodies should nevertheless not depend on public action alone.

The centre's conclusion is that public bodies do not have a strategy that covers all public actions in knowledge clusters. An exception is the Regional Innov- ation and Technologies Transfer Strategies and Infrastructures programme (RITTS) of the province of North-Holland, co-financed by the EU, which aims to establish such a strategy. Its ultimate goal is to improve the innovative capacity of SMEs in the province by improving intermediary public advisory agencies and encouraging collaboration between other public bodies. This requires organizational skills and consensus on regional policy goals. The programme is divided into research, strategy development and implementation phases, but focuses on best practice. So far, it has been relatively *ad hoc* in encouraging the transfer of technology and knowledge ('let all flowers bloom').

Regional initiatives supporting knowledge clusters include:

- Amsterdam Knowledge Network: Seeks to enhance exchange of knowledge between firms. Organizes workshops, publication of knowledge guides, newsletters, working visits.
- ATO (Associatie Technologie Overdracht – Association of Technology Transfer): Aims to improve knowledge transfer between public organizations and the private sector.
- Centre for Life Sciences and Technology: Encourages the development of Amsterdam into a significant biometric/technical centre.
- KCLT (Kennis Centrum Logistiek en Telematica): Seeks to improve

SME knowledge of logistics and telematics, including traffic and flow of goods in the context of congestion in Amsterdam.

- TCIJ (Technologie Centrum IJmond): Encourages the technological development of SMEs in the IJmond area, around the Hoogovens steel mills.
- RTC (Regionaal Technologie Centrum – Regional Technology Centre Alkmaar): Incubator centre for approximately twenty-five knowledge-intensive firms. Provides support for management and other business services.
- POKB (Platform onderwijs Kennis en Bedrijfsleven – Foundation for Education and Knowledge in the Business Community): Aims to improve knowledge transfer between public organizations and the private sector.
- TIFAN (Technologie en Industriefonds voor Amsterdam en Noord-Holland): Participating company in technical start-ups. Provides venture capital up to Fl 0.5 million.
- Vlechtwerken ('Interlacement'): Initiates cooperative networks between SMEs in specific sectors. KIS must be involved.

Such initiatives are typically public–private partnerships, concerned with the transfer of information and knowledge. They mostly aim at the manufacturing sector, but also involve KIS to a certain extent. The last initiative, Vlechtwerken (Interlacement), aims solely at the transfer of KIS expertise to manufacturing companies. It employs consultants of the 'InnovatieCentrum' (a national agency with regional offices to stimulate innovation by SMEs) to select companies for cooperation. Such cooperation always requires at least one KIS firm (usually an engineering consultancy).

The InnovatieCentra, ultimately aiming at innovation in manufacturing firms, face a problem in this region, since the Amsterdam economy is largely service based. The only remaining large manufacturing industries are found in old-established food processing and printing. Fifteen per cent of the centres' enquiries are from service companies (considered to be high). Innovation in services is important for the region, like the impact of KIS for services in general. It was recognized that innovation concepts from manufacturing cannot be transferred to services, and a different innovation/knowledge transfer approach by consultants is required from that undertaken by the InnovatieCentrum.

National R&D levels in The Netherlands are falling behind, but economic policy in relation to the labour market is regarded as successful. At the regional level, especially in Amsterdam, where the service sector is predominant, innovation levels are perceived as being below average. This may be a consequence of the conceptual and policy neglect of service innovation. There is almost certainly an element of 'hidden innovation' within service activities. Recent surveys of innovation, however, suggest that there is a high level of innovation in KIS in general, although it appears lower in Amsterdam.

It must be assumed that innovation in the service sector needs a different conceptual approach to that applied to innovation as an industrial phenomenon. So far, policy has not addressed such an approach.

Innovative expertise and the interaction of KIS supply and demand

Table 6.2 shows that the manufacturing sector employs economic consultancy and research institutions to a relatively large extent, especially in high- and medium-technology markets. Banks, insurance companies, trade and wholesale use relatively more computer consultancies. Business services also use relatively more engineering, architects and technical design agencies. But do these interactions between service companies, as well as between manufacturing and services, contribute to innovation? Do the stronger client firms require more inputs, implying a mutual causal relationship?

Two prevailing hypotheses, arising from current debates about innovation systems and policies, are relevant to discussion of the innovative influence of KIS. These are:

1   That the proximity of firms, especially SMEs and including KIS, enhances innovation.
2   That, within any innovation system, KIS firms contribute distinctively to innovation in other firms.

The first of these propositions has been given most attention. In general, economic geography has developed the theory that innovation and proximity are positively related. This is elaborated in several concepts: the incubation hypothesis, the filter-down hypothesis, and the qualities of the so-called 'milieu-of-innovation'. The economic recession in the early 1980s directed more attention to these ideas and they have since often been discussed in economic literature. Feldman (1994: 27) suggests why the association of innovation and proximity might have become more important: 'The increased complexity of innovation suggests that other sources of information such as related industry presence as specialised business services are key to innovative success.'

In economics this synergy is denoted as 'positive external effects'. Proximity causes positive external effects, so that the cost of information and knowledge (including tacit knowledge) decreases and the chance of innovation rises. In the United States, Jaffe (1989) found evidence for such positive external effects ('spillover'), with the presence of universities and the laboratories of large companies having a positive effect on the level of patent applications in the states where these institutions were located. However, he found, ' ... only weak evidence that spillovers are facilitated by geographic coincidence of universities and research labs within the state' (Jaffe 1989: 968). Acs *et al.* (1992: 366) focused on product innovations actually introduced to the market and obtained more positive results, concluding that spillovers from geographical proximity between innovative firms, universities and research laboratories are more important than Jaffe found. Feldman found the same: state-level innovation is enhanced by the presence of academic and company research laboratories and also related industries and specialized business services. He concluded: 'Specialized business services (the variable) is statistically significant in relationship to innovative output at the state level' (Feldman 1994: 68).

Conclusions at the level of the state or region may nevertheless not apply to inter-firm relationships. Stöhr (1986) stated that proximity in itself contributes nothing to innovation. At this level, various studies often show contradictory results. Vaessen (1993) found many capable, innovating firms in suburban and peripheral regions in The Netherlands. Leus and Pellenbarg (1991) found that many production subcontractors are situated far away, although they expected them to be located near each other. Vlessert and Bartels (1985) put the importance of proximity into perspective concluding that, in the case of highly qualified knowledge transfer, distance did not matter, even for the smallest dynamic innovative firms.

Regional policy makers have similar perceptions of the impacts of proximity: that is, it depends on the kind of inputs involved. Companies such as Microsoft may need proximity to specialist labour markets, but proximity to clients is clearly not important. Field experience suggests to policy makers that proximity is important primarily in the early, conceptual phases of the product cycle (Manshanden 1996). It is often required for scarce inputs. Within this overall pattern of external links, the need for local access to KIS is equally varied. In relation to service inputs, however, the psychological benefits of proximity may still be decisive, especially for SMEs not used to involving KIS in their businesses (Manshanden 1996).

A case of 'KISful' thinking?

Brouwer and Kleinknecht have studied the determinants of innovation, defined in terms of market impacts, for manufacturing and service firms of all sizes in The Netherlands. They found that the most important influences are firm size, sector and export rate. Consultation with the Dutch InnovatieCentrum also proved to contribute significantly to innovation, although other information sources did not show significant influence (Brouwer and Kleinknecht 1996). In fact, this research surprisingly found that the use of external sources of knowledge does not enhance innovation: 'Against all expectations, we found very little evidence that firms which collaborate on R&D or acquire external technological knowledge have a higher innovation output' (ibid.: 118).

This is remarkable, since many policy programmes are targeted towards encouraging just such collaboration. One possible explanation arises from Teece's (1988) argument that firms do not want to be dependent on others for crucial assets such as innovation and technological renewal. Such findings of course do not appear to support the proposition that KIS enhance innovation. Brouwer and Kleinknecht (1996) may not have specified all such external sources, but respondents were offered sixty-six different possibilities. Their results certainly suggest that external sources in general, including KIS, are not of *decisive* strategic value for innovation.

On the other hand, this evidence does not mean that the expertise KIS convey has no influence on innovation. It is inconceivable that firms can be innovative without having some access to outside technical or market knowledge. If this is not done through collaboration, it must come through in-house R&D, linked to market knowledge on the basis of organizational and human resources development skills. In these processes, the widespread and growing use of KIS, at the very least, reflects a broader trend towards increased inter-organizational collaboration. Like other aspects of organizational change, innovation thus now depends more on interdependent relations, if not the dependent relations defined by Brouwer and Kleinknecht (1996). It is also widely accepted that innovative success is influenced by more than R&D or technological knowledge. It depends on organizational, marketing and human resources expertise as well, and it should further be borne in mind that large and multinational firms are the main users of many types of consultancies. Since formal R&D is concentrated in such large firms, it may be that innovation is enhanced more by KIS expertise among such large firms.

Knowledge-intensive services in inter-firm networks

The nature of services requires their simultaneous production and consumption, and excess production capacity cannot be stored. As a result, they have higher levels of market uncertainty for both suppliers and clients compared to that of goods, because their impacts depend on the nature of the transaction and cannot be established until after the event. Manshanden (1996) showed that regional markets, whether in manufacturing or service companies, are relatively uncertain and of low value added. On the other hand, national and international markets are more certain and have a higher value added. This suggests a weak relationship between local KIS use and innovation. Long-distance KIS trade, in contrast, may be associated with more specialized and innovative services.

Recent evidence from a sample of SMEs in the Amsterdam region, including food processing and metal processing firms, machinery manufacturers and engineering consultancies, examined inter-firm relations and innovativeness, measured by the market introduction of new products, processes and designs (Manshanden 1996). The external relations examined included contacts made in developing new products, within each market sector (for example through other manufacturers or business services); in different types of market (*ad hoc* project markets or more planned, series markets); and through the level of exports. Consultation with clients about possible innovations occurred more in series markets than in project markets, although equally for export and home markets. Proximity was not decisive in supporting the effect of consultancy on innovation, but the stability of the producer–client relationship was important. Engineering consultancy firms had a larger share of project markets than manufacturing firms, reflecting the unstandardized nature of the production of knowledge in such KIS.

In general, engineering consultancy firms and manufacturing firms show the same pattern. Exporting firms innovate more, have a larger share of series markets and have more contact with the client on innovation. In discussion, respondents stressed that innovation requires close producer–user relationships, but that this did not require close proximity. Most critical was that the client and the consultant should be at similar competence levels, and this relationship is also influenced by business and technical culture. This is important not only for SMEs but also for large firms, even when KIS expertise comes from in-house sources.

'Culture' is important, indicating that a certain level of organizational competence and experience is required to enhance the transfer of knowledge, keeping transaction costs low. One problem for intermediate organizations, for example in agriculture, is the so-called 'closed shop' mentality under which well-organized branches of industry set their own standards, suppressing competition through tacit agreement, and thus obstructing technological development. In other words, organizational forms may result in conservatism. Competition in a branch of industry needs to be maintained through openness to outside influences. Even though KIS are usually expected to work within established client cultural norms, their growing use might therefore be expected to enhance competitiveness and innovation, by exposing clients to outside ideas from wider, even international practice. Innovation is, after all, a strategy to compete.

Conclusion

The demand from firms for knowledge to innovate is influenced by their size, reflecting the scale of internal resources; sector, reflecting their technological base; and export level, reflecting the degree of market competition. Each of these sustains different types of external knowledge-intensive requirements for expertise, technological and market knowledge. The certainty, stability and predictability of markets also contribute to a firm's innovation level. The same factors that determine the demand for knowledge and innovation also determine the demand for KIS. There are problems in The Netherlands, including the small home market and the small size of many KIS firms.

Two conclusions may be drawn concerning the impacts of KIS on innovation. The first concerns specific evidence for the unique circumstances of The Netherlands. The second concerns evidence from The Netherlands that may also apply to other countries and areas.

The small home market of The Netherlands, for manufacturing for example, is potentially disadvantageous for KIS, especially for potential exporters such as engineering consultancies. Thus firms have to reap the benefits from specific circumstances or 'benefits by chance'. For example, Dutch KIS perform well if they build on the natural advantages of The Netherlands, such as wet infrastructure, trading experience or international language proficiency. Employment growth in The Netherlands is fast, but KIS have grown even faster. Nevertheless, sub-sectors that work in international markets are growing more slowly. In The Netherlands, KIS firms are not especially located in the centres of the main cities, but tend to disperse around them. KIS mainly support commercial services, although research and economic consultancy are especially employed by manufacturing sectors.

Nevertheless, national R&D expenditure has not grown as fast as intended. Policy initiatives have been undertaken to change this, but KIS do not play a predominant role in these. At the regional level, the dominance of services in the regional economy limits the scope of current forms of innovation policy. Structural innovation policies involving KIS are lacking. In the province of North-Holland, and in Amsterdam, much effort is being put into improving technological and knowledge intensity.

Dutch evidence suggests that proximity between collaborating firms is not a decisive factor for innovation. The demand for external knowledge (and therefore innovation) relates to the branch, size and export rate of sectors, influenced by the relative certainty, stability and predictability of markets. Large firms provide the main markets for knowledge, and thus for KIS. However, evidence for the impact of KIS on innovation in SMEs is too limited to draw more definite conclusions. Therefore, further research is needed on the benefits of the use of KIS by SMEs.

References

Acs, Z.J., Audretsch, D.B. and Feldman, M.P. (1992) 'Real effects of academic research: comment', *American Economic Review* 82(1): 363–7.

Bartels, C.P.A. (1996) *Kennis als strategische kracht: een beleidsvisie voor Amsterdam*, Assen/Utrecht: Bureau Bartels.

Baumol, W.J. (1967) 'Macroeconomics of unbalanced growth: the anatomy of urban crisis', *American Economic Review* 57: 414–26.

Brouwer, E. and Kleinknecht, A.H. (1994) *Innovatie in de Nederlandse industrie en dienstverlening (1992): een enquête-onderzoek*, The Hague: Ministerie van Economische Zaken (Ministry of Economic Affairs) (Directie Algemeen Technologie Beleid (beleidsstudie No. 27) (Department of General Technology Policy (Policy Report no. 27)).

—— (1995) 'An innovation survey in services: the experience with the CIS questionnaire in The Netherlands', *STI Review* 16, Paris: OECD, 141–8.

—— (1996) 'Determinants of innovation: a micro-econometric analysis of three alternative innovation output indicators', in Kleinknecht, A.H. (ed.) *Determinants of Innovation and Diffusion*, London and Basingstoke: Macmillan Press, 99–124.

Budil-Nadvornikova, H. and Kleinknecht, A.H. (1993) *De regionale spreiding van produktinnovaties in Nederland*, Amsterdam: Stichting voor Economisch Onderzoek der Universiteit van Amsterdam.

CBS (Centraal Bureau voor de Statistiek – Central Bureau of Statistics) (1995) *Onder loep genomen: zakelijke dienstverlening (A Close Look: Business Services)*, Voorburg and Heerlen: CBS.

—— (1996) *Employment Statistics*, Voorburg: CBS.

Centraal Plan Bureau (1997) *Centraal Economisch Plan*, The Hague: Sdu Uitgevers.

Clark, C. (1940) *The Conditions of Economic Progress*, London: Macmillan.

Elfring, T. (1988) *Service Employment in Advanced Economies: A Comparative Analysis of its Implications for Economic Growth*, Aldershot: Gower.

Feldman, M. (1994) *The Geography of Innovation*, Dordrecht: Kluwer Academic Publishers.

Fourastié, J. (1949) *Le grand espoir du XXe siècle*, Paris: Presses Universitaires de France.

Fuchs, V.R. (1964) *Productivity Trends in the Goods and Service Sectors, 1929–1961: A Preliminary Survey*, New York: National Bureau of Economic Research (Occasional paper 89).

Hessels, M. (1992) *Locational Dynamics of Business Services: An Intrametropolitan Study on the Randstad Holland*, Utrecht: Rijksuniversiteit Utrecht.

Jaffe, A.B. (1989) 'Real effects of academic research', *American Economic Review* 79: 957–70.

Kamers van Koophandel en Fabrieken, NV Databank (1995) *Adressen, Bedrijfs-informatie en Tellingen*, Woerden: Kamers van Koophandel en Fabrieken (Dutch Chambers of Commerce and Industry).

Kleinknecht, A.H. (1994) 'Heeft Nederland een loongolf nodig? Een neo-Schumpeteriaans verhaal over bedrijfswinsten, werkgelegenheid en export', *Tijdschrift voor politieke economie* 17(2): 5–24.

Leus, E.H.B.M. and Pellenbarg, P.H. (1991) 'Production subcontracting in The Netherlands: a survey of developments', in Smidt, de, M. and Wever, E. (eds) *Complexes, Formations and Networks*, Utrecht and Nijmegen: Royal Dutch Geographical Society, 103–10.

Manshanden, W.J.J. (1996) *Zakelijke diensten en regionaal-economische ontwikkeling: de economie van nabijheid* (doctoral dissertation, University of Amsterdam), Utrecht and Amsterdam: Elinkwijk.

Ministry of Economic Affairs (Ministerie van Economische Zaken) (1995) *Knowledge in Action*, The Hague: Sdu Uitgevers.

Ministeries van VROM, EZ, LNV en V&W (1997) *Nota milieu en economie: op weg naar een duurzame economie*, The Hague: Sdu Uitgevers. (Ministeries van Volkshuis-vesting, Ruimtelijke Ordening en Milieubeheer (VROM), Economische Zaken (EZ), Landbouw, Natuur en Visserij (LNV), Verkeer & Waterstaat (V&W): Ministries of Housing, Spatial Planning and Environment, Economic Affairs, Agriculture, Nature and Fisheries, Traffic and Water Management.)

OECD (1998) *OECD Economic Surveys 1997–1998: Netherlands*, Paris: OECD.

Stöhr, W.B. (1986) 'Regional innovation complexes', *Papers of the Regional Science Association* 59: 29–44.

Teece, D. (1988) 'Technological change and the nature of the firm', in Dosi, G., Teece, J.T. and Chytry, J. (eds) *Technical Change and Economic Theory*, London: Pinter, 256–81.

Tordoir, P.P. (1993) 'The professional knowledge economy: the management and integra-tion of professional services in business organizations', PhD thesis, University of Amsterdam.

Vaessen, P. (1993) *Small Business Growth in Contrasting Environments*, Nijmegen: Department of Economic Geography/Catholic University of Nijmegen.

Vegt, van der, C. and Manshanden, W.J.J. (1996) *Steden en stadsgewesten: Economische ontwikkelingen 1970–2015*, Amsterdam: Stichting voor Economisch Onderzoek der Universiteit van Amsterdam.

Vlessert, H.H. and Bartels, C.P.A. (1985) *Kenniscentra als elementen van het regionale produktiemilieu*, Oudemolen: Buro Bartels.

7 The United Kingdom

Knowledge-intensive services and a restructuring economy

Peter Wood

The rapid growth of commercial knowledge-intensive services (KIS) in the UK since the 1970s, as elsewhere, reflects the changing needs of the rest of the economy, in the private and the public sectors. The response of the KIS sector itself has also been influential, developing specialized technical and management consultancies, and offering legal, financial, human resources, real estate, marketing and much other expertise. As we shall see, these demand trends and emerging patterns of KIS supply have generally been well documented in recent years. The operational basis for KIS success, however, is less well documented, although it is now attracting more commercial and academic interest. The more general impacts of their growing use, however, especially on business development and innovation, have hardly been explored, particularly in relation to the parallel activities of other agencies, including clients.

This chapter will summarize the demand and supply patterns of KIS in Britain. Much of their influence on change, however, also depends on the outcomes of the process of consultancy, the active engagement of KIS expertise by clients to serve their needs. The conditions governing this interaction will also be reviewed, together with UK evidence for its outcomes, especially in relation to strategic organizational change and technology transfer. Some attention will be given to the best-documented activities: management, engineering and computer consultancies.

The implications of this evidence for UK innovation policy is, at best, indirect. Such policy has concentrated on support for technical research and development (R&D) in manufacturing, within a broadly *laissez-faire* commercial regime. This has actively involved private consultancy firms only to a very limited extent. In the 1990s, further policies sought to foster change and innovation among small–medium enterprises (SMEs), and in economically lagging regions and cities. In Britain, as elsewhere in Europe, this has encouraged the involvement of public and other non-market business advisory agencies, usually supported by the Department of Trade and Industry (for England), the Scottish and Welsh Offices (before political devolution in 1999), local government authorities, trade associations and chambers of commerce. Again small commercial consultancies have been engaged only in limited and controlled circumstances. In general, these interventions have had marginal impact on national patterns of innovation, dominated by the main-stream corporate and public sectors, which increasingly engage both major consultancies and many smaller, more specialist counterparts.

Supply: overview of KIS developments in Great Britain[1]

The 1980s

A good picture of changing British patterns of KIS supply since 1981 is available from publicly available data on employment and numbers of firms. By 1989 there were over a million workers in the measured 'management-related' professional and technical, commercial, computer and business services (Wood 1996a), operating primarily on the basis of consultancy. Many consultancy services were also offered by firms in other sectors, such as the legal and financial professions. Measured KIS employment had more than doubled between 1981 and 1989, creating over half a million extra jobs (Table 2.3a). Computer services and other business services, including consultancy, achieved sustained annual growth rates of around 15–16 per cent. Growth slowed after this, to 7–9 per cent per annum in the early 1990s as the recession of that time deepened. Professional and technical services were most affected by this, through the slump in construction and manufacturing demand. This appears, however, to have been the precursor of a more sustained contraction during the 1990s, as this chapter will demonstrate later.

Another perspective on the significance of KIS for the British economy is seen in Table 7.1. This shows the national share of key groups of skilled workers, defined as occupational units in the 1991 census, employed in broadly categorized 'business services' firms, rather than in all other sectors. Such firms dominated the supply of legal professionals (occupation unit 24), and provided well over half of business and financial professionals, including accountants and management consultants, and architecture/planning professionals (25 and 26). They also controlled over 20 per cent of national technical expertise in business and financial methods (36), surveying and mapping (31), and computer analysis and programming (32), and specialist management skills in finance, purchasing, marketing, personnel and advertising (12). Business services firms thus command a significant share and variety of key UK knowledge-based expertise.

The location of this expertise is also highly concentrated. The management-related business services are defined to include services ancillary to finance; professional and technical services; advertising and computer services; and other business services, such as management consultancies, market research and public relations, employment agencies and copying services. In 1989, over half of their employment was based in London and the South East, one-third in London itself (Wood 1996a). This mirrors the wider regional distribution of knowledge-based skills, although in a more concentrated form

Table 7.1 Great Britain: proportions of selected occupational units employed by business service organizations, 1991

Occupational unit ranked by % share		Business services		% share of all industries	
		Male	Female	Male	Female
24	Legal professionals	5,103	1,818	85.7	80.0
25	Business and financial professionals	8,000	1,910	61.1	54.2
26	Architects, town planners, surveyors	5,076	439	55.3	50.6
36	Business and financial associate professionals	7,031	2,101	40.3	28.0
31	Draughtspersons, surveyors	3,511	263	32.7	29.4
32	Computer analysts/programmers	3,432	785	29.9	25.8
12	Specialist managers	8,297	5,475	18.6	23.4
41	Numerical clerks and cashiers	5,075	9,780	17.8	12.9
38	Literary, artistic and sports professionals	3,513	1,800	17.1	15.0
All SIC business services		90,104	74,333	6.1	6.7

Source: 1991 Census of Population.

Notes
Data based on 10 per cent sample data; classification according to the 1980 Standard Industrial Classification (SIC).
Under SIC Class 83, 'Business services' includes: *8310/20, Activities ancillary to banking and insurance; 8340, Housing and estate agents; 8350, Legal services; 8360, Accountants, auditors, tax experts; *8370 Professional and technical services; *8380, Advertising; *8394, Computer services; 8395, *Business services not elsewhere specified; 8396, Central offices not elsewhere allocable. (* 'Management-related services' which predominantly serve business clients (see Table 7.3). The other activities also serve appreciable consumer markets).

(Hepworth *et al.* 1987; Porat 1977). The patterns nevertheless vary between activities. Advertising and financial support services were particularly concentrated in London, while computer services were best represented in the rest of the South East outside London. Professional and technical services, although concentrated to some degree in the South East, were more dispersed across the industrial regions of the midlands, west and north of Britain. In the early 1980s most activities had become further concentrated in the South East and adjacent areas of East Anglia, the South West and the East Midlands. After 1987, however, other regions appeared to perform more strongly, although this now appears to have been recession related, as London and the South East led the economy into the early 1990s slump. The national market nevertheless remained dominated by the 'expertise-rich' greater South East, combining

Table 7.2 British regions: selected business services, percentage share of employees by establishment size, 1991

	Ancil. to finance (8310)		Prof. and technical (8370)		Advertising (8380)		Computer services (8394)		Business services (8395)	
	Establishment size (numbers of employees)									
	<10	>100	<10	>100	<10	>100	<10	>100	<10	>100
London	9	67	28	30	18	42	26	34	17	39
South East	25	51	41	22	42	10	31	33	22	30
East Anglia	71	0	45	11	45	0	33	20	24	30
South West	52	13	42	10	46	0	32	28	22	25
West Midlands	29	39	33	22	43	4	24	40	19	41
East Midlands	51	0	40	18	39	0	33	25	21	33
Yorkshire and Humberside	34	36	42	8	44	5	35	10	21	29
North West	37	28	39	22	35	6	26	35	19	36
North	45	40	42	17	36	0	26	26	20	38
Wales	28	42	33	19	66	0	44	30	27	28
Scotland	18	47	41	13	32	0	32	20	18	42

Source: 1991 Census of Employment (unpublished tables, Department of Employment).

Note
The figures in brackets are the 1980 Standard Industrial Classification (SIC) code numbers.
Business services (8395) are 'Business services not elsewhere specified.

rationalizing corporate control functions with the burgeoning specialist KIS (Howells 1988: Ch.3; Coffey and Bailly 1992).

The characteristic KIS combination of small-scale provision, through professional individuals and partnerships, with growing corporate domination, is demonstrated by workplace size evidence from the 1991 Census of Employment (see Table 7.2). Over 80 per cent of KIS workplaces in most activities employed fewer than ten members of staff. In contrast, KIS employment and market shares were dominated by larger firms and establishments. For example, the top twenty management consultancy firms in 1994 accounted for over 93 per cent of the market share of the top sixty-four, and the largest three held 40 per cent of this market. In terms of establishments, London dominated the large workplaces of several key KIS activities, reflecting their higher corporate status and orientation towards international as well as national markets. Around 60 per cent of London's workers in 'services ancillary to finance', for example, were in offices with over 100 employees. This share was 42 per cent for advertising and (not shown in the table) 57 per cent for accountancy and 44 per cent for legal services. Although in professional and technical services the share in large offices was only 30 per cent, as for all these activities the capital still possessed the greatest regional proportion of such workers. Large offices were less dominant among London's thriving management consultancy, market research and other business services. They were also relatively more important in these sectors in Scotland and the West Midlands, mainly as branches of national firms. Computer services also showed a wider regional spread of large establishments.

The 1990s

The move to the 1992 Standard Industrial Classification, based on the European Union standard, NACE (Nomenclature des Activités Communauté Européen), broke the continuity of monitoring at a critical time during the early 1990s recession, when secular and cyclical influences in KIS development seemed to come together. The biennial sample census of employment in the UK was also replaced after 1993 by a new annual survey of employment, using a tax payment database. Both of these changes make the comparison of detailed trends before 1991 (see Wood 1996a) and after 1993 unreliable. Table 7.3 presents national employment data for 1995, the first year for which both data changes can be combined, as the basis for examining sectoral trends to 1998. The table distinguishes those 'core' KIS that are predominantly management related (computer services, market research) from others that either serve both business and consumer markets (real estate, legal and accountancy), or offer more routine support functions (recruitment, security, cleaning). Financial service trends are also shown for comparison. The largest defined group of 'core' KIS in 1995 was architectural and engineering consultancy, with 325,000 employees. As in computer services (265,000 in total), two-thirds of the workforce were men, mainly in full-time jobs. Other business services employed more women, many in clerical and part-time positions as indicated in Table 7.1. These included services ancillary to finance (179,000), business and management consultancy (150,000) and advertising (73,000), with market research (38,000) employing the highest proportion of female workers. The business-orientated component of real estate, legal and accountancy employment is difficult to estimate, but if we assumed it to be half of the total, 325,000 KIS workers would be added to specialist KIS. Recruitment agencies employed a further 352,000, many working temporarily for other organizations. These data thus suggest a total of around 2 million workers in independent firms engaged in business service provision, often based on consultancy-type relationships with their clients. This compares with about 800,000 in the financial services.

The three years to 1998 demonstrate the continued growth of KIS in the UK through the mid- to late 1990s, but also some significant shifts in the structure of provision compared with the 1980s. The main change was the continued rapid development of computer consultancy and services, especially data-processing activities, often through the outsourcing of corporate functions. By 1998, these accounted for over 400,000 jobs, still two-thirds occupied by men. Market research and advertising also expanded rapidly, in contrast to their earlier relative stagnation. On the other hand, employment growth in business and management consultancy appeared to falter. This may reflect some retrenchment of earlier rapid growth through mergers and rationalization, and greater efficiency in workforce use. Statistically more significant, however, is the focusing (and thus probably reclassification) of much consultancy activity into computer-related work. Evidence later in this chapter will show that there is little sign of stagnation in management consultancy demand, but a strong move towards support for IT- and computer-driven change, often subsuming more traditional management consultancy advice.

The most startling decline, however, was in technical consultancy, dominated by architectural and engineering work. The first signs of this during the late 1980s recession seem to have heralded a persistent trend (also seen in France, see Chapter 4). A switch of much of this work to computer-orientated consultancy may again explain some of the decline, but it may also be due to changes in the organization of construction and major engineering projects. Large construction companies, for example, have increasingly integrated technical and design functions into project management, reducing the role of independent architectural, design and engineering consultancies. Organizational shifts may thus not always favour KIS growth; 'internalization' is possible as well as 'externalization'. Whatever the causes, the result was an apparent national loss almost 50,000 KIS jobs, 15.3 per cent in three years, with growth only in part-time positions.

Among the other service activities shown, the high rate of corporate consolidation in UK business is reflected in the 68.1 per cent growth of the specialist headquarters functions of holding companies. This itself is a significant generator of demand for consultancy, whose major markets continue to focus around corporate headquarters. Recruitment agencies also showed a remarkable apparent expansion of over 60 per cent, 233,000 jobs. These are primarily contracted to other organizations, however, so that this growth is not directly comparable with changes in other KIS. It is, however, a striking general indication of changes in modern employment relations.

Increasing employment flexibility during the 1990s is also reflected in the growing proportion of part-time staff (working less than thirty hours per week) in business services. The data suggest that this continued earlier trends, the proportion of part-time staff rising from 29 per cent to 32 per cent over three years for all business services, and from 16 per cent to 23 per cent for core KIS activities. In Table 7.3, to illustrate the potential significance of the

Table 7.3 Great Britain: business service employment trends, 1995–98

SIC Code/Description[a]	1995				1995–98	
	×1,000				% change	
	Male	Female	Total	FTE[b]	Total	FTE[b]
Management-related KIS						
67: Ancillary to finance	94	85	179	171	6.8	7.8
7210/20: Computer consultancy	137	60	197	187	32.4	34.2
7230/40: Database/ analysis, etc.	46	22	68	65	107.7	110.6
73: Research	53	32	84	81	17.1	17.4
7413: Market research	9	28	38	26	30.1	43.6
7414: Business/ management consultancy	75	74	150	132	−1.2	−13.7
7420: Architecture/ engineering consultancy	221	104	325	302	−15.3	−26.8
7430: Testing	16	12	28	25	6.8	6.6
7440: Advertising	34	39	73	65	20.7	8.1
Total Management- related KIS	685	456	1,142	1,051	12.3	7.8
Other business services						
70: Real estate	124	163	286	248	3.9	3.7
7411: Legal	60	154	213	190	−3.5	−8.8
7412: Accountancy	62	89	152	136	13.6	6.4
7415: Holding company HQs	32	29	60	56	68.1	49.6
7450: Recruitment	171	182	352	308	60.2	58.4
7460: Security/ investigative	74	18	92	82	36.1	43.7
7470: Cleaning	111	298	409	256	−19.4	−9.9
748: Other business services	120	140	259	218	19.9	12.0
Total other business services	754	1,073	1,823	1,493	—	—
Total all business services	1,439	1,529	2,965	2,544	15.0	12.6

(continued on next page)

Table 7.3 (cont.)

SIC Code/Description[a]	1995				1995–98	
	×1,000				% change	
	Male	*Female*	*Total*	*FTE*[b]	*Total*	*FTE*[b]
Financial services						
65: Banking (excl. central)	236	351	586	538	–4.6	–2.0
66: Insurance	106	103	209	197	14.8	16.8
Total financial services	342	454	795	735	0.6	3.0
All services	7,116	9,368	16,481	13,637	7.2	7.8

Source: Annual Surveys of Employment, 1995, 1998, Office for National Statistics (unpublished tables).

Notes
a SIC codes and descriptions according to the 1992 Standard Industrial Classification.
b FTE: 'Full-time equivalents', counting part-time jobs as 0.5 full-time; rounded to nearest 100. Column and row totals may not sum, due to rounding error.

trend, the 'full-time equivalent' (FTE) of part-time work is estimated as 0.5 of a full-time job at both dates. The growing share of part-time work thus reduces core KIS growth from 12.3 per cent in gross numbers, to an estimated 7.8 per cent in FTE equivalents. The comparison has little effect in some sectors, such as computer services, research and testing, and in real estate or security, where total and FTE changes were very similar. Some activities, such as market research and cleaning services, with already high levels of part-time work, even reduced their dependence on it significantly. The sectors most affected by growing part-time work, in which FTE-based changes were significantly lower than the total employment trend, were advertising, accountancy and holding company headquarters. The apparently declining professional activities also became more dependent on part-time work, including business/management consultancy, legal services and architectural/engineering services. In the last case, the shift to part-time work accentuated decline, with an estimated loss of FTE labour resources of over one-quarter from 1995 to 1998.

Notes to Table 7.4 (opposite)

a SIC Codes and descriptions according to the 1992 Standard Industrial Classification.
b GB: Great Britain. RoSE: Rest of South East. SW/Mid/EA: South West, West Midlands, East Midlands, East Anglia. N&W: North West, Yorkshire and Humberside, North, Wales, Scotland. Column and row totals may not sum, due to rounding error.

Table 7.4 Great Britain: regional distribution of business services, 1998

SIC Code/Description[a]	GB[b]	London		RoSE		SW/Mid/ EA	N&W
	1,000	1,000	%GB	1,000	%GB	1,000	1,000
Management-related KIS							
67: Ancillary to finance	191	80	42	33	17	37	39
7210/20: Computer consultancy	260	54	21	96	37	60	46
7230/40: Database/ analysis, etc.	141	37	26	41	29	34	27
73: Research	98	14	14	44	45	21	22
7413: Market research	49	23	47	14	29	6	6
7414: Business/ management consultancy	148	51	34	41	28	28	30
7420: Architecture/ engineering consultancy	274	49	18	62	23	70	100
7430: Testing	30	2	7	8	27	9	15
740: Advertising	88	36	41	23	26	13	17
Total Management-related KIS	1,289	346	27	361	28	276	301
Other business services							
70: Real estate	297	76	26	65	22	74	81
7411: Legal	206	60	29	29	14	46	71
7412: Accountancy	172	56	33	30	17	39	47
415: Holding company HQs	101	29	29	17	17	25	29
7450: Recruitment	585	144	25	130	22	148	134
7460: Security/ investigative	124	31	25	22	18	25	43
7470: Cleaning	329	68	21	72	22	76	122
748: Other business services	310	91	29	70	23	71	76
Total all business services	3,405	903	27	796	23	783	899
Financial services							
65: Banking (excl. Central)	560	186	33	87	16	118	169
66: Insurance	240	43	18	64	27	58	72
Total financial services	800	229	29	151	19	176	241

Source: Annual Survey of Employment, 1998, Office for National Statistics (unpublished tables).

Table 7.5 British regions: percentage changes in management-related KIS, 1995–98

	Total	FTE	Difference (Total minus FTE)
Great Britain	12.3	7.8	4.5
London	15.6	10.3	5.3
Rest of South East	19.4	15.7	3.7
East Anglia	20.2	14.7	5.5
South West	11.1	5.2	5.9
West Midlands	10.6	7.9	2.7
East Midlands	14.7	6.4	8.3
Yorkshire and Humberside	13.8	5.9	7.9
North West	−3.0	−6.7	3.7
North	10.2	2.4	7.8
Wales	−5.3	−9.2	3.9
Scotland	−2.9	−5.0	2.1

Source: Annual Surveys of Employment, 1995–98, Office for National Statistics (unpublished tables).

Notes
Total = all employees; 'Management-related KIS': see Table 7.4.
FTE: estimated 'full-time equivalents', counting part-time jobs as 0.5 full-time.

Overall, the employment growth dynamic so characteristic of KIS in the 1980s seems to have become focused during the 1990s into computer-based services and revived market research and other research activities. Support services, such as recruitment agencies and security, have also grown. In contrast, there has been a shift away from more traditional business and technical consultancy, with rationalization and restructuring towards IT and computer-driven work. Decline in the legal sector has also mirrored what is occurring in banking. In FTE terms, other sectors showed only limited growth, including services ancillary to finance, testing, advertising, real estate and accountancy. The possibility of KIS market contraction is raised by the experience of architectural and technical services. Another pervasive trend is growing diversity in the forms of employment offered. This was reflected not only in the growth of part-time work, including that undertaken by men, in business and technical consultancy, advertising, legal and accountancy services. It also emerges in the extent of employment agencies' activities, assigning over 580,000 workers in 1998 to mobile and temporary work across the economy.

Regional trends

Developments since the early 1990s suggest that the apparent regional dispersal of KIS away from the South East noted in the late 1980s (Wood 1996a) was recession related and therefore temporary. The dominance of the southern regions in the growth of most types of KIS was revived after 1993, including the Rest of the South East (RoSE), the East Midlands and the South West. There also appeared to be some reconcentration of growth in London, especially in services supporting finance, real estate, advertising, market research and other business services.

Table 7.4 traces the distribution of the business-related services in 1998, between London, the RoSE, and two groups of other regions, the southern and midlands regions, and the more peripheral north and west regions, including Scotland and Wales. The management-related KIS remained heavily concentrated in London and the South East. Three-quarters of national market research and over two-thirds of advertising employment were located there, along with over half of each of the other activities except architectural/ engineering consultancy and testing. About half of most other business-related and financial services are also based in and around the capital. Within the South East, London itself dominates services ancillary to finance (42 per cent of national employment), market research (47 per cent) and advertising (41 per cent) and, to a lesser extent, business and management consultancy (34 per cent) and accountancy (33 per cent). The RoSE was particularly important for research (45 per cent) and computer consultancy (37 per cent), while only architectural and engineering consultancy, generally in decline, and testing were well represented in other regions.

More significant, Table 7.5 suggests that this concentration of the core KIS was reinforced between 1995 and 1998. Most regions still showed healthy aggregate employment expansion, but this was led by East Anglia and the South East, and there was actually a decline in KIS in Wales, the North West and Scotland. When allowance is made for the variable increase in part-time work between regions, this polarization appears to be even more marked. The KIS base of Wales (with an estimated FTE loss in the three years of almost 10 per cent), the North West and Scotland was further eroded. In the East Midlands, Yorkshire and Humberside and the North, KIS growth relied more than average on part-time work, indicated by the larger differences between the total and estimated FTE growth rates. The estimated FTE growth of KIS in London, the South East and East Anglia thus appears even more dominant. The widespread pattern of UK KIS growth established during the 1980s was thus fully sustained only in the core regions around London during the 1990s.

Full-time work for men in KIS also fell by 17,000 FTE jobs between 1995 and 1998, concentrated especially in Wales, the North West, Yorkshire and Humberside, and the North. To compensate, full-time employment of women grew by 43,000 FTE jobs, spread across most regions except Wales and Scotland. The major shift, however, remained to part-time work. For women, this expanded further by 33,500 FTE jobs between 1995 and 1998, almost half, although Scotland did not benefit even from this growth. The fastest and most widespread change, however, was to part-time work for men, which more than doubled in three years, adding 22,700 FTE jobs.

As well as more modest growth compared with the 1980s, the mid- to late 1990s showed a reintensified regional polarization of KIS supply. Outside London, the South East and adjacent regions, KIS employment appears less skilled and secure, with a greater reliance on part-time employment. It is also more dependent on branch offices and small firms offering more routine, including consumer and contract, services. These comparisons confirm what survey research evidence suggested during the 1990s (O'Farrell *et al.* 1992; O'Farrell *et al.* 1996). KIS employment is generally of a different quality in southern Britain, especially in London and the South East, from that elsewhere, especially in northern England, Wales and Scotland. The southern nexus of information-rich, innovation-orientated services, working in close association with high-level client corporate and innovative activities, may be expanding into adjacent regions. But of greater significance from a national perspective is the growing southern concentration in the 1990s of both core KIS employment opportunities and access to the knowledge-intensive support they offer other organizations.

Demand: KIS and the restructuring of the UK economy

The growth of KIS in the UK since 1980 reflects a series of significant changes in the wider UK economy favouring their use. Demand has been driven, first, by the often radical restructuring of client sectors, based on the rationalization and modernization of manufacturing and energy production (including North Sea oil and gas developments) and growth and change in the private services. These trends have been supported by predominantly equity-based and internationally liquid capital markets, promoting active takeovers, mergers and corporate financial restructuring. The liberalization of labour markets has also removed traditional barriers to corporate adaptation, as well as supporting new forms of work relations, including subcontracting, outsourcing and consultancy. Thus, one consequence of restructuring since 1980 has been an increase of over 50 per cent, 1.3 million, in the number of firms employing fewer than 250 workers, which now account for over half of private sector employment (Curran and Blackburn 2000).

There has been an equally radical change in the volume and mode of public sector activity. Privatization during the 1980s and early 1990s linked many important sectors formerly dependent on public finances to private capital markets, often under statutory regulation. These included water, gas and electricity production and distribution, state steel and motor vehicle interests, much public housing, British Airways, the British Airport Authority, British

Telecommunications, the remains of the coal industry and British Rail. Fundamental business cultural changes were required, especially in the utilities and transport. The privatization of British Rail in 1995–97 was estimated to have involved £800 million in consultancy fees. Corporate restructuring has continued in the post-privatization period as market deregulation has increased competition.

Much other government administration has also become subject to quasi-independent forms of management. These are often based on commercial efficiency norms monitored by the National Audit Office. Consultancy advice has also been increasingly sought in such public services as defence management and the health and education services. Both private and public sector agencies have been encouraged to focus on core functions, externalizing other tasks to cut running costs, reduce overheads and respond to unpredictable change. Externalization also has the potential to achieve more than cost cutting, allowing improved marketing, planning for growth and innovation, and human resources management.

Consultancy use has also been encouraged by the growth, within a liberal regulatory regime, of national information and computer technology (ICT) capacity. Despite an inherently scarce supply of ICT expertise and applications experience, many market and regulatory changes have boosted demand. One of the main examples was financial sector deregulation in the mid-1980s, encouraging the City of London to adopt ICT to sustain international competitiveness. Regulatory changes have also supported ICT-related restructuring in domestic markets, including financial services, retailing, transport and the utilities. Extensive process, as well as product innovation, much of it supported by communications and computer-based change, has been led by the major innovative sectors, including electronics, information technology, pharmaceuticals, aerospace, automotive engineering, financial services and retailing. Mobile telephones and Internet-based services also began to transform both organizational processes and market expectations in the 1990s. Some specific policies have encouraged the employment of consultancies. These include the 'private finance initiative', attracting private investment into health, transport and education schemes. The formulation of such initiatives often requires specialist expertise, as does the management of complex defence and other procurement policies. To a more limited degree, consultancy use has been encouraged by some aspects of industrial and regional policy, especially by SMEs.

At the local level in the UK, the autonomy of local and regional government agencies for independent action was generally reduced in the UK during the 1980s. Regional aid was cut, with some diversion of funds towards urban policy initiatives, although these were accompanied by reductions in local authority spending powers. The emphasis of assistance shifted to 'centralized localism', with government ministries targeting local problem areas, including programmes such as Urban Development Corporations and Enterprise Zones, or schemes focusing on areas of mining or industrial decline which also attracted EU funds. Local government thus employed consultancies largely in response to national government policies, especially to control operating costs and privatize services. Planning and environmental consultancy has also become increasingly important. The urban development agencies established in the 1980s with limited in-house professional capacity depended heavily on consultancies for much of their specialist development work. With regional inequality persisting, a new phase of regionalism was entered in 1999–2000, with elected assemblies established in Scotland and Wales, an elected mayor/assembly in London, and regional development agencies in the English regions, opening up further opportunities for specialist consultancy work.

In addition to these UK changes, European markets and projects have become a fertile breeding ground for UK-based consultancy work. KIS markets are becoming increasingly open internationally, with growing proportions of UK KIS firms operating throughout the EU (O'Farrell *et al.* 1996). Exchange is also being internalized within the increasingly integrated European operations of global consultancies. The main driving force for these trends has been the scope and complexity of the European market, raising the expertise requirements associated with international market development, and offering expanding KIS potential across Europe.

In the past two decades the UK economy has experienced a plethora of institutional and technical innovations. This has created unprecedented demands for specialist expertise, at a time when private and public agencies have sought to reduce the overhead costs of employing such expertise in-house. This is why consultancy has thrived. It is also evident that any innovative role possessed by consultancies cannot be confined to technical matters. The impacts, most obviously, of the ICT their specialists promote are much more general. Also, in the UK institutional or workforce innovations have been important in recent years, providing the conditions under which technical innovation may succeed. In practice, therefore, distinctions between technical (especially computer-related) and other forms of innovation, or even between innovation and other forms of change, are increasingly difficult to draw. The trends already noted in the 1990s, for computer and business consultancies to merge and overlap, are one practical consequence of this.

Consultancy–client relationships

UK institutional traditions and recent changes

The focus of the consultancy process, which determines its ability to assist client innovation, is the consultancy–client exchange. This is not a new process in the UK, although the number and range of consultancy options available to potential clients have expanded greatly in the past twenty years. The move from in-house to consultancy sources of specialist expertise has nevertheless extended a long-established UK tradition of independent, self-regulated external business advice. Under this, the consultancy–client relationship is based on specific areas of expertise, codes of professional practice and individual reputation. These form the basis of legal services, the various engineering specialists, architecture and town planning, and accountancy (cf. Table 7.1). Such externalization is legally required only for business accountancy. Specialisms enjoy market protection by professional bodies who control access through the accreditation of training, qualifications and service quality.

As we have seen, many firms are small, sometimes with only a few professionals. Larger KIS firms have formed from partnerships of experienced professionals rather than as limited liability companies. Those with more than twenty professionals tend more to serve business and government clients. In general, the professions are widely distributed geographically, serving a combined consumer and business demand. The larger firms are concentrated in major urban centres, especially London, which sustains the head offices of the major national business and public sector specialists.

Within this tradition of independence, professional status is attached to individuals rather than firms. Many private and public organizations thus also employ their own professionally qualified staff. Consultant–client interaction is thus common between outside specialists and parallel client experts, for example over financial and legal regulation. It has historically also been active within UK industrial and commercial regions, for example in the engineering-based midlands regions, the textile regions of the north, or in port centres. The decline of such regions and the growth of new forms of consultancy have favoured the South East, where national and international consultancy experience can be combined with related forms of corporate business expertise.

More recently developed activities, including management, computer-related, product design and marketing-related consultancies, possess less clearly defined accreditation requirements and looser professional control. In a more open and competitive environment than in traditional professions, accreditation is exerted mainly through voluntary codes of practice. Success depends on individual and company reputation and referrals. The concentration of these activities in London and the South East reflects their dependence on private and public corporate demand.

The rapid development of consultancy in the UK has taken place in an environment which imposes few commercial limitations on the establishment or operation of firms. This has supported an accommodation between the traditions of individual professionalism and the drive to corporate consolidation. Even the traditional professions have seen deregulation removing barriers between specialist functions, encouraging a more diversified, competitive, consultancy-style of business. Large, often internationally active firms now appear to dominate many markets, especially to serve large clients. These include financial and management consultancy, advertising, commercial law and real estate firms, offering clients linked expertise in financial, organizational, legal, property management and ICT.

In some cases, especially where financial and consultancy services are offered by the same major firms, anxieties among clients have begun to have an impact on consultancy practice. The impartiality of some advice has also been called into question, especially when diverse expertise is brought in. Major clients have become more proactive in combining consultancies, by including smaller, more specialist firms on major projects (for case studies, see Appendix in Chapter 3, page 85). The growth of the consultancy market, outstripping the traditional accountancy base of the largest firms, also led them to disengage their accountancy and consultancy activities in the late 1990s. The most public case was the legal battle for the Arthur Andersen name, resulting in the renaming of Andersen Consulting as Accenture in 2001. In some sectors, the traditional partnership structures of consultancy firms are also being replaced by limited liability arrangements. This is a response to international competition and the increasing risk of conflict of interest or liability claims and legislation. Whatever future forms of consultancy–client relations emerge from these developments, however, the need for close and semi-permanent relations with major clients will continue to favour companies operating in London and the South East.

Consultancy–client relations in the UK

The success of all specialist consultancy firms, large or small, depends on the quality of staff and company reputation in a relatively sophisticated, quasi-competitive corporate market. Their most important competitive asset is the quality of their technical, managerial, IT and computing expertise, based on staff training and work experience. Their innovative contribution is inherently difficult to separate from the wider networks of inter- and intra-firm exchange within which they operate. Such relations are also mediated by many other personal and behavioural qualities, especially when significant change or

Table 7.6a UK firms: main types of recent corporate change

Changes (change types ranked as 'highly significant' or 'significant')	Number of firms	Percentage
Management structures and processes	95	83
Organizational culture	93	81
People	82	72
Organizational performance	68	60
Tasks and activities	49	43
Image of the organization	48	43
Set-up of the organization	47	42
Technology used	22	20

Source: Survey of 124 major UK companies (Wood, 1996a 1996b).

Table 7.6b UK firms: reasons for using consultancies on corporate change projects

Reasons for using consultancies	Number of firms	% of all projects
Required special knowledge and skill	68	58
For an impartial, outside viewpoint	48	41
Required intensive temporary help	44	38
Timescale did not allow development of internal skills	18	15
Good past experience with consultancies	14	12
To confirm an internal management decision	11	9
Unable to recruit appropriate staff	2	2

Table 7.6c UK firms: use made of consultancies in corporate change projects

Use made of consultancies	Number of firms	% of all projects
Worked in partnership with in-house staff	50	43
Carried out specific elements of project	37	32
Provided specialist technical advice	36	31
Provided overall 'blueprint' for change	30	26
Trainer/educator role	24	21
Involved directly with implementation of change	17	15
Provided specialist human resource management advice	11	10
Provided specialist market research advice	9	8

Table 7.6d UK firms: types of consultancy used in corporate change projects

Types of consultancy	Yes	No	% Yes
Large, multifunctional firms	53	34	61
Small and medium firms	37	50	43
Specialist technical firms	24	62	29
Sole practitioners	19	68	22
'Consulting professors' (e.g. business schools, etc.)	15	72	17
Consultants from banks	14	73	16
Proprietary advice from suppliers of equipment, services, etc.	4	83	5

Source of above three tables: Survey of 124 major UK companies (Wood, 1996a, 1996b).

innovation is the outcome. One consequence of this is the continuing diversity of consultancy activity and the ability of small consultancies to compete effectively with the dominant firms.

This diversity and the interdependence of consultancy influence with that of other actors have been demonstrated in a study of the management of strategic change by over 120 major UK companies (Wood 1996a, 1996b). Some of the results are summarized in Tables 7.6a–d. 'Top 500' *Financial Times* UK companies were included, with a range of manufacturing and service sectors, size groups, locations and types of organization. Very broadly, KIS consultancy use was greatest among large, growing, service companies in southern Britain, compared with smaller, slower growing manufacturing firms based in the North. In about one-quarter of the change projects examined, firms had not employed consultancies at all.

The strategic changes studied were concerned largely with management and organizational reforms, including associated human resources developments (Table 7.6a). Technological change was seldom dominant in its own right. In general, the study indicated that consultancies are not employed simply for their specific technical or managerial expertise. Their impartial perspective on the deliberations of client experts and managers is also often valued, as is their ability to offer support during periods of intensive specialist need (Table 7.6b). The most common requirement of consultancies in supporting strategic innovation is that they work closely and cooperatively with client staff (Table 7.6c). Strategic consultancy expertise is most employed by clients that already possess significant related staff experience, usually based on past projects. Lack of client expertise is recognized only when more specialist technical or market research consultancies are engaged. Much consultancy work is based on joint production with client staff, each contributing information, ideas and team personnel, and monitoring the outcomes. At these strategic levels, therefore, the distinction between internalized and externalized functions is difficult to sustain. They are essentially interdependent.

The consultancy–client relationship may best be characterized in terms of a division of labour between (1) the wider experience in combining expertise to manage change of the consultancy, based on work with many clients; and (2) the specific expertise, experience and responsibility for a particular organization, of client staff. Consultancies can tap international 'state-of-the-art' thinking about change and innovation, as well as bring to bear outside perspectives on change processes, but their value can be realized only by effective implementation on the part of the client. The main outcome of consultancy provision is often the transfer of expertise between staff and training, building on established client skills.

This evidence suggests the importance of close interpersonal and reciprocal relations and sensitivity to the strategic, political and cultural context of innovation. Project management commonly draws on a continuous and evolving requirement for expertise, itself negotiated through client–consultancy interaction. Operationally, consultancies and clients emphasize the need to be clear about their respective roles and responsibilities. When the successful outcome of a complex project is examined, however, these are often difficult to separate. The outcome is also determined by the client's capability to change and respond to the quality of this interchange.

It was also notable that the large clients in the sample employed large multifunctional (including international) consultancies in less than two-thirds of the cases (see Table 7.6d), and only 17 per cent employed them solely on the project. Most commonly, smaller, specialist consultancies were also engaged for some of the work, and large consultancies were not used at all on almost 40 per cent of the projects. In these cases clients preferred to select and combine various types of smaller consultancy, including individual experts, under their own direction.

There are wide variations in how consultancy expertise is integrated into company operations. For manufacturing firms undergoing critical restructuring, for example, the most important consultancy role may be to support the emergence of a new set of cultural norms, orientated towards the more effective exploitation of innovation, perhaps moving from a 'production-' to a 'market-' orientated mode of working. However, many companies, especially large service corporations, routinely employ specialist consultancies as extensions of their own capabilities, drawing on them regularly as a controlled resource of expertise and advice. In this they may support either technical or operational improvements and enable senior managers to plan innovation strategically.

This indicates why it is so difficult to assess the effectiveness of KIS support on innovation. Various aspects of client operations may be affected, and more than one consultancy may be engaged. Contract negotiations can define the required outcome for a specific project assessment, but more subjective responses often dominate the general evaluation of consultancies, coloured by social and interpersonal relations. Box 7.1, summarizing some of the comments from interviews with UK-based representatives of international management consultancies, gives a flavour of how they operate.

Box 7.1 UK interviews 1994–5: a sample of major consultancy comments on innovation

---

A major multidisciplinary consultancy

- Innovation is defined as enabling clients to 'go beyond the competition'. For example, financial service innovation comes from analysis of the total service delivered to the public, integrating various functions in new ways to create coherent products for target market segments (e.g. A/B social class).

- This often adapts methods successful in one sector to other sectors. Major sources of innovation thus often lie in marketing and organization, for example, employing supermarket principles in other increasingly competitive market, such as financial services, banking, telecommuni-cations, utilities.
- As a result, consultancy is increasingly organized by competence (e.g. marketing, technology, process) rather than sectors. These approaches arise from the need to assist clients with established brands (e.g. in food retailing) to adapt their strengths to new product areas, in response to deregulation and increasing competition in core markets.
- Emphasis of KIS organisation on teamwork; combining expertise on strategy – change – technology – process. Need to focus on outcomes; successful implementation of strategic change is crucial, and many client companies have little experience of this.

Accountancy-based consultancy I

- Wide range of clients, but 30–40% manufacturing. Not generally interested in SMEs. Management consulting is directed mainly at 'operating consultancy' and process innovation, rather than strategy: especially adapting general change processes to specific client needs. These roles are often linked to auditing functions, growing out of financial management issues, including cost management or shareholder value.
- 'Support clients in achieving strategic goals they have set themselves', especially through development of computer and information systems e.g. SAP software and systems applications in financial management, business processes, facilities management; integrated supply-chain management. Some consultancies, such as McKinsey, specialize in advising on strategy.
- But becoming more concerned with strategic considerations, following growing demand from major clients in top 2,000 global MNCs. Encouraging clients to think ahead on technology, marketing, competition, management methods, etc.
- Consultancy draws on 'a global currency of ideas', developing from a global database of best practice. But local teams/companies are strong, and may contribute to global activities. Strong in UK and Spain; growing in France, Germany and Eastern Europe. The working culture is moving more towards client–consultancy partnerships, based on integration of IT and business culture issues.
- Public sector work and privatization is important but less innovative, mainly domestic, and offering lower margins.

Accountancy-based consultancy II

- Audit and tax basis of traditional work has extended through advice on corporate finance and recovery into management consulting in support of major business transformations. Some interest in the 'middle market' organizations, but mainly among medium-sized firms with good growth prospects. Work much more with human resources and organizational issues than ten years ago.
- The partnership has grown as an amalgamation of formerly local groups of partners. A confederation, offering common training, recruitment methods, international links, while also enjoying benefits of local knowledge and client links.
- Work focuses more on adaptation and improvement than innovation in the traditional sense. Consultancy acts a facilitator of client change. Also widely engaged in ICT systems implementation, for example of SAP software.

The interaction of supply and demand: key consultancy sectors

Management consultancy

Although information about markets for KIS consultancy is diffuse and elusive, the activities of management consultancies are regularly reviewed for the UK in the journal, *Management Consultancy* (Management Consultancy 2000). The estimates for 1999 suggest that the fee income of the top 100 management consultancy firms was £4.3 billion, ten times what it had been in the late 1980s. In continuing boom conditions, there had been a 16 per cent increase during 1998, equivalent to up to 5,000 extra consultant jobs in a year. About two-thirds of this growth was estimated to be due to expanding demand, the rest to increasing charge-out prices. The journal's figures include all services sold by the management consultancy companies, including IT support, computer software and systems development and outsourcing, now intrinsic to much consultancy. Table 7.7a indicates the top ten management consultancies by total fees, and the major client markets served. These firms account for almost two-thirds of the UK management consultancy market, and the top five for 40 per cent of the total.

Client relations are dominated by the financial sector, which accounts for over one-third of management consultancy business. In comparison, manufacturing firms accounted for almost 10 per cent, and the public sector and the utilities, each just over 8 per cent. Among the dominant consultancies in 1999, Andersen, KPMG and the FI Group relied most on financial service clients, while Cap Gemini was most orientated towards manufacturing. The ICL

Table 7.7a Top ten UK management consultancies: 1999 fee income, by top four markets served (£m)

	Total fee income	Financial services	Manu-facturing	Public sector	Utilities
Andersen Consulting	609	286e	31e	49e	85e
PricewaterhouseCoopers	540	157	27	43	40
KPMG Consulting	282	144	21	25	4
Cap Gemini	276e	97e	74e	—	52
FI Group/OSI	238e	117	2	16	55
ICL Group	212e	58e	40e	60e	19e
PA Consulting Group	176	38	19	22	10
Ernst & Young	174e	59e	13e	4e	17e
Deloitte Consulting	162	21	22	32	13
McKinsey & Co	135e	47e	26e	13e	—

Source: Management Consultancy 2000.

Notes to Tables 7.7a–c
e   estimate.
—   negligible.

Group undertook more public sector work than others. Although many smaller consultancies thrive by specializing in particular sectors and clients, most sectors are dominated by these large firms.

By 1999, IT expertise accounted for over 40 per cent of the value of services provided to clients, about 40 per cent of this from outsourcing financial management, including data processing (Table 7.7b). E-commerce consultancy and other Internet-related activities, although still small, were growing rapidly. Advice to clients on corporate strategy also grew by 25 per cent during 1998–99. On the other hand, *Management Consultancy* noted that demand for more traditional consultancy tasks such as project management, business process re-engineering and change management was being absorbed into IT or other strategic programmes, and these were declining as distinct activities. This is in line with the evidence presented earlier of national employment trends (see Table 7.3). As client needs for specialist advice change, so does the nature of consultancy, and it becomes increasingly difficult to distinguish between overlapping old and new requirements. The major firms claim varying service profiles. Cap Gemini, for example, focuses on project management expertise, while McKinsey continues to exploit its traditional strength in corporate strategy. The work of both is nevertheless driven by the impacts of ICT. Outside the top ten, other top twenty-five firms are significant in some markets (Table 7.7c): CMG in IT work; Sema and CSC in IT and project

Table 7.7b Top ten UK management consultancies: 1999 fee income, by major product activities (£m)

	IT	Project manage-ment	Corporate strategy	Business process re-eng.	Change manage-ment	Finance and admin systems	Human resources
Andersen Consulting	468e	44	40e	12	—	24	—
Pricewaterhouse Coopers	261	4	80	54	44	39	—
KPMG Consulting	70	36	55	74	2	—	2
Cap Gemini	31	210	3e	14e	—	17	—
FI Group/OSI	172	46	6	3	3	4	—
ICL Group	164e	4e	6	4e	7e	1	13
PA Consulting Group	61	48	15	7	2	—	11
Ernst & Young	14e	30	12	29e	17e	12	—
Deloitte Consulting	85	25	13	17	10	—	—
McKinsey & Co	—	—	135e	—	—	—	—

Source: Management Consultancy 2000.

management; Bain, Gemini and Boston Consulting in corporate strategy, but the brevity of this list confirms the general dominance of the top ten.

Much management consultancy work is based on a core expertise in financial monitoring and management procedures, but the impacts of IT are increasingly evident. The predominant markets are in the private and public services, including retailing, the utilities, the media, communications and transport. Their significance for manufacturing-based technical innovation and competitiveness may therefore be primarily in the encouragement of more dynamic management in other sectors. Management consultancies have fostered widespread IT-based process and organizational innovation. Such market-driven process change perhaps provides more commercially secure guidance to technical innovation than technological speculation alone. The restructuring of manufacturing still accounts for the second largest sector of management consultancy demand. In these cases there are probably more direct relationships to core manufacturing innovation, especially where management consultancy work overlaps with that of more specialist technical and computer consultancies.

Table 7.7c Top UK management consultancy firms in each activity not in top ten (overall rank), 2000

	IT	Project manage- ment	Corporate strategy	Business process re-eng.	Change manage- ment	Finance and admin systems	Human resources
CMG UK (11)	88	—	—	—	—	—	—
Sema (12)	79	30e	—	5e	—	—	—
CSC (16)	57e	13e	—	—	9	—	—
Bain & Co (UK) (22)	—	—	43	—	—	—	—
Gemini Consulting (14)	—	—	36	—	—	—	—
Boston Consulting (23)	—	—	35e	—	—	—	—
IBM Management Consulting (17)	—	—	6e	—	—	—	—
Towers Perrin (18)	—	—	—	—	9	4e	36

Source: Management Consultancy 2000.

Technical, including computer consultancy

Consulting engineering services (CES) remain quite distinct from the pervasive influence of management consultancy. About 40 per cent of CES in Europe serve manufacturing markets, with a further 40 per cent supporting construction and 20 per cent infrastructure (Sema Group 1996). A study based on 1992 data suggested that, of an estimated European market of ECU 52.6 billion, about 42 per cent was supplied by professional specialists, with 30 per cent coming from in-house sources and 27 per cent from other types of organization. The UK market was estimated at ECU 6 billion (over £4 billion), with over 17,000 CES firms, many relatively small, and 168,000 employees. The level of specialist CES professionalization in the UK was among the highest in the European Union, at around 50 per cent of the market (about £2 billion), especially in construction-related activities. The UK also had the largest export market, much of it outside Europe, about 40 per cent of turnover compared with a European average of 24 per cent.

The Sema Group study for the European Commission specifically discussed the innovative role of CES companies, emphasizing their generally low-risk approach in the advice they offered clients. Most consultancies were portrayed as employing tried and tested means, with development work mainly subcontracted. They may simply apply existing solutions or adapt them to different needs. They may also seek out and adapt a solution from another country or sector. Only rarely do they develop truly innovative approaches.

Some types of CES nevertheless do engage in product or process development and R&D, usually in close and semi-permanent relationships with key clients in particular sectors. These 'producers without production lines' operate through project-based development work, relying on financial support from one or more clients. Like consultancies in general, they may thus not be independently innovative, but support innovation which would not otherwise occur. Some earn income from patents and licensing. They are certainly able to engage innovative staff on a wider variety of projects than can individual clients, so offering important sectoral economies of scale and scope. Given the pace of change in many sectors, 'tried and tested' methods may be increasingly inappropriate, emphasizing the importance of innovativeness for CES firms willing to share the risks of new developments with clients.

The methodologies of many CES companies have also been profoundly affected by computer-aided design and manufacturing (CAD and CAM) techniques. These may enable assessments of many more technical alternatives to be offered at lower cost than in the past. International 'virtual' design networks are being developed to tap global expertise in design, simulation and testing, and for the redesign of operational methods. Collaborative arrangements are also becoming more common, including between innovative equipment suppliers, universities and other research agencies.

The dominant clients for CES are large private and public organizations in technology-intensive sectors, such as oil and gas, chemicals and pharmaceuticals, energy, metallurgy, construction and food and drink. These seek specialist skills from consultancies on building, utilities and production design, equipment selection, testing of specifications and the coordination of complex projects. The projects require regular and close contact with clients but, unlike management consultancy, the detailed work is often undertaken in the consultancies' offices. New requirements, such as environmental expertise (for the construction, energy or automotive sectors, for example) are generating demands for CES expertise. CES companies are not extensively employed in some sectors, such as aeronautics in the UK, because work is carried out largely in-house or by specialist quasi-monopolistic CES companies. The apparent decline in UK technical consultancy employment during the 1990s suggests that such approaches may be becoming more widespread, for example in the construction sector. The computer basis of much technical innovation (including CAD) is also affecting the traditional structure of CES supply, with more firms focusing on this technology, and being classified as such in official statistics.

The universal importance of computer-related change suggests that the burgeoning computer consultancy sector should be more innovative than many other KIS. Also, as hardware costs have fallen, more emphasis has been placed on software and systems development and even more on applications consultancy. Consultancies have therefore proved adaptable as conduits for the diffusion of changing computer-based technology to many users. Some may be highly innovative but, like the CES, many also valuably support the spread of established practice rather than solutions that are 'new to the world'. The primary problem for clients is thus to chose the right consultancy.

The dominant players in UK IT services are the major management consultancies, Andersen Consulting, PricewaterhouseCoopers, and the ICL and FI Groups. As we have seen, they primarily serve large-scale corporate needs. However, many other agencies link technical developments in computing to other client organizational requirements. These include hardware and software systems specialists, and independent consultancies that may break off from large telecommunications, energy and even public sector organizations (see Chapter 2). Smaller, specialist consultancies also continue to offer either innovative business-orientated software products, which may be marketed by other consultancies, or subcontracted programming expertise. Even though the large computer and management consultancies appear increasingly similar, the evidence on technology transfer presented in Chapter 3 suggests that consultancies' origins still influence what they offer clients. Thus clients have a wide range of consultancy options to consider. As in all consultancy, however, success depends essentially on the quality of client–consultancy interaction. Such choices must reflect clients' own technical capabilities, and their ability to chose and manage outside expertise.

Policy responses to KIS growth in the UK

Sectoral policies

The UK, perhaps even more than Europe as a whole, has traditional problems of transforming its high level of basic R&D and scientific excellence into practical economic benefits. A high proportion of R&D is government sponsored, especially in defence and related industries or in the universities which generally still focus on longer-term goals. Since 1980 stronger incentives have been offered to encourage closer links between industry and publicly funded (including university) research, for example through the Technology Foresight programme. By international standards, the UK also has persistently low levels of private R&D, although its national share has risen since the mid-1980s mainly because of the decline in public sector contribution and the influence of inward investment. Some sectors, such as pharmaceuticals, are more successful than others, such as much engineering. A strategic approach to R&D is generally lacking in UK corporate manufacturing, although some service sectors are innovative, especially in the implementation of ICT. The UK nevertheless lacks both the collaborative public–private research institutes and the link between fundamental and applied research commonly found, for example, in Germany (Bennett and McCoshan 1993: 55).

National innovation policy in the UK is implemented by the Department of Trade and Industry (DTI) and contains sectoral and generic, including SME and regional, components. The overarching framework is set by the Technology Foresight programme, established in 1993, which aims to overcome traditional barriers to linking scientific R&D and commercial practice through industrial and research-based initiatives, and to guide support for publicly funded science and technology. The DTI operates in this area through an Innovation Unit, directed by a team of seconded senior business managers. This reviews best practice to guide officials in the promotion of business innovation, linking business with universities and other research and technology providers, and sources of finance. It also supports panels of experts, whose remit is often based on sector trade associations, encouraging collaboration between major companies, for example in motor manufacturing, defence/aerospace, pharmaceuticals or biotechnology. The panels review major issues of common interest such as supply-chain management, public policy developments, education and training, and competitiveness and innovation in their sectors. Major consultancy firms support these activities through advice to their clients. In general, however, much UK innovativeness depends essentially on relationships between major companies and government procurement and research agencies, and recent policy has been based on a relatively *laissez-faire* approach to market processes, competition and regulation.

Area-based policies and SMEs

As in other countries in recent decades, the perceived need to encourage the development of SMEs in the UK and the regionally uneven impacts of economic change have focused much innovation policy on the encouragement of enterprise, often at a local level. Regional and local economic development policies before the 1980s were primarily directed towards high-unemployment areas and depended on sustaining traditional, usually publicly owned industrial sectors, or attracting 'branch plant' investment by large companies. From the late 1970s, and increasingly through the 1980s, emphasis switched to the privatization and rationalization of the traditional sectors, support for SMEs, the encouragement of international manufacturing competitiveness, and major inward investment projects. The last of these has contributed significantly to regional restructuring, for example promoting electronics in central Scotland and south Wales, and modern motor vehicle production in the North East and the East Midlands. The old development areas of high unemployment in the north and west regions remained as EU Objective 2 areas, often also qualifying for aid as a result of the closures in coal mining and steel making (Sadler 1992).

For much of the 1980s local development policies emphasized the encouragement of private land development, for example in a series of urban development corporations and twenty-two enterprise zones. Over twenty science parks were established through a variety of private and public initiatives, although their success in encouraging innovation and in creating employment was limited, especially in the high-unemployment regions (Massey *et al.* 1992). A plethora of other localized policies was attempted, affecting inner cities, port areas and enterprise development more generally. These included the provision of pump-priming finance and attempts to coordinate government agencies. The most successful local schemes during the 1980s were generally those which could benefit from service-based land developments, often in urban areas, based on office, retailing, tourist and housing schemes.

By the early 1990s it was widely recognized that this mixed inheritance of agencies, development purposes, area coverage and old-established and new programmes was serving no clear development purpose. Bennett and McCoshan (1993) analysed the situation in detail. They argued that the traditional UK policy emphasis on macro-economic policies and micro-economic programmes directed towards training, enterprise, education and innovation made little policy provision for improving the capabilities of firms and public agencies to absorb the consequent changes. Policy was thus required to develop these capabilities in the complex processes of innovation, market development, organizational adjustment and human resources development. At this time, of course, consultancies were already thriving in the corporate sector by offering just such services.

At regional and local levels in the UK, however, there was a chaotic confusion of policy institutions supporting SMEs and local economic development. Bennett and McCoshan describe the fragmented state of local economic service delivery, including significant overlap and competition between agencies, gaps in support, and a focus often on provider interests rather than client needs. Even without the separate arrangements for Scotland, and for some functions Wales, enterprise support involved nine major agencies, and finance from four government departments (Bennett and McCoshan 1993: 74). Sustained partnership arrangements were needed which could bring together the many private, government and voluntary/community agencies into effective economic development organizations.

It also became accepted during the 1990s, almost without question, that the main basis for developing such arrangements should be the local business environment, within which business–business links and business–community relations can be fostered (Bennett and McCoshan 1993: 96–7). The emphasis was on business networking among SMEs and the social and community context of development. This was promoted by EU regional innovation strategy, reinforced by 'new regionalism' thinking, emphasizing local enterprise, networks, 'learning regions' and clusters. Such approaches have traditionally been neglected in the centralized UK national policy regime, especially in England. It has thus proved difficult to forge effective locally active institutions out of the disparate inheritance of ministries, agencies and responsibilities.

A major attempt to coordinate and deliver support by the DTI for SMEs through local agencies was begun in 1993. Business Link (BL) aimed to develop a partnership framework for promoting enterprise, including all agencies, especially training and enterprise councils (TECs), chambers of commerce and enterprise agencies. The lead agencies were to be the local TECs, established in 1988, developing their enterprise-support responsibilities. The focus was on a 'one-stop-shop' agency to assist companies to identify their business objectives, assess business opportunities, gain access to information and advice, and devise appropriate individual development plans. The key to success was the recruitment of 600 personal business advisers (PBAs), supported by other exporting, design, innovation and technology specialists, to offer a single point of high-quality service and information. The main focus was on established SMEs with growth potential who would be assisted in finding specialist advice, either from BL or from outside consultancies. In the late 1980s, the enterprise initiative scheme had shown that subsidized support to selected consultants could make a measurable improvement on SME performance, especially in marketing (Segal, Quince, Wickstead, Ltd 1991).

BL was expected to develop over a period of years as best practice developed. Funds were available to support successful firms, including EU co-financing in Objectives 1, 2 and 5B areas, although schemes were intended to become self-supporting. It was to be more than a repackaging of current services, however, requiring the development of a collaborative approach, creating radical changes of attitude towards innovation, change and local networking with established companies, banks, accountants, solicitors and trade associations. In 1996, BL also became the main point of contact for the DTI's investment support services to SMEs, including a loan guarantee scheme, an enterprise investment scheme, and support for venture capital trusts. There were other measures to broaden the scheme's impact, by engaging other areas of government support, including Technology Foresight (DTI 1997).

By the time of the change of national government in 1997, BL had still not made the hoped-for impact (Curran and Blackburn 2000). In 1998, over 200 separate SME support initiatives were still active. Most of the pre-existing agencies remained, along with their institutional rivals, including other ministries, such as the DTI and the Department of the Environment, Transport and the Regions (expanded in 1997). Surveys of small businesses suggested that take-up was low even among target firms, and the schemes offered little incentive to broaden the base of clientele. Areas already well served by effective chambers of commerce or enterprise agencies were in the best position to develop any new collaborative arrangements. Although PBAs were useful in channelling general information to SMEs, the cost and intensiveness of the PBAs' work limited their impact, and their recruitment and training also proved to be difficult.

The 1998 White Paper, *Our Competitive Future: Building the Knowledge-driven Economy*, reiterated the importance of entrepreneurship and innovation at the local level (DTI 1998). The Small Business Service was set up in April 2000, to prioritize SME support at national government level, combining government and private sector expertise. It consolidated the 1996 broadening of measures and repackaged established schemes, although hardly streamlined them. BL continues as the basis of local advice delivery, like its equivalents in

Scotland, Wales and Northern Ireland. Meanwhile, regional development agencies (RDAs) were also established in 1998–99, with outline regional economic strategies developed in some haste with the government offices for the regions. Each RDA is built on the basis of very different prior experience of regional institutional collaboration. The drafting of the strategies also revealed continuing tensions between regional aspirations and regulatory and funding constraints imposed by central government. They represent the latest phase in UK efforts to allay central government suspicion of devolved institutional power and to encourage regional institutional structures, including new approaches to promoting entrepreneurship and innovation.

Conclusion: KIS and innovative expertise – research needs

The characteristic pattern of UK innovation in the twentieth century was of high basic scientific inventiveness but weak ability to translate this into commercially competitive products. Formerly strong sectors, such as engineering, machine tools and vehicles production also failed to adopt modern process innovations. These conditions have been associated with poor management attitudes and training. More general cultural and institutional explanations, however, include the legacy of old industries and production practices, a historical tradition of protected international markets, an inadequate vocational education system, chronically poor labour relations before the 1980s, alleged pressures from equity markets for short-term returns, and competition for capital investment from British overseas interests. In sum, an inability to take the strategic risks required for innovation has been reinforced by the generally low expectations of UK investment opportunities.

UK economic reforms during the 1980s were driven by a particular political view of how these deficiencies might be overcome, primarily by supporting international private investment interests, rather than those of traditional UK-based capital or labour. Manufacturing innovation has been stimulated most by this encouragement of foreign investment, importing US-based, European and Japanese production methods either directly or as a result of takeovers of UK firms. Much of this innovation, however, is process- rather than product-based. UK-based production may now be more competitive, but the fundamental and longer-term problems of UK-based innovativeness, related for example to the education system or risk-taking attitudes of capital, have not been directly resolved.

It has also come to be accepted that public sector ownership, control and regulation restrict the availability of capital, undermining competitiveness and innovation. There remains political debate over the degree to which this is the case, but radical challenges to traditional UK business attitudes have been stimulated by privatization and the deregulation of consumer markets, including retailing, the media, finance and property development. Their effects, at least in the short term, have been most evident in the vitality of UK consumer service activities.

We have seen that these transformations of private and public management methods have provided fertile ground for consultancies. They have supported privatization and public sector restructuring, as well as widespread changes in services, utilities and even manufacturing process innovation, and associated management methods. They have encouraged a more strategic view by client managements, especially in relation to international production and marketing trends. They have offered improvements in management training, and their growth is itself a symptom of more flexible and innovative approaches to management and change. They appear to have been particularly influential in the adaptation of ICT to many end-uses.

Questions still remain over the part consultancies may play in promoting the core innovativeness of clients, through the stimulation of ideas, the choice of innovation paths, the establishment of contacts, the linking of innovative ideas to the needs of potential markets and the managerial changes required to deliver them. Processes of innovation are fundamentally complex, making the influence of particular agencies difficult to isolate within national, industrial or regional innovation systems. UK evidence suggests that full understanding of the significance of consultancy in innovation will require closer examination of the major sectors and innovative systems. These include aerospace; medical research and health; information technology and computing; distribution and logistics; engineering, architecture and urban planning; commercial and industrial design; financial and business services; consumer services; banking; tourism and travel; media and cultural activities; defence; oil and other primary sectors; and agriculture/food/biotechnology. These are the key technological, national and regional innovation systems, influenced by corporate agencies (such as multinational corporations), public sector agencies (for example in defence and health), and perhaps also some groups of innovative SMEs, including consultancies.

The major national and regional innovation priorities are concentrated in these systems. So is the expertise required to support them from various international and national sources, including consultancies. At present, the consultancy role can be identified only through the declared activities of the major consultancies, with little systematic evaluative evidence. More case studies are required of key national and regional innovation projects, including their use of various types of consultancy in relation to other sources of expertise.

The questions that would need to be addressed by such studies, for which this chapter has offered only indirect evidence, may therefore be summarized as:

1   What are the impacts of consultancies on national innovation/diffusion systems? Are large consultancy firms more or less effective? How do they affect different sectors, types of firm, forms of interaction, regional/local needs?

Consultancies seem to have had a major impact on the diffusion of new, especially ICT-based technical and organizational methods in the UK over the past twenty years. They have achieved this by responding to the wider challenges faced by clients, especially in the more technically and commercially innovative sectors. Without consultancies, such adaptation would have been slower and less effective. Large consultancies dominate the most comprehensive range of expertise, but small consultancies are effective by specializing and working with clients with a clear sense of their own requirements. Outside this nexus, consultancy remains relatively inaccessible, for example to SMEs.

2   Are some regions better or worse off because of the innovative activities of consultancy firms?

A clear division has emerged between the 'knowledge environment' provided by innovative (including corporate) clients and consultancies in southern Britain, and that in other parts of the country. The major consultancies operate through branch offices in most regions, but these focus on corporate and public sector demands. These regional trends seemed to intensify during the 1990s. From the point of view of smaller potential clients, the pool of consultancy choice is smaller in peripheral regions, its quality is lower and the needs of such sectors are being increasingly bypassed by the pace of change in the corporate economy, based in core regions. It is not clear how far this adds to the disadvantages of operating competitively in these regions.

3   With the internationalization of consultancy, can local/small consultancy firms compete directly (for example on price) or by niche adaptation? Can multinational consultancies serve the needs of SME clients?

Many local consultancies focus on relatively routine support to familiar clients. Small/local consultancies can make an innovative impact by niche specialization, the exploitation of 'local knowledge', and the pursuit of complementary strategies to the major consultancy firms (see Chapter 3). There is good evidence that successful specialist consultancies in the UK are also developing active international markets themselves, often following work with multinational clients, or through networking arrangements within overseas partners.

4   What are the main points of leverage for possible policy? Can particular aspects of policy be identified that might be adapted to take account of the growing influence of KIS firms?

Consultancies are well established and influential in mainstream corporate and public sector innovation and change programmes in the UK. Policies have attempted to provide public business advisory services where access to commercial advice is clearly deficient, among SMEs or in peripheral regions, or to support consultancy advice under regulated conditions. These policies have achieved limited success, and new regionally based support arrangements seem unlikely to respond effectively to the pace of innovative change to which, of course, consultancies contribute. Knowledge internationalization is widening the gap between the capacities of conventional policy and the needs of successful entrepreneurs. One facet of a new policy might well be to promote access to private consultancy, recognizing that it has grown as rapidly as the activities of SMEs since the 1970s when much SME policy was initiated, transforming the commercial environment (Curran and Blackburn 2000). SME policies need arrangements better attuned to the circumstances of the 2000s, recognizing the limited success of traditional approaches, especially under highly centralized UK policies. The first point of departure might then be to tap the expertise of the consultancy sector. This would provide a suitable challenge not only to government, but also to the consultancy sector itself.

Note

1   The data analysis in this chapter is for Great Britain, and does not include Northern Ireland.

References

Bennett, R.J. and McCoshan, A. (1993) *Enterprise and Human Resource Development*, London: Paul Chapman.

Coffey, W. and Bailly, A.S. (1992) 'Producer services and systems of flexible production', *Urban Studies* 29: 857–68.

Curran, J. and Blackburn, R.A. (2000) 'Panacea or white elephant? A critical examination of the proposed new small business service and response to the DTI consultancy paper', *Regional Studies* 34(2): 181–9.

DTI (1997) *Report on Business Link and Business Link Regional Supply Network Management, July 1995 to September 1996*, London: DTI.

—— (1998) *Our Competitive Future: Building the Knowledge-Driven Economy*, London: HMSO.

Hepworth, M., Green, A.E. and Gillespie, A. (1987) 'The spatial division of information labour in Great Britain', *Environment and Planning A* 19: 793–806.

Howells, J. (1988) *Economic, Technological and Locational Trends in European Services*, Aldershot: Avebury.

Management Consultancy (2000) *Trends in Management Consultancy, 1999: Annual Survey* (July/August), London: VNU Publications.

Massey, D., Quintas, P. and Wield, D. (1992) *High Tech Fantasies: Science Parks in Society, Science and Space*, London: Routledge.

O'Farrell, P.N., Hitchens, D.M. and Moffat, L.A.R. (1992) 'The competitiveness of business service firms: a matched comparison between Scotland and the SE of England' *Regional Studies* 26: 519–34.

O'Farrell, P.N., Wood, P.A. and Zheng, J. (1996) 'Internationalisation by business services: an inter-regional analysis', *Regional Studies* 30: 101–18.

Porat, M. (1977) 'The information economy: definitions and measurement', Special publication 77–12(1), Washington, DC: Office of Telecommunications, US Department of Commerce.

Sadler, D (1992) *The Global Region: Production, State Policies and Uneven Development*, London: Pergamon.

Segal, Quince, Wickstead, Ltd (1991) *Evaluation of the Consultancy Initiatives: Third Stage*, London: Department of Trade and Industry, HMSO.

Sema Group (1996) *The Role of Consulting Engineering Services in Innovation*, EIMS (European Innovation Monitoring System) Publication No. 17, Brussels: European Commission, D-G XIII.

Wood, P.A. (1996a) 'Business services, the management of change and regional development in the UK: a corporate client perspective', *Transactions, Institute of British Geographers* NS 21: 649–65.

—— (1996b) 'An "expert labour" approach to business service change', *Papers, Regional Science Association* 75(3): 325–49.

8 Italy

The influence of regional demand and institutions on the role of KIS

Lucia Cavola and Flavia Martinelli[1]

Introduction

This chapter reviews the development of knowledge-intensive services (KIS) in Italy throughout the 1980s and early 1990s. Three main themes underlie the commentary.

1 The close link between the characteristics of KIS demand and supply in explaining their national development and geography.
2 The influence of the socio-institutional context in explaining KIS development and regional trajectories of change, including the structures of and relationships among firms, labour market characteristics and government institutions.
3 The need to adopt a historical and evolutionary perspective on the socio-economic processes that may foster or hamper the development of KIS.

In the first section the Italian productive structure and institutional environment will be sketched, outlining the major demand influences on KIS. The second section will detail the supply structure and geography of business services and KIS. Section three will illustrate how the geographical diversity of local institutional and policy contexts influences regional trajectories, through three regional cases.

National demand and the regulatory environment

The industrial structure of Italy in the 1980s: a stylized picture

To highlight the influence of demand on the growth of KIS in Italy, the specific characteristics of the country's productive basis need to be sketched. The Italian industrial structure in the early 1980s may be summarized as a tripartite system with regard to both firm characteristics and geography, although these do not strictly correspond. In relation to firm characteristics, a few oligopolistic financial/industrial groups operate in a number of basic and modern industries, such as steel, chemicals, electronics, automobiles, electromechanical equipment, telecommunications, food and pharmaceuticals, and in the financial and distribution services. These may be further divided between the private sector, with an important presence of foreign multinationals, and a state-controlled system. The latter was traditionally quite important in the Italian economy although it has experienced a significant degree of privatization. There is also a large, diverse small–medium enterprise (SME) sector, primarily in the so-called 'traditional' industries of food, textiles, clothing, footwear and furniture, but also in some modern and highly innovative sectors, including pharmaceuticals, electronics and machinery, as well as business and consumer services.

The geography of these three industrial segments was quite distinct. Among the large national corporations, private groups were mostly located in the North West, in the so-called 'industrial triangle', made up of Lombardy, Piedmont and Liguria, with headquarters mainly located in Milan and Turin. The headquarters of state-controlled groups and foreign multinationals were mainly located in Rome and, to a lesser extent, in Milan. Their production plants were also mostly located in the North, although with a strong presence also in the South, as a consequence of the industrialization policy of the 1960s and early 1970s. SMEs were present in all regions but were particularly strong in the so-called 'Third Italy' (the North East–Centre).

Three main regional models, with distinctive urban systems, thus emerged. First, the North West, the old industrial core of the Italian economy, based on large firms, but also on a dynamic system of SMEs, some of which supplied large firms while others possessed more autonomous market outlets. In this region the metropolitan area of Milan emerged as the main headquarters and service axis of the Italian economy, followed by the Turin area, the original cradle of Fiat and Olivetti.

Second, there was the so-called 'Third Italy', which included the North East and the Centre (NEC), with an industrial base almost exclusively made up of SMEs. In a number of regions (Emilia Romagna, Tuscany, Veneto) such SMEs were organized as true 'industrial districts': territorially and economically integrated, self-supporting productive systems, often strongly export orientated.[2] These regions were characterized by a well-articulated urban system, with several medium-sized cities acting as provincial or regional service centres for their hinterlands. Within central Italy, Lazio is a special case, with the metropolitan area of Rome hosting the central state apparatus, the headquarters of several state corporations and foreign multinationals, as well as international organizations and institutions.

Finally, there was the South or Mezzogiorno (including Sicily and Sardinia), which covers about 40 per cent of the national territory and includes about 37 per cent of its population. This was historically, and still is, the lagging region of Italy. It experienced significant industrialization during the 1960s and early 1970s, mostly through inward investment by large national and foreign corporations, but was hard hit by the economic crisis of the mid-1970s. By the early 1980s its industrial structure was still rather dualistic, with several 'branch plants' of large (especially state) corporations and a scattered population of often very small SMEs, which organizationally and economically lagged behind their NEC counterparts (Martinelli 1985, 1989; Del Monte and Martinelli 1988; Giannola 1982, 1987; Giunta 1992). They mostly operated in poorer market sectors, with few interlinkages, or as 'captive' subcontractors for large companies. Many belonged to the 'underground', unregulated economy, thriving on illegal labour, and fiscal and regulatory evasion.

Demand factors affecting KIS growth: economic restructuring, large firms and SMEs

The dramatic growth of KIS recorded in all Western economies during the 1980s must be primarily related to the booming demand arising from the profound restructuring of the international productive system. The opportunities offered by the new information technology (IT) merged with the needs of technological and organizational restructuring associated with the decline of 'Fordist' methods since the 1970s economic crisis and the internationalization of competition, including that within the Single European Market. In Italy, as elsewhere, the oligopolistic sector was the first to engineer a transformation of its organizational and technological structure. This created a major focus for KIS demand throughout the 1980s.[3] Public administration lagged behind, although a few central government institutions did take the lead. Most, especially peripheral region authorities, started reorganizing their information and management systems only in the 1990s (AIPA 1997; Epifani 1996; Assinform-Nomos Ricerca 1996; Anasin-Nomos Ricerca 1994).

The majority of large Italian firms engineered their organizational transformation by coupling internal know-how with external consultancy services from large hardware producers and/or IT consultants, mostly foreign multinationals. An exception was Olivetti, which undertook its restructuring on the basis of internal resources and then adapted this acquired knowledge for the outside market by establishing an independent consultancy (cf. Moulaert *et al.* 1990).

On the other hand, one of the major features of IT development was its accessibility to small firms. Indeed, another relevant focus of Italian demand for IT-related services during the late 1980s was SMEs. The latter, however, were far from uniform, either sectorally or geographically, and this also affected their use of business services and KIS. At least three types of SMEs can be identified. Some were independent firms operating in innovative segments/sectors, with strong internal know-how and a highly 'targeted' demand for KIS. Others were in the so-called industrial districts of the NEC region, operating in traditional sectors, which developed an increasing demand for both traditional and sector-specific KIS, often with the help of local institutions. Finally, there were the territorially and/or sectorally isolated SMEs which showed little demand for innovation. These were organizationally weak and mostly orientated towards local markets, or dependent on larger firms' subcontracting.

Regional trajectories in restructuring

Because of their different economic structures, industrial restructuring in the 1980s had different features in the three main geographical areas of the country (cf. also Del Monte and Giannola 1997). In the North West the crisis of the old industrial apparatus was more acute and the losses in manufacturing employment higher, especially in large establishments. However, such trends were partly compensated for by the tertiarization of the economy and by the vitality of SMEs. In the NEC region the de-industrialization process was less acute and the losses in manufacturing employment limited. SMEs actually recorded positive employment performance, as did both producer and consumer services.

In the Mezzogiorno the crisis of Fordism further weakened the already fragile economic structure of the region, jeopardizing progress towards internal cohesion and reducing the aggregate competitiveness of the regional economic system. After over forty years of development policy, the per capita gross regional product of the Mezzogiorno in 1994 was 67 per cent of the national average, having reached a peak of over 70 per cent in the mid-1970s. More significant, the unemployment problem worsened dramatically, with southern unemployed making up 61 per cent of the national total in 1991 compared with 49 per cent in 1980. The official unemployment rate in the South in 1994 was 19 per cent, compared to 8 per cent nationally, and youth unemployment reached 52 per cent compared to 18 per cent in the North and Centre (Del Monte and Giannola 1997). Many branch plants established in the 1960s and early 1970s closed down or were radically rationalized (cf. Giunta and Martinelli 1995). SMEs recorded a better performance, especially in the smallest size class (up to twenty employees). But in this region the restructuring was not coupled with a significant tertiarization process. Although business services recorded above-average rates of growth, their weight remained largely below the national average (see further below).

The national institutional context and the regulatory environment

Besides factors related to the demand for KIS, such as the structure of and changes in the industrial system, factors related to the broader national institutional and regulatory environment must also be considered. In the Italian case at least four specifics must be mentioned.

First, among the effects of the increased internationalization of markets and, more particularly, of the accelerated integration of Italy into the European market, the intensification of national regulation of business operations has been significant, including accounting and filing procedures, technical standards and requirements, quality and safety certification of products, and health and safety norms in the workplace. The impact of this regulatory intensification, which Gallouj (1992) has called 'regulatory entropy', is clearly seen in the significant growth of such business services and KIS as veterinary, testing and certification, technical, and accounting and fiscal consultancy services.

A second institutional feature of Italy is reflected in national policies in support of innovation. With regard to technological innovation, Italy is basically a 'latecomer' (Papagni 1995; Graziani 1997). A gap also exists between the South and the rest of the country (Istat 1990; Formez-Istituto Tagliacarne 1994; Silvani and Prisco 1993). This position explains the limited progress of research and development (R&D)-related services in the private sector[4] and also the low and late engagement of national policy in support of technological innovation and R&D compared to other European countries. Indeed, it was only in the late 1980s that policies and programmes in this area picked up momentum (see below).

The third factor is the process of partial administrative decentralization from national to regional governments that began in the 1970s and became fully operational in the 1980s, although with different results across the country.[5] The newly established regional governments did not have authority over industrial and development policy, which remained a national government prerogative. However, they did have legislative power in the area of 'handicraft firms' (small firms), which provided a cover, employed by many north–central regional governments, for strong SME policies during the 1980s and 1990s, especially in relation to business services.

Finally, a long-established feature of the Italian regulatory structure that has affected KIS development is the former national policy for the development of the Mezzogiorno. From 1950 to 1991 such action was centrally regulated through special legislation known as the 'Intervento Straordinario' for the Mezzogiorno and its local institutions. Policies, tools and funding evolved throughout this period, but were mostly centred on public works (physical infrastructure) and on financial incentives to productive investment. Only from 1986 did central regional policy include measures in support of business services and R&D activities in the South. With the termination of the centralized Intervento Straordinario in 1991, development policies and programmes for the Mezzogiorno have become a matter of 'ordinary' public spending procedures, mostly delegated to regional and local authorities, within the EU Structural Funds framework for 'depressed areas'. In practice, lagging regions of the Mezzogiorno (EU Objective 1 regions) now have to compete with other less depressed areas of the country (Objectives 2 and 5B) for reduced levels of development aid.

The role of local institutions

As will be illustrated below, the actions of a number of regional and local institutions have had a significant influence on the development of KIS and on innovation diffusion, especially among SMEs. Many Italian regional governments, as well as local institutions associated with provinces, municipalities, and chambers of commerce have had a crucial catalysing role in supporting local economic development and innovation. However, a major cleavage exists in Italy between north–central and southern regions. Even though regional governments have been established throughout the country for more than twenty years, there remains a clear differentiation in their planning and management capabilities, reinforcing the dualism of the country.

There are several reasons, deeply rooted in history, for these divergent trajectories (cf. Putnam 1993). Two general views can be contrasted. On the one hand there is the prevailing neo-liberal view, which blames the Intervento Straordinario for all the evils of the Mezzogiorno. These include high public expenditure as well as corruption and mismanagement, which have, it is argued, prevented the development of a market economy, and produced a dependent and parasitic culture. Only by eliminating the distortions introduced by such centralized public spending and by decentralizing responsibility can conditions for self-reliance and entrepreneurship be restored in the South (Lizzeri 1983; Pontarollo 1982; Trigilia 1992; Bodo and Sestito 1991). Such views, reinforced by federalist sentiments in the North East, have legitimized the demise of national policy for the Mezzogiorno and paved the way for deregulation.

On the other hand, a more articulated assessment of the links between class structure, interest groups and institutions, both at the national level and in the Mezzogiorno, still survives, which re-emphasizes the need for a reformed, but active, national development policy for the South (Del Monte and Giannola 1989, 1997; Wolleb and Wolleb 1990; Giunta and Martinelli 1995).[6] Martinelli (1998) argues that the development of modern, democratic and efficient local institutions in the South, especially regional governments, was hampered by the establishment of a national social bloc increasingly based on a client-focused system of public expenditure in the South. With the economic crisis of the mid-1970s, political consensus increasingly replaced the original developmental aims of regional policy. This interpretation explains how formally similar institutions such as regional governments can behave in very different ways. In the North and Centre there is a strong cohesion between markets and institutions. In the South the development of market mechanisms has been interrupted by the economic crisis, while local institutions have become passive recipients of declining transfer flows, mostly redistributing money and jobs to maintain political consensus. It also explains how institutions supporting change, including KIS, operate more effectively in the cohesive environment of the North and Centre than in the poorly integrated and disjointed institutional conditions which still dominate the South.

The supply of KIS in Italy: trends, structures and geography[7]

Despite some difficulties involved in comparing the 1981 and 1991 censuses it is clear that, between 1981 and 1991, business services were the fastest growing sector of the Italian economy, contributing the largest share of new employment. As shown in Table 8.1, employment in the Italian tertiary sector increased by 15 per cent in that period (1,427,000 new jobs). Nearly half of this increase (688,000) was created in the business service sub-sector, broadly defined, which grew by 61 per cent. These trends are particularly striking when compared with the manufacturing sector, which lost about the same amount of employees during the period (604,307, a fall of 10.4 per cent). From a 12 per cent share of the tertiary sector in 1981, the business service share grew to 17 per cent by 1991, of which 7.1 per cent were KIS.

The KIS, including IT services, R&D and professional and technical consultancy, almost doubled employment, by 88 per cent (360,000 new jobs). The most spectacular growth, as might be expected, was in information technology services, which more than trebled employment in the 1980s (130,000 new jobs). Within the professional and technical consultancy services, the fastest growth (more than fourfold) was in the small group of testing and technical certification services (over 324 per cent), but the largest absolute increase, of 130,000 jobs (over 70 per cent), was in professional consultancy. Growth in R&D activities, although appreciable, was the slowest (over 48 per cent). These data clearly show that the Italian economy has experienced a profound shift from manufacturing to services and that 'free-standing' business services were a leading growth sector throughout the 1980s.

The growth of managerial and clerical employment within non-service sectors, or 'in-house' service employment, also showed a significant tertiarization of production. Nationally, white-collar employment (managerial, technical and clerical employees) in manufacturing recorded an increase of 9 per cent, partially compensating for a 17 per cent decline in the numbers of manual workers (Table 8.2). This trend was most marked in the North East and Centre, and was strong also in the Mezzogiorno, but not in the North West, where in-house service employment was already high.

These aggregate trends hide differences in the performances of individual regions. From 1981 to 1991 it appears that business service growth rates were higher in more peripheral regions, whether in North–Central Italy (Val d'Aosta, Trentino Alto Adige, Veneto, Umbria), or in the South (Abruzzi, Molise, Basilicata, Campania, Sardinia), whereas the traditional centres of KIS supply (Lombardy, Lazio and Piedmont) slowed down (Table 8.1). Some 'spreading' thus seems to have taken place. On the other hand, when absolute employment figures are considered, Lombardy, Piedmont and Lazio still accounted for 42 per cent of new business service jobs. Adding the 'newcomer' regions of Emilia Romagna, Tuscany and Veneto, these six regions accounted for as much as two-thirds (68 per cent) of new business service employment.

The geography of business services and KIS in 1991

Having summarized growth trends, we can now focus on the 1991 situation, using, among other things, a fuller and more detailed database for KIS. The

Table 8.1 Italy: trends in business services employment, 1981–91 (absolute values and growth index)

Service activities	NW			NEC			Mezzogiorno			Italy		
	1981	1991	Index[a]	1981	1991	Index	1981	1991	Index	1981	1991	Index
KIS	160,517	288,693	179.9	174,168	334,604	192.1	77,467	152,472	196.8	412,152	775,769	188.2
Information technology services (72)	22,414	69,683	310.9	23,579	84,065	356.5	5,889	27,204	461.9	51,882	180,952	348.8
Research and development (73)[b]	9,380	13,468	143.6	15,102	22,356	148.0	4,948	7,634	154.3	29,430	43,458	147.7
Professional consulting and spec. services (74)	128,723	205,542	159.7	135,487	228,183	168.4	66,630	117,634	176.5	330,840	551,359	166.7
Professional consulting services (74.1)	68,143	111,024	162.9	76,603	132,835	173.4	39,947	70,340	176.1	184,693	314,199	170.1
Technical services (archit. and engin.) (74.2)	47,499	68,386	144.0	50,401	77,877	154.5	24,232	42,409	175.0	122,132	188,672	154.5
Testing and technical certification (74.3)	673	3,929	583.8	739	2,736	370.2	407	1,055	259.2	1,819	7,720	424.4
Advertising (74.4)	12,408	22,203	178.9	7,744	14,735	190.3	2,044	3,830	187.4	22,196	40,768	183.7
Employment and training services[c] (74.5)	n.d.	1,471	—	n.d.	720	—	n.d.	1,480	—	n.d.	3,671	—

Other business services	259,544	367,436	141.6	322,982	464,774	143.9	140,265	211,637	150.9	722,791	1,043,847	144.4
Banking and finance (65 + 67.1)	124,469	152,806	122.8	146,402	185,852	126.9	62,659	84,873	135.5	333,530	423,531	127.0
Insurance (66 + 67.2)	40,304	54,833	136.0	47,265	63,805	135.0	18,413	27,366	148.6	105,982	146,004	137.8
Real estate (70)	18,575	36,396	195.9	23,613	39,811	168.6	3,711	6,982	188.1	45,899	83,189	181.2
Rental of productive equipment (71, less 71.1)	3,489	2,492	71.4	3,856	2,903	75.3	1,251	1,420	113.5	8,596	6,815	79.3
Security services (74.6)	11,520	14,473	125.6	12,107	16,280	134.5	9,734	15,877	163.1	33,361	46,630	139.8
Cleaning services (74.7)	23,898	50,558	211.6	31,763	77,032	242.5	15,011	35,925	239.3	70,672	163,515	231.4
Other business services n.e.c. (74.8)	24,256	40,477	166.9	30,426	46,045	151.3	19,127	26,940	140.8	73,809	113,462	153.7
Business associations (91.1)	13,033	15,401	118.2	27,550	33,046	119.9	10,359	12,254	118.3	50,942	60,701	119.2
Total business services	420,061	657,600	156.5	497,150	800,098	160.9	217,732	365,589	167.9	1,134,943	1,823,287	160.7
Total services	2,781,138	3,166,269	113.8	4,049,137	4,662,761	115.2	2,686,126	3,114,004	115.9	9,516,401	10,943,034	115.0

Source: Istat 1995a.

Notes a Index 1981=100 b Figures in brackets are Istat industrial classification codes. c In 1991 'Employment and training services' were classified elsewhere. They are therefore not included in the totals for 1991.
n.d. no data.

regional endowment of such activities must be assessed against some indicator of demand (cf. Martinelli 1984). Regional 'tertiarization quotients' (TQs) have thus been calculated, which measure regional employment in given business services against total regional value added, and compare it to the national average.[8] As Table 8.3 shows, there was a strong geographical polarization in the supply of business services in Italy in 1991. Lombardy, Piedmont, Lazio, Emilia Romagna and Tuscany exhibit a high concentration of these activities, compared to their productive basis. All other regions, especially in the South, show a business service endowment below the national average.

In 1991, therefore, despite the lower growth rates of the established business service regions and the faster growth in peripheral regions, large portions of the national territory (especially the Mezzogiorno) were still excluded from the supply of business services (and of KIS in particular). On the other hand, the geography of polarization changed significantly between 1981 and 1991 (Tables 8.2 and 8.3). Lombardy, Piedmont and Lazio maintained their business services primacy, whereas Liguria fell behind. They were joined by the 'newcomers' Emilia Romagna, Tuscany and Friuli.

Furthermore, different specializations within KIS emerged among regions. Piedmont, Lombardy and Lazio dominate the IT services, especially in the most strategic activities of installation and software, and consultancy. Most regions of the 'Third Italy' exhibit TQs close to the national average, but they appear to concentrate more on the 'operational' IT services, such as data processing, data banks and maintenance. In the Mezzogiorno the under-supply of this group of KIS, despite higher growth rates during the 1980s, appears dramatic. Aggregate TQs were below 0.70 in all regions, with the sole exception of Basilicata. In the most strategic components the regional endowment was barely half the national average.

With regard to R&D services, the most endowed regions, with TQs above 1.10, are Piedmont, Lazio, Liguria and, surprisingly, Basilicata and Sardinia, in the South. In these regions there is, indeed, a strong presence of public R&D centres. Lombardy, on the other hand, exhibits a significant under-endowment of specialist R&D activities compared to its productive basis, most likely because R&D in this region is mainly carried out in-house and therefore not counted separately by the census. Emilia Romagna, Tuscany and Friuli follow, again with TQs just below the national average. Most of the South lags behind, with TQs below 0.70.

With regard to professional consultancy and specialized business services, regional contrasts appear less sharp. Piedmont and Lombardy lead, with TQs above 1.10, closely followed by Emilia Romagna and Tuscany. Lombardy is most diverse, especially in strategic services such as market studies and polls, advertising, testing and technical certification, employment and training services, management consultancy, and accounting and fiscal consultancy. Piedmont appears more narrowly focused on advertising, congress services and technical consultancy. Emilia Romagna and Tuscany also specialize in accounting and fiscal services, congress services, and testing and technical

Table 8.2 Italy: occupations in selected industries, 1981–91

Industries	Entre- preneurs	Manag. /clerical	Manual	Total	Entre- preneurs	Manag. /clerical	Manual	Total	Entre- preneurs	Manag. /clerical	Manual	Total all occupations
	1981 %				1991 %				1981–91 % change			
Italy												
Manufacturing/extr.	14	17	69	100	16	21	64	100	4	9	–17	–10
Energy, gas, water	1	44	55	100	1	52	47	100	–10	–3	–31	–19
Constructions	31	7	62	100	33	10	56	100	19	65	2	12
North West												
Manufacturing/extr.	11	21	68	100	13	25	62	100	0	0	–24	–16
Energy, gas, water	1	46	52	100	1	54	44	100	–23	–14	–38	–27
Construction	35	8	57	100	36	11	53	100	21	61	9	17
North East-Centre												
Manufacturing/extr.	15	15	70	100	17	19	65	100	3	19	–12	–5
Energy, gas, water	1	42	57	100	1	52	47	100	–6	9	–27	–12
Construction	34	7	59	100	36	11	53	100	10	56	–8	2
Mezzogiorno												
Manufacturing/extr.	16	12	72	100	19	15	66	100	13	15	–11	–4
Energy, gas, water	1	43	56	100	1	49	50	100	4	–7	–27	–18
Construction	23	6	71	100	26	10	64	100	41	89	10	22

Source: Istat 1995.

certification and, in the case of Emilia Romagna, advertising. Most of the remaining north–central regions, including Lazio, exhibit TQs just below 1.00. Lazio concentrates on only a few activities, including congress services, market studies and polls, and legal services. With the notable exception of Basilicata, the South shows aggregate regional TQs below 0.90, and even lower for market studies and polls, advertising, testing and technical certification. These had TQs below or barely above 0.50. The only activities significantly concentrated in the South are legal and employment and training services.

In-house service employment confirms the same geographical polarization (cf. Table 8.2). In the North West, the tertiarization of production is higher than in central and north-eastern Italy, and much more than in the South. The dominance of the traditional centres of business services is even more striking for higher-level service employment. The 1991 census identifies employment in holding companies, the headquarters and most strategic in-house service activities of large corporations. Table 8.3 shows that these were overwhelmingly concentrated in Lazio, Piedmont and Lombardy (with TQs of, respectively, 3.80, 2.40, 1.63). These regions alone contained almost all national holding company employment in Italy (92 per cent).

The structure of supply

The structure of supply of KIS in Italy is quite fragmented, characterized by a pre-eminence of small firms and establishments. In 1991 the smallest sizes of firms, with averages of fewer than three employees per firm, were found in professional consultancy and specialized business services, especially legal, accounting and fiscal, management, and technical consultancy. A higher average firm size, of over five employees, was found in both IT services and R&D activities. In professional consultancy and specialized business services, about three-quarters of firms had only one or two employees. In IT services, this share was about half. Once again, significant differences emerge between regions. Lombardy, Piedmont and Lazio stand out quite clearly as having more large firms, whereas all the other regions, including those in the 'Third Italy', show a distribution biased more towards the smaller sizes.

The prevailing small size of Italian KIS firms is confirmed by data about their enterprise structure. Over 90 per cent of firms were single-office enterprises. Multi-establishment firms were significant only in R&D activities (4.5 per cent of firms and 46 per cent of employment) and in IT services (6.2 per cent of firms and 34 per cent of employment). The latter also possessed the greatest share of establishments outside the firm's home region. Among professional consultancy services the shares of multi-locational firms were generally lower. Once again, they were mostly concentrated in Lombardy, Lazio and Piedmont.

Table 8.3 Italy: tertiarization quotients for KIS in selected regions, 1991

Services	Lomb.	Lazio	Pied	E.R	Ven	Tusc.	NW	NEC	STH	Italy
All business services	1.2	1.2	1.1	1.1	0.9	1.1	1.1	1.1	0.8	1.0
Knowledge- intensive services	1.2	1.2	1.1	1.0	0.9	1.0	1.2	1.0	0.8	1.0
Information technology services	1.3	1.6	1.3	1.0	1.0	0.9	1.2	1.1	0.6	1.0
Installation (72.1)	1.1	1.9	1.7	0.9	1.3	0.7	1.2	1.1	0.5	1.0
Software and consultancy (72.2)	1.3	2.3	1.4	0.9	0.7	0.7	1.3	1.1	0.5	1.0
Information processing (72.3)	1.3	1.0	1.0	1.1	1.2	1.3	1.1	1.1	0.6	1.0
Data banks (72.4)	0.8	2.4	0.9	1.1	1.7	1.0	0.8	1.5	0.5	1.0
Maintenance and repair (72.5)	1.1	1.2	1.0	1.1	1.0	1.0	1.1	1.1	0.8	1.0
Other (72.6)	1.4	1.9	1.3	0.6	0.6	0.7	1.3	1.0	0.7	1.0
Research and Development (73)	0.7	2.8	1.3	0.9	0.4	0.9	1.0	1.2	0.7	1.0
Prof. consultancy and specialized services	1.2	0.9	1.1	1.1	0.9	1.1	1.2	1.0	0.8	1.0
Legal (74.11)	0.9	1.1	0.8	0.9	0.8	1.1	0.9	1.0	1.1	1.0
Accounting and fiscal (74.12)	1.1	0.8	1.0	1.2	0.9	1.4	1.1	1.0	0.8	1.0
Market studies and polls (74.13)	2.3	1.2	0.8	0.9	0.6	0.9	1.7	0.8	0.4	1.0
Management consultancy (74.14)	1.3	1.0	1.1	1.1	1.1	1.0	1.2	1.0	0.7	1.0
Technical (arch., engin., etc.) (74.2)	1.1	0.8	1.2	1.1	1.0	1.0	1.1	1.0	0.9	1.0
Testing and technical certification (74.3)	1.7	0.4	1.1	1.2	0.9	1.2	1.6	0.8	0.5	1.0
Advertising (74.4)	2.0	0.8	1.5	1.1	0.9	0.8	1.7	0.9	0.4	1.0
Employment and training (74.5)	1.7	0.4	0.5	0.6	0.4	0.5	1.2	0.5	1.6	1.0
Congress, fairs, and exhibit (74.83)	1.1	1.2	1.0	1.4	0.9	1.2	1.1	1.1	0.7	1.0
Holdings (74.15)	1.6	3.8	2.4	0.3	0.3	0.2	1.7	1.1	0.0	1.0

(continued on next page)

Table 8.3 (cont.)

Services	Lomb.	Lazio	Pied	E.R	Ven	Tusc.	NW	NEC	STH	Italy
Finance, insurance and real estate	1.2	1.2	1.0	1.1	0.9	1.1	1.2	1.1	0.7	1.0
Banking and finance (65+67.1)	1.2	1.3	1.0	1.1	0.8	1.1	1.1	1.0	0.8	1.0
Insurance (66+67.2)	1.2	1.3	1.1	0.9	1.0	1.0	1.2	1.0	0.7	1.0
Real estate (70)	1.4	0.9	1.2	1.3	1.2	1.6	1.4	1.1	0.3	1.0
Other Business Services	1.0	1.1	0.9	1.3	1.0	1.1	0.9	1.1	0.9	1.0

Source: Istat 1995.

Note
Figures in brackets are Istat industrial classification codes.

Recent trends in selected KIS: slowdown and growing foreign influence

In the early 1990s, the national economic recession brought the growth of business services to a halt. Growth picked up again only in the second half of the decade, although at a slower pace (Istat 1998). Between 1993 and 1997 business service employment in Italy increased by 13 per cent. Growth was highest in the North East (17 per cent) and lowest in the South (10 per cent). Further information is available for two relevant KIS sub-sectors: IT consultancy and management consultancy. These data also show increasing Italian dependence on foreign suppliers.

For new IT services the growth of the 1980s seemed to continue into the early 1990s, although at a slower pace and with some structural changes (see Table 8.4). In 1994–95 the number of firms with twenty or more employees in computer services was growing rapidly (at 13.6 per cent; cf. Istat 1997), but the average size of firms was falling. Aggregate sales also increased by 11.1 per cent, but employment declined, by 6.7 per cent. The geography of such trends strongly penalized the South. The extra firms were located almost entirely in north–central Italy. Moreover, the decline in employment was mostly confined to the South.

In this market, the penetration of foreign multinationals is quite marked. Only two large national firms succeeded as free-standing suppliers during the 1980s, but their growth in the 1990s took place under increasing constraints. Indeed, data collected by Nomos (1997) on software and computer consultancy services show that in 1996 almost 50 per cent of the Italian market was supplied by fifteen large firms, only two of which were Italian. These were Olivetti, which was established principally as a hardware supplier, but also provides integrated software, management consultancy

Table 8.4 Italy: firms with more than 20 employees in IT services, 1994–95

Regions	1994			1995		
	Number	Empl.	Average	Number	Empl.	Average
NW	430	26,978	63	483	26,325	55
NE	238	13,599	57	246	13,222	54
Centre	235	19,610	83	284	22,129	78
Mezzogiorno	164	16,941	103	158	10,297	65
Italy	1,067	77,128	72	1,171	71,973	61

Source: Istat 1997.

and training, especially to the banking and insurance sectors, and Finsiel, which owed its growth mostly to demand from central government administration.

In management consultancy, trends towards concentration and dependence on foreign suppliers are even greater. According to Assoconsult (1996),[9] the growth of the Italian market for organizational and management consultancies slowed down during the early 1990s. In the recession conditions of 1994–95 aggregate sales increased by only 5 per cent. Moreover, there was a clear process of concentration, with large firms gaining greater control of the market. The top twenty firms (in terms of sales) increased their share from 63 per cent of total national sales in 1991 to 74 per cent in 1995. Foreign multinationals (mostly Anglo-American) also became more dominant. The top five foreign firms increased their share of the national market from 31 per cent in 1991 to 40 per cent in 1995, while the top five Italian firms continued at around 18 per cent. Management consultancy firms operating in Italy also work overwhelmingly for the national market. In 1995, 95 per cent of their operation was in the national territory, with no change compared to 1994.

The implications of KIS supply characteristics

Despite a marginal process of KIS dispersion to less central regions during the 1980s, the geography of supply in Italy remains significantly polarized. The historical centres of KIS development (Lombardy, Lazio and Piedmont) still dominate, especially in the most strategic and innovative services. But 'newcomer' regions, such as Emilia Romagna and Tuscany, have also emerged, lessening the primacy of the traditional centres, although their KIS are provided mainly by small firms and orientated towards the regional market. Within this more detailed picture, the Mezzogiorno remains highly marginal. Yet Italian territorial polarization appears less emphatic than in other

European countries, such as the United Kingdom, France and Spain, where, despite some processes of decentralization, the supply of KIS remains overwhelmingly concentrated in one dominant region.

Italian KIS firms, especially those providing professional consultancy services, also appear mostly to consist of small, single-location firms, orientated towards local markets. Although comparisons with other European countries are difficult to make, it seems that on average in Italy KIS providers are more fragmented and local-market orientated, with few large national firms. Medium-sized and large firms, some with multi-establishment structures, are significant only in computer installation services, data banks, testing and certification, market studies and polls, advertising, employment and training services, and, to a lesser extent, in management consultancy services. Such firms are mostly located in the established centres of KIS supply. In information technology and management consultancy, the Italian market appears dominated by large foreign firms, whereas Italian export capacity remains negligible.

The limited development of large national firms is certainly related to the characteristics of the Italian market, dominated by a few large industrial and service firms, a developed system of SMEs, and penetrated by foreign multinational suppliers. In Italy, compared to other 'latecomer' European countries such as Greece, Portugal and Spain, the penetration of foreign multinationals appears low, but is still appreciable.

The fragmentation of Italian KIS supply, however, is not necessarily a negative characteristic. KIS may be based on highly specialized know-how, which can be provided efficiently by small firms, especially when demand is also overwhelmingly from SMEs. The development of a more diffuse basis of KIS, mostly orientated towards the local market, might actually be considered to be a positive national attribute, although its capacity further to develop and improve remains to be tested.

The role of policy in KIS development

No coherent policy approach has emerged in response to the significant growth of Italian business services and KIS during the 1980 and 1990s. It took longer for awareness of the crucial role of KIS, in both restructuring and increasing the competitiveness of the national productive structure, to translate to the policy arena in Italy than it did in some other European countries. The few national measures enacted to support KIS have been indirect, with KIS included as only one facet of wider regional or industrial policies. More direct action is evident in the policy approaches of some regional and local governments (for a taxonomy of policies directly or indirectly supporting KIS and a basic glossary, see Tables 8.5a and 8.5b). The significant influence of EU regional policy must also be mentioned. Thus, the increasing complexity of the interaction of national, European and regional actions makes it difficult to present a thorough assessment of the full policy regime. Nevertheless, the various policies will be summarized in the following section.

National policy

National Law 46/1982 for technological innovation

The first important national measure which indirectly supported KIS, and was later used as a co-funding device for access to EU funds, was National Law 46/1982 which provided for a programme of aid for the development and diffusion of technological innovation. The programme provides financial incentives, in the form of subsidized loans and grants, to firms or consortia of firms for the development of strategic and risky research projects (through the FRA – Fund for Applied Research), and for the development of relevant product and process innovations (through the FIT – Revolving Fund for Technological Innovation). These financial incentives are mostly directed towards innovative manufacturing firms, thereby only indirectly stimulating the supply of related services. Although a significant portion of funding is reserved to firms operating in the South, it is mostly north–central firms that have benefited from these incentives over the years.

National Law 317/1991 for the development and innovation of SMEs

Another measure that indirectly supports KIS is National Law 317/1991, which established a detailed aid framework specifically geared towards SMEs, especially in 'industrial districts'. The main objectives of the law are (1) to support SMEs' competitiveness, through subsidies to new technology investment, business service acquisition and R&D projects; (2) to promote the development of new financial tools; (3) to promote membership organizations and consortia geared to provide business services. The programme was slow to take off, especially in services and R&D. Most of the funding has been used for investment in fixed capital and only in 1995 did funding start to be granted to R&D projects (Svimez 1999).

National Law 488/1992: financial incentives to firms' investment

National Law 488/1992 is the main tool to support both manufacturing and service investment in depressed areas. However, only up to 5 per cent of total available funds is eligible for service firms, and only to those organized as partnerships or joint-stock companies. Business services eligible for support include IT services and a range of professional consultancies. The programme, run by the Ministry of Industry, became operational only in 1995 and the first round of approved projects was published at the end of

Table 8.5a Italy: a taxonomy of policies directly or indirectly supporting KIS

1 Domain of intervention	Technological innovation
	research and development
	business services
	training
	SMEs
2 Form of support	Direct supply of
	information
	infrastructure
	business services (KIS)
	Financial support to
	supply
	demand
3 Target recipient	Individual firms
	Groups and associations of firms/ institutions
	Groups of firms (or sectors)

Table 8.5b European/Italian regional policy glossary

SF	Structural Funds = SF/Fondi Strutturali
ERDF	European Regional Development Fund = FESR/Fondo Europeo Sviluppo Regionale
ESF	European Social Fund = FSE/Fondo Sociale Europeo
EAGGF	European Agricultural Guidance and Guarantee Fund = FEOGA/ Fondo Europeo Orientamento e Garanzia Agricoltura
FIFG	Financial Instrument of Fisheries Guidance = SFOP/Strumento Finanziario Orientamento Pesca
CSF	Community Action Programme = PAC/Programma di Azione Comunitaria
GG	Global Grant = SG/Sovvenzione Globale
PA	Priority Action = AP/Azione Prioritaria
CIP	Community Initiative Programme = PIC/Programma di Iniziativa Comunitaria
IMP	Integrated Mediterranean Projects = PIM/Programmi Integrati Mediterranei
Objective 1	(lagging regions): the eight administrative regions of the Mezzogiorno
Objective 2	(declining industrial regions): various sub-areas of north–central Italy
Objective 5b	(vulnerable rural regions): various sub-areas of north–central Italy

1996. Only 1.1 per cent of total funding had been granted to service firms (Svimez 1997).

Science parks, technology parks, and business innovation centres

Science parks (SPs), technology parks (TPs), and business innovation centres (BICs)[10] are, perhaps, the most direct tools for supporting KIS and innovation transfer. Generally speaking, SPs, TPs and BICs operate on three main supply levels, in varying proportions:

1   Logistical infrastructure (real estate development/redevelopment, physical infrastructure).
2   Information technology infrastructure (hardware infrastructure, telematic networks).
3   Service infrastructure (basic business services but also KIS).

With regard to host activities they can promote:

1   Research activities (university laboratories, public and private R&D centres).
2   Productive activities (new or old firms operating in innovative sectors).

In Italy, national policy in support of SPs, TPs and BICs was developed much later than in other European countries. Many initiatives started autonomously and then joined national and European programmes. Indeed, the many fragmented and diverse initiatives, launched in different periods, with different aims and structures,[11] reflect the lack both of a coherent national strategy and of a clear reference model throughout the 1980s (Marinazzo 1995). A number of Italian studies provide useful classification methods, case studies and/or analyses of the complex logic for promoting institutions, financial sources, support services and their use (Cappellin and Tosi 1993; Campo dall'Orto and Roveda 1989; Wolleb 1991; Artusi and Romano 1990). Nevertheless, no consistent recent survey of the various initiatives has been published. From the sparse literature it appears that only a few have become truly operational. Many are still in the form of programmes awaiting funding. Moreover, the excessive paper proliferation of such initiatives may reflect a rush on the part of local institutions to obtain resources, regardless of the actual conditions for effective start-up.

The influence of the European Union

As already mentioned, the actions of the European Union through its Structural Funds have now largely replaced national policies for depressed regions. They have also somewhat influenced the focus of Italian policies, especially in respect of support for business services and innovation, and among SMEs.

Within the Community Support Framework (CSF) 1994–99 the shift towards 'economic' infrastructure, rather than physical infrastructure is quite evident. The relevant priority measures are:

1   The Industry, Handicraft and Business Services measure, where funding was significantly increased and specific provision was made for business services.
2   The Infrastructure in Support of Economic Activities measure and, more specifically the Research, Development and Innovation measure, where EU funding was also significantly increased. Action in this area is covered by two sub-measures relating to (1) scientific and university infrastructure; (2) research and development, technology transfer and research personnel training (European Commission 1995a, 1995b).

Also significant are programmes established outside the national CSFs and within the European Regional Development Fund, called the Community Initiative Programmes. For the 1994–99 period, in line with the Commission's *Green Book* of June 1993, seven themes and thirteen initiatives were established, some of which are relevant for KIS. Among these the one that most closely concerns KIS is industrial restructuring with initiatives such as Adapt, Retex and PMI, all geared to promote inter-firm cooperation in the area of services, organizational innovation and R&D.

Regional and local action in support of KIS

Over the last twenty or thirty years, and within the slow process of institutional decentralization, many regional and local institutions in Italy have taken direct initiatives in support of local economic development. Here we present three case studies: Lombardy, Emilia Romagna and Puglia. These are by no means representative of the whole of Italy but exemplify three different territorial contexts. The Lombardy region covers a clearly established area, comprising corporate headquarters (Milan) with a strong industrial base, including large firms and very active and diversified SMEs. Emilia Romagna, among the most cited in national and international literature, includes the advanced frontier of the so-called 'industrial districts', increasingly considered as a model of alternative industrialization. Finally, the Puglia region is one of the most industrialized and developed regional economies of the Mezzogiorno, including both older industrial centres, now undergoing restructuring, and a diverse source of SMEs.

Regional policy approaches

LOMBARDY

Since the early 1980s the Lombardy Regional Government has developed an active industrial policy in support of SMEs (cf. Lassini 1985), with particular emphasis on services and innovation.[12] Unlike the Emilia Romagna Regional Government, which used its regional finance corporation as the main tool for direct participation in a number of projects, Lombardy mostly used its legislative power to enact financial programmes in support of individual projects prepared by local institutions and firms.

EMILIA ROMAGNA

The Emilia Romagna Regional Government's action in the area of industrial policy and local development is largely coincident with that of its regional finance corporation, Ente Regionale per la Valorizzazione Economica del Territorio (Ervet).[13] Established in 1973, with the regional government as the main shareholder and the participation of a number of regional banks and the regional union of chambers of commerce, Ervet is the executive arm of regional government planning, with the task of formulating strategies, as well as directly participating as a shareholder in specific actions and projects (cf. Bellini and Pasquini 1996).

PUGLIA

The action of the Puglia Regional Government differs significantly from the other two cases.[14] The Lombardy and Emilia Romagna strategies were both strongly orientated towards the promotion of services to SMEs, as a way to strengthen the competitiveness of their productive basis. In Puglia, despite the fact that the productive structure includes a significant presence of SMEs, the regional government has only recently expressed an interest in supporting business services. The prevailing strategy throughout the 1980s was that of strengthening the regional R&D supply structure, first within the legislation for the Intervento Straordinario in the Mezzogiorno (especially National Law 64/1986), and then within the framework of EU programmes for Objective 1 regions.

Service centres

Despite the very rich experience developed over the last twenty years, no comprehensive recent study is available on the numerous business service centres for SMEs established in the various Italian regions, their success and failure rate, their evolution, or their institutional and operational characteristics. The last survey took place at the end of the 1980s by Nomisma (1988), and identified seventy-five such initiatives in Italy, twenty of which were located in the Mezzogiorno.

The form and structure of business service centres varied significantly. They included joint-stock consortia, joint-stock companies, consortia and membership associations, but were generally based on some sort of association between public and private institutions, including local authorities (regional government, provincial governments or municipalities); local business institutions and associations (such as chambers of commerce); research and training institutions (universities, the National Research Council); banks; and private firms. Funding came from partners or associates, but also from national and European sources.

The Nomisma study highlighted the different characteristics of initiatives in north–central Italy and the South, the latter proving more orientated towards R&D activities rather than supplying support services. Here, we update this analysis for the three regional cases, based on information from regional governments and service centres themselves.

LOMBARDY

The regional pattern of service centres in Lombardy is quite diverse, in terms of their social form, promoters, participants, funding sources, service portfolios and markets. Five major categories have been identified (Lassini 1994):

1   Sectoral service centres.
2   Area service centres.
3   Membership associations service centres.
4   Chambers of commerce service centres.
5   Research and technological innovation service centres.

Quite important within the regional government action is the Centro Lombardo per lo Sviluppo Tecnologico e Produttivo delle PMI (Cestec), the regional agency for technological development and diffusion which, besides providing services to SMEs, also represents the region at the European level and carries out the evaluation of projects submitted to the regional government for financing under its various programmes.[15]

Service centres successfully operating in Lombardy at present are (Cespri-Università Bocconi 1997):

- Associazione Tessile, Como: textiles;
- Agenzia Lumetel (Lumezzane Telematia), Lumezzane;
- AQM (Centro Assistenza Qualificazione Meccanica), Brescia: mechanics;
- CESAP (Centro Europeo di Sviluppo Applicazioni Plastiche), Zingonia: plastics;
- Centro Servizi Calza, Castell Goffredo: textiles;
- Centro Servizi alle Imprese, Vigevano;

- SECAS, Darfo;
- Agenzia di Innovazione e Sviluppo, Sermide;
- CR&S (Centro Ricerche e Sviluppo per il Mobile e l'Arredamento), Lissone: furniture;
- CRSLP (Centro Ricerche e Servizi per il Legno), Sustinente: lumber;
- CLAC (Centro Legno e Arredo), Cantù: lumber and furniture;
- Isval (Istituto di Promozione Attività della Valtellina e Valchiavenna), Sondrio;
- Centrocot (Centro Tessile Cotoniero e Abbigliamento), Busto Arsizio: textiles and garments.

EMILIA ROMAGNA

As in Lombardy, the experience of Emilia Romagna in terms of service centres is quite rich. Since the early 1980s, the regional government, through its regional finance corporation Ervet, has directly promoted a number of centres (cf. Nomisma 1985), as well as a regional agency for technological innovation transfer, the Agenzia per lo Sviluppo Tecnologico dell'Emilia Romagna (ASTER).

Including ASTER, the Ervet system controls eight sectoral service centres and four area centres operating in Emilia Romagna. These are:

- ASTER (Agenzia per lo Sviluppo Tecnologico dell'Emilia Romagna): technology;
- CITER (Centro di Informazione Tessile per l'Emilia Romagna): textiles and garments;
- CESMA (Centro Servizi Meccanica Agricola): agricultural machinery;
- Cercal (Centro Servizi settore Calzaturiero): footwear;
- Quasco (Centro Servizi per la Qualificazione delle Costruzionio): building industry;
- CERMET: laboratory analysis, testing, and quality certification;
- Democenter: industrial automation;
- Centro Ceramico: ceramics;
- Promo (Società per la Promozione dell'Economia Modenese), Modena area;
- Sipro (Società Interventi Produttivi), Ferrara area;
- Soprae (Società di Promozione Attività Economiche), Piacenza area;
- Soprip (Società Provinciale Insediamenti Produttivi), Parma area.

Among the most successful Emilia Romagna service centres, most cited in the international literature, is CITER, one of the first sectoral centres established to support the textile and garment district of Carpi.[16]

On the basis of information made available by individual centres, a few studies carried out at the end of the 1980s (Nomisma 1988; Cles-Fnaa/Cna

1988), and information gathered from local policy experts, the main components of Emilia Romagna success in their service centres are as follows.

1 The implementation of a close interaction between the supply and the demand for business services, that is, between promoting institutions, supply structures and users, which has helped identify and supply appropriate services and, at the same time, has made the latter more accessible to SMEs.
2 The existence of geographically clustered and economically integrated systems of SMEs, often featuring sectorally homogeneous industrial districts, which has made it easier to achieve economies of scale in service supply and a greater impact on the regional productive structure. This strength/opportunity should not be underrated, since it partly explains the difficulties involved in setting up service centres in regions where such clusters and integration are not present.
3 A strong commitment of the public sector and its ability to train and attract technically and organizationally competent staff (Ligabue 1992).

Among the positive effects of the service centres' strategy are:

1 The increasing capacity of centres to self-finance their activities through membership fees and/or service fees (although public support remains important).
2 The growing diffusion of an awareness and sensitivity to organizational and technological innovativeness among SMEs.

PUGLIA

There has been no significant development of service centres in Puglia. National and regional government action has been geared mainly towards supporting research activities, mostly public in character. More recent developments, and projects launched within European Union programmes, show a greater attention to KIS, but it is too early to make any evaluation. The delivery of some KIS has been carried out under the CSATA-Tecnopolis structure, in the form of assistance to new firm formation within its Technology Park Incubator programme near Bari.[17] But this activity is only a marginal part of the Tecnopolis project, which remains mainly orientated towards publicly supported R&D projects.

Planning lessons from diverse regions

Significant differences emerge between the Mezzogiorno and the rest of the country as to the characteristics and impact of these various initiatives. There is a major contrast between the roles of the respective regional governments,

which have proved much more active in north–central Italy in developing procedures and tools to support service centres and local development.

Policies implemented by the regional governments of Lombardy and
Emilia Romagna in support of SMEs, particularly in activities concerning KIS,
have several common features. Both regions have, directly or indirectly, supported the creation of service centres for SMEs, especially by means of the
regional finance corporations, although through different institutional agencies. In both regions the majority of such initiatives have been successful,
because they were able to spark a beneficial interaction between suppliers and
clients. Both regional governments, because of the pressures of international
competition, now believe that a new policy phase is necessary, based on better
coordination of initiatives, upgrading of the services provided and greater
international integration.

The experience of the Puglia Regional Government is significantly different. Its strategy has been that of strengthening publicly supported R&D activities, rather than linking with the regional productive basis and supporting
SMEs. The former is a long-term strategy, highly dependent on public funding, without immediate effects on the regional economy. Despite the existence of a number of systems of SMEs and a quite well-articulated productive
structure, no efforts have been made to mobilize the associative capacity of
local firms and institutions, or support the establishment of service centres.
Finally, from the sparse available evidence, including evidence of significant
delays in the management of funding and implementation of programmes,
and the transfer outside of design and monitoring functions, it would seem
that the regional administrative apparatus lacks the flexibility and/or expertise
needed to carry out strategic planning tasks.

The interaction between KIS supply, demand and the institutional context for innovation

The relationship between demand and supply of KIS is interactive and cumulative. However, although this interaction is crucial in explaining the 'take-off'
of KIS development, the subsequent unfolding of growth and innovation
processes is also strongly affected by the socio-institutional environment
within which these processes are set.

The role of corporate and market structures in national, regional and sectoral trajectories

During the restructuring of the Italian industrial system in the 1980s, both
the large corporate sector and SMEs responded to the increasingly competitive pressures of the post-Fordist transition. The consequent increasing use of
KIS explains the latter's strong growth throughout the decade. In comparison
with other European states such as France, Germany or the UK, Italy does not
possess a strong indigenous corporate structure. Its productive basis is largely

made up of SMEs. Many of these are quite dynamic on the international scene, often integrated into flexible systems of production, and may also have innovative peaks. They nevertheless lack the market, financial and, especially, knowledge strength of large hierarchical corporate systems. In terms of technological innovation Italy is, in fact, a 'latecomer' and a net importer.

These national structural features partially explain why, although fast growing, KIS have not reached the pre-eminence they have in other countries such as the US, France and the UK. They also help explain why Italian KIS are overwhelmingly provided by small firms and orientated towards local markets. With the exception of a few firms in the information technology services, there are no large indigenous national or multinational KIS corporations in Italy, as there are in other European states. In the absence of such 'national champions', Italy has actually become a key market for foreign suppliers. Compared to the position in other European countries, the penetration of foreign KIS multinationals into Italy appears to be inferior only to the situation in Greece and Portugal.

Regional trajectories: three territorial models

Three main territorial models reflect the different regional productive structures and institutional contexts already summarized.

1   The 'old' service centres

These are the three regions hosting the traditional tertiary metropolitan areas of Milan, Turin and Rome, which maintain strong KIS specialization (although with different characteristics) and also are home to the largest KIS firms. Lombardy and Piedmont belong to the 'industrial triangle', the cradle of Italian industrialization, where most large national firms are still based. These two regions alone include 53 per cent of national employment in headquarter offices. Milan is also the financial capital of Italy and includes the marketing departments of most foreign multinationals in the country, explaining its broader specialization compared to Turin. Piedmont maintains a strong specialization in production-orientated KIS, such as information technology services and R&D, because of the presence of Olivetti. On the other hand, according to Ciciotti *et al.* (1994) the competitive position of Milan as an advanced KIS centre is quite fragile compared to other European capital cities. Milan is a gateway to the Italian market, rather than a potential export platform.

Rome, in contrast, is the seat of the main national and international government institutions, as well as of most state-enterprise holdings and multinational corporations. Lazio alone hosts 39 per cent of national employment in holding companies. This explains its specialization in R&D activities and information technology services. Many National Research Council agencies, as well as international R&D centres, are based in the Rome metropolitan area. There is also a strong contribution of information technology services, based on the presence of Finsiel, the other national giant, with Olivetti, in this sector, which mostly caters for the central state administration. The KIS structure of these three regions is thus orientated mainly towards large firms and institutions, but it also reverberates throughout the whole of their regional productive systems.

2 The 'newcomers'

Parallel to the strengthening of traditional service centres, a significant growth of KIS has been recorded in most north-eastern and central regions. The most dynamic have been Emilia Romagna, Veneto and Tuscany, where manufacturing industrial district systems of firms are strongest. Emilia Romagna and Tuscany have actually become specialized in a number of KIS activities and represent secondary centres of national supply. However, this pattern of supply is different from the previous model. It is more fragmented and territorially diffuse, with several medium-sized cities acting as local supply centres. It is also based on smaller firms, which are closely orientated towards local markets. The explanation for this success lies in the vitality of SMEs and industrial districts since the 1970s, and also in the particular socio-institutional context which couples cooperative attitudes between firms with a proactive strategy in support of KIS from local institutions.

3 The Mezzogiorno

In contrast to the previous models and despite recent above-average rates of KIS growth, the Mezzogiorno exhibits a persistent underdevelopment of these activities, especially in the most advanced and strategic functions. The reasons for this poor performance lie in the structural weaknesses of the region's productive system, which worsened during the 1980s and early 1990s. These include a shrinking and dualistic manufacturing base and a weak organizational structure of alternative sectors such as tourism and agriculture. The large branch plants of outside corporations, which constituted the bulk of industrial investment in the 1960s and early 1970s, were dramatically rationalized during the 1980s. Among the apparent benefits of the National Law 64/1986 for Intervention in the Mezzogiorno, is the reconversion by some firms of old facilities or new investments in R&D. However, the durability and the impact of such ventures on the local economy remain questionable. SMEs, despite isolated examples of sectoral and territorial clustering, generally exhibit a highly fragmented and organizationally weak structure in all sectors. Public administration is lagging in the introduction of new technology and advanced management systems. Altogether, these characteristics explain the low level of aggregate demand for KIS which does not allow sustained development of local supply. These structural market weaknesses are further reinforced by the socio-institutional characteristics of the region.

The socio-institutional and regulatory context

Territory-based social, cultural and institutional conditions further affect economic performance and policy effectiveness. In our view, such environmental conditions are historically determined and influenced by both internal and external relations. When the internal cohesion of local systems is strong, their external relations with, and integration in, the wider national and international context may further stimulate internal dynamics. When internal cohesion is low, dependent relations with stronger outside systems undermine internal linkages, further increasing external dependence.

Despite significant territorial differentiation of the national productive structure, the geography of its institutional context still reflects the old dualism between north–central Italy and the South. In north–central regions, an almost 'organic' interaction between the institutional environment and the market has developed. Although with limited powers, regional governments have been active in designing and implementing strategies to support local economies, even in 'latecomer' regions such as the North East and Centre. Policy has been grounded in the local context and programmes have supported restructuring and innovation within the local economic system, including the direct provision of KIS. Some such service centres are actually among the most successful policy innovations in support of SMEs.

Among the factors maximizing the impact of these initiatives are (1) the technical and managerial capacities of administrative staff in designing policy; (2) the close interaction between private and public institutions; and (3) the existence and/or development of strong internal cohesion based on the presence of territorially clustered and often sectorally integrated systems of firms. A beneficial interaction arises from the progressive diffusion of 'innovation consciousness', the build-up of technical and strategic know-how among both suppliers and clients, an increased capacity to face competitive pressures and develop an international perspective and, ultimately, a further strengthening of internal integration, with good evolutionary capacities.

In the Mezzogiorno, both the low level of internal economic cohesion and the historically dependent integration with the stronger economy of the North have been major obstacles to bridging the socio-economic gap, also affecting the evolution of local institutions. Two major periods can be identified. First, from 1950 until the late 1970s, when the 'southern question' was considered a national issue and national development policy was geared towards sustaining and developing the productive basis of the region through investment. Second, the 1980s and early 1990s, when national economic difficulties diverted attention from the South, national regional policy lost strategic perspective, and public action shifted to the mere support of income. Once productive investment stopped and the restructuring of the established industrial base further weakened the Mezzogiorno, regional governments and local institutions proved unable to sustain a strategic role, or to design and implement policies geared towards the strengthening of internal cohesion. They have remained captives of the old 'public transfer' philosophy. Whether this failure can be blamed exclusively on southern institutions is a matter for debate, but it certainly represents a powerful brake to policy effectiveness.

Future scenarios and issues for further research[18]

Trends in supply and demand

Since the surge in KIS demand in the 1980s, linked to the industrial restructuring process and the introduction of new IT, a slackening of growth has been under way. Nevertheless, there may still be continuing growth potential, based in the late modernization of much Italian public administration (Assinform-Nomos Ricerca 1997) and lags in demand, especially for information technology services by more marginal SMEs, sectors and regions. Foreign demand, reflected in an increase in KIS exports, does not appear to be a viable outlet, given the dependent nature of the national market *vis-à-vis* foreign multinationals. Although a few highly specialized small suppliers (notably a number of north–central service centres) are successfully developing some exports, such a component is hardly significant in aggregate terms.

With the slowing of demand, the KIS sector will probably experience significant restructuring. Larger firms, both national and foreign, are already undergoing such a rationalization process. At the end of the 1980s many large national and multinational new IT suppliers held an optimistic view, especially about the SME market, and were planning the opening of new regional branches (cf. Moulaert *et al.* 1990). By the second half of the 1990s, however, many regional offices were being closed or were lightly staffed, and many contracts were being supported by staff mobility from headquarters. Moreover, the SME market is being abandoned in favour of the consolidation of relationships with large clients.

As for smaller KIS suppliers, there are indications that a process of market selection is taking place, based on performance and competitiveness. It is also possible that greater cooperation will develop among small and medium-sized KIS suppliers of different specializations, as well as between large and small KIS suppliers. Alternatively, some sector experts believe that the slackening of demand may encourage the further fragmentation of supply. The large reserve of educated unemployed, together with the early retirement of specialized personnel from large firms, may boost the supply of KIS in the form of self-employed professionals competing for shrinking markets. In the absence of a broadening of demand, it is possible to forecast a deepening of supply, with the further specialization of KIS towards greater knowledge content and more integrated service portfolios. In north–central regions, interviews with representatives of both local government and service centres clearly suggest that KIS policy must enter a new phase, strengthening integration and networking processes and increasing the quality of both supply and demand, including through training programmes.

Issues in designing and implementing KIS-based policy

Designing efficient policy in the area of KIS and innovation transfer is not easy and its implementation even less so. KIS, as the most advanced and strategic form of business services, are very complex, and thus quite difficult to understand and support. They are of an intangible nature and highly customized for specific organizations and aims. Moreover, the high level of knowledge required is varied and specialized within the wide range of firms' functions, from the mobilization of resources (finance, R&D, training), throughout the production process (management, engineering, quality), and on to marketing (distribution, promotion).

This complexity is partially reflected in the large number of programmes that have been experimented with to influence KIS use and innovation diffusion at all institutional levels, including supply- and demand-side policies; programmes to support R&D; programmes for the diffusion of given innovations among firms; 'pilot' targeted projects; and financial incentives available to all. Although apparently positive, with a policy to suit each possible need, this proliferation of measures may actually constitute a form of institutional overload, which discourages the weakest segments of the market.

Not surprisingly, in Italy the most successful policies have been implemented in regions where both market and institutional forces were strongest, where active interaction between suppliers and users, as promoting institutions and recipients, could be built upon, and action most clearly targeted at the actual and potential demand of local businesses. A sufficient 'mass' of demand was thus territorially aggregated and initiatives were fast, efficient and shared clear demonstrable effects.

Reaching marginal regions and firms

It would thus appear, at least from Italian experience, that KIS-based policy works only where market and institutional conditions already exhibit some level of internal cohesion. This means that such policies cannot easily reach marginal regions, institutions or firms. In the current context, policy action in the Mezzogiorno, for example, should be guided by the following principles.

1   The idea of a national strategy for the development of the Mezzogiorno should be revived and a policy geared towards supporting productive investment relaunched.
2   In particular, the idea of a 'solidaristic federalism', within a national recovery strategy, should be developed, rather than conceding to northern pressures for *laissez-faire*. This collaborative approach is already being witnessed in a number of bilateral agreements between, for example, Lombardy and Calabria, or Emilia Romagna and Puglia, to establish common policy initiatives. Such an approach, if effectively carried out,

could help southern institutions in the process of knowledge building, capitalizing on existing experience.

3   Greater emphasis should be placed on human resources. Targeted national training programmes are required, geared towards raising the planning and technical capabilities of southern authorities, business and professional associations.

4   Universities and research centres operating in the South, which have the qualified personnel and are increasingly called upon to provide expertise, should also develop a more rooted role within the local system. Some policy experts believe that the university system may become a strong competitor for the private KIS sector in relation to some demands from the productive system because of its lower costs. This is already happening in the North.

5   Policies and projects should be geared towards stimulating local associative capacities and identifying local needs, not just in the manufacturing sectors. Individual SMEs are not sufficiently strong to act independently and many policy initiatives are based precisely on the SMEs' technical ability to formulate projects. Demonstration projects and campaigns should thus be emphasized.

Notes

1   This chapter is the product of research work jointly carried out by Lucia Cavola and Flavia Martinelli, with the support of the staff and facilities of Iter SrL, the Naples-based planning consultancy. Lucia Cavola is responsible for the section on policies, whereas Flavia Martinelli is responsible for the sections on demand and supply.

2   This model emerged throughout the 1970s and was formally recognized towards the end of the decade (cf. Bagnasco 1977; Brusco 1982). It came to the fore in the 1980s when, in the wake of the crisis of large firms, its flexible attributes proved competitive. For a period it was even considered as an alternative model to Fordist development.

3   Among the earliest sectors to restructure were automobiles, electronics, electromechanical engineering, telecommunications and pharmaceuticals, together with a number of service industries, such as banking and insurance, parts of the hospital system, and transport and distribution.

4   According to Istat (1998), in 1995 Italy lagged in 17th position (behind all major industrialized countries, but also behind Spain, South Korea, Ireland, Finland) in terms of share of private R&D expenditure over total national R&D expenditure.

5   Although the establishment of regional governments (Regioni a Statuto Ordinario) was written into the Italian Constitution of 1948, only a few regional governments were immediately created, with a special status, in the most politically problematic regions of Sicily, Sardinia, Val d'Aosta and Trentino Alto Adige. It was only in 1970 that the remaining regional governments were actually established, with limited legislative authority: urban and territorial planning, handicraft firms, agriculture, tourism, forestry, quarries. In the last few years, the strong separatist pressures from north-eastern regions (primarily Lombardy and Veneto) have brought about major institutional reform, still in process, towards a 'federalist' structure.

6   This view stresses how the centralized Intervento Straordinario began deteriorating when the national development strategy, based on productive investment in the South, was

abandoned at the end of the 1970s in favour of a mere 'income support' strategy, including subsidies and transfers to households and special employment and training measures.

7   The section on the structure and geography of service supply in Italy mostly relies on research by Martinelli (Martinelli and Gadrey 2000).

8   The formula is: $\dfrac{\text{Employees in sector } i \text{ of region } j}{\text{Total value added of region } j} \bigg/ \dfrac{\text{Employees in sector } i \text{ of Italy}}{\text{Total value added in Italy}}$

9   Assoconsult is the Italian association of management and organizational consultancy firms. In 1995, it had 276 associated firms, covering 76 per cent of the sector's total Italian sales. Eighty-seven per cent of the associated firms are based in north–central Italy.

10   Of approximately thirty-five SPs and TPs promoted in Italy during the 1990s, twenty-two were supported within the EU Sprint programme between 1990 and 1993, mostly in north–central Italy. Another thirteen were promoted in the Mezzogiorno, started in 1994 by the Ministry of University and Scientific and Technological Research. However, by 1997 no financing had yet been approved for the latter (Marinazzo 1995). BICs, established under Article 10 of the European Regional Development Fund (ERDF) regulations, are in Italy called CISI (Centri Integrati di Sviluppo dell'Imprenditorialità) and are managed by the SPI (Società per la Promozione e lo Sviluppo Imprenditoriale), with EU support. About twenty BICs have been programmed, half in the South.

11   For example, an S&TP established in Milan (Pirelli-Bicocca) stems from a private investment project by Pirelli and is geared towards reviving abandoned industrial areas by attracting the research and production activities of public and private firms. In contrast, the two S&TPs established in Bari (Tecnopolis) and Trieste (Area), are part of the national programme to support research and development, designed to attract research by public and private institutions.

12   In the 1980s, policies supporting SME innovation financed up to 40 per cent of R&D and technological innovation projects, and supported training and service assistance for organizational and technological innovation (Lombardy Regional Law 34/1985). Later, Regional Law 41/1990 for the development of quality systems financed projects for quality improvement, and cooperative initiatives in quality certification. Policies in the 1990s addressed issues of SME 'systemic' strengthening and internationalization, including harmonizing existing legislation and supporting projects proposed by 'aggregations' of local institutions and SMEs (Regional Law 7/1993). New measures also financed up to 50 per cent of projects from local institutions and firm associations for business services support, local systems integration and participation in European and international R&D programmes (Regional Law 35/1996). More recently, emphasis has been placed on coordination of programmes through a 'regional information network', to identify new needs; integrate the regional government, promoting institutions and users; and improve the link with European policy (Lassini 1994).

13   In the 1970s Ervet was geared towards supporting the regional system of SMEs by establishing industrial areas, improving the financial system and supporting basic services. Ervet's more distinctive projects were launched in the 1980s, supporting organizational and technological innovation among SMEs through the promotion of service centres. Its strategy was again reformed in the early 1990s, responding to economic and political changes and new needs, including demands for higher-level services, growing market internationalization, increasing local autonomy of the regional government, and the need to coordinate institutional activities (Bellini and Pasquini 1996). As in Lombardy, the new strategy focused on innovation, finance and internationalization (cf. Mazzonis 1996–97; Ervet 1995), upgrading services, integrating service suppliers into a regional network system, rationalizing regional funding, developing new financial tools, and strengthening Ervet as a development agency of European standing.

14   The first R&D centre established in Puglia was the Centro Studi e Applicazioni in Tecnologie Avanzate (CSATA) in the early 1970s. In the late 1980s, this became a technology park, Tecnopolis-CSATA Novus Hortus, providing research, training and

advanced business services. Other research initiatives were the Cittadella della Ricerca in the Province of Brindisi, and the Centro Nazionale per la Ricerca e lo Sviluppo dei Materiali (CNRSM) (Law 64/1986). Only in the Puglia Regional Programme 1994–99 (under the EU CSF) did the regional government begin to promote business services in support of SME innovation. This contracted out the management of the Research, Development, and Innovation programme to an *ad hoc* Regional Inter-university Consortium, emphasizing research activities rather than services, in three sub-programmes addressed to (1) promotion of technological innovation for SME systems, providing up to 50 per cent of eligible costs to SMEs and to consortia including SMEs, research centres and national or international firms planning to establish facilities in the region; (2) transfer of research and innovation to the system of SMEs, with a feasibility study to develop an innovation transfer development plan, including pilot projects (1994–96), followed by implementation; (3) technological mediation for innovation support among SMEs, supporting service supply, especially private advanced business service, with aid for up to 50 per cent of cost. The Industry, Handicraft, and Business Services initiative, to support SME internationalization and the transnational cooperation was also contracted out. Since 1991, a 'global grant' programme has funded the 'BIC Puglia Sprind', a European centre for business and innovation. During 1991–95, 9,507 firms were contacted by mail, 1,794 field visits were made, and 420 firms assisted. No evaluation of the innovative impact of this scheme has yet been made.

15 Lombardy service centres include Centro Lombardo per lo Sviluppo Tecnologico e Produttivo delle PMI (Cestec: Lombardy Centre for the Technological and Productive Development of SMEs), operational since 1981, a joint-stock corporation, with 50 per cent of capital from the regional government, the rest from regional business associations and the regional union of chambers of commerce. It provides services to all firms and sectors, including promotional activities, technical and managerial training, customized consultancy, sectoral consultancy, assistance in project preparation, internationalization consultancy, and institutional services such as screening and evaluation of projects (for the regional government). Demand has especially grown from firms in the preparation of project applications and customized consultancy (Cestec 1997). With six other regional institutions (the Milan Chamber of Commerce and the University of Milan, and local research centres), it has been recognized by the EU as Innovation Relay Centre for Lombardy (Lombardy Area Innovation Centre: Larice). Centro Tessile Cotoniero e Abbigliamento (Centrocot: Centre for Cotton Textiles and Garments) is one of the most successful service centres in Lombardy. It was established in 1987 in Busto Arsizio near Varese, the centre of an extensive textile area, as a non-profit joint-stock corporation, and is jointly owned by the regional government, the Province of Varese, seven municipalities, chambers of commerce, business associations, unions and banking institutions. Centrocot's seven divisions are concerned with quality certification, training, promotion, consultancy, international projects, R&D and technology monitoring. By 1996 it had sixty-two employees and a support team of about a hundred outside experts; a turnover of L 4.8 billion; and 3,500 client firms, including 30 per cent outside the region and 5 per cent abroad (Centrocot 1997).

16 Emilia Romagna service centres include: Agenzia per lo Sviluppo Tecnologico dell'Emilia Romagna (ASTER), established in 1983 as a joint-stock cooperative between Ervet and a number of sectoral service centres, joined subsequently by several business associations. ASTER supports the technological modernization of SMEs in Emilia Romagna, through broadly based projects and specialized support to firms. It offers customized diagnosis of technological innovation needs/projects; quality information/documentation; information and assistance on EU funding; support for networking; and a data bank and information teller. It employs about thirty permanent technical staff but develops most of its programmes and services in cooperation with business associations and chambers of commerce. ASTER also acts as a link for EU projects between the regional government, the European Commission and the firms. Centro di Informazione Tessile dell'Emilia Romagna (CITER – Centre for Textiles Information of Emilia Romagna) was established in 1980 by Ervet with local business associations. CITER stems from a mid-1970s European Social Fund (ESF)-funded

training programme promoted by local and business institutions in the textile district of Carpi. After the initial four-year programme, which trained about 800 entrepreneurs and managerial staff for about 350 SMEs, a more permanent service structure was established, with a membership of over 500 firms, including some outside the Carpi area. CITER supplies information, and has progressively extended its range. Initially it focused on fashion trends; then, during the 1980s, product and process technology. This included the design and development of a graphic (CAD) workstation (CITERA). In the later 1980s, its remit extended to market research, and CITER also started to provide technical assistance and training on quality management and market internationalization.

17   Tecnopolis-CSATA Novus Hortus. The Tecnopolis experiment originated in 1969, supported financially by the Cassa per il Mezzogiorno, through a 'special project' to support R&D in the South. It was established as CSAT, a non-profit association between the University of Bari, the Cassa per il Mezzogiorno, the Formez (a training institution for the Mezzogiorno) and the Pignone Sud (an industrial group). In 1978 it became a consortium, adding the Bari Chamber of Commerce, a regional banking institution and some private firms. The technology park project was launched in 1982 with European Union support, to integrate research, training and advanced services. In 1984 the first embryo of 'Tecnopolis-CSATA Novus Hortus' was established in Valenzano, near Bari. In its first phase, Tecnopolis worked mostly for large national private and state firms (Fiat, Olivetti, Italsiel, Telettra), some with branches in the area. After the late 1980s its activities were directed more towards transferring technological innovation into regional and local administration and, to some extent, the local productive system, drawing on European projects. Support was also given to the creation of new firms. More than 400 projects were evaluated, and fifteen new firms established in the Park Incubator, with fifteen established firms. Since 1995 Tecnopolis has participated in European projects for the valorization of research. It is also responsible for the EU Regional Innovation Strategy Programme in Puglia (Ris Puglia Innova, 1997–98). Today, Tecnopolis undertakes research and development (40 per cent of turnover), training (29 per cent) and provides services (21 per cent). Between 95 and 100 per cent of its activities depend on European, national, or local public funding (cf. Bozzo 1995).

18   This final section integrates many inputs gathered from sector and policy experts at the KISINN National Workshop, held in Naples, 16 May 1997.

References

AIPA (Autorità per l'Informatica nella Pubblica Amministrazione) (1997) *Lo stato dell'informatica nella pubblica amministrazione*, Rome: Istituto Poligrafico e Zecca dello Stato.

Anasin-Nomos Ricerca (1994) *Survey sulla spesa informatica della pubblica amministrazione*, Rome: Anasin (mimeo).

Artusi, C. and Romano, A. (1990) *Linee metodologiche e orientamenti programmatici per il progetto strategico 'Promozione di una rete di parchi tecnologici nel Mezzogiorno d'Italia'*, report to the Ministry for the Mezzogiorno, Rome: Ministero per il Mezzogiorno (mimeo).

Assinform-Nomos Ricerca (1997) *Rapporto 1996 sull'informatica e le telecomunicazioni*, Milan: Assinform.

Assoconsult (1996) unpublished data, Rome: Assoconsult.

Bagnasco, A. (1977) *Tre Italie: la problematica territoriale dello sviluppo economico italiano*, Bologna: Il Mulino.

Bellini, N. and Pasquini, F. (1996) *Il caso dell'Ervet in Emilia Romagna: un'agenzia di sviluppo di seconda generazione*, Bologna: Ervet (mimeo).

Bodo, G. and Sestito, P. (1991) *Le vie dello sviluppo*, Bologna: Il Mulino.

Bozzo, U. (1995) 'Tecnopolis: esperienze fra sistema locale e contesto nazionale ed europeo', paper presented at the meeting on RTD Potential in the Mezzogiorno of Italy: The Role of Science Parks in a European Perspective, promoted by the Consorzio Mario Nagri Sud, S. Maria Imbaro, 21–22 September 1995.

Brusco, S. (1982) 'The Emilian model: productive decentralisation and social integration', *Cambridge Journal of Economics* 6.

Campo dall'Orto, S. and Roveda, C. (eds) (1989) *Parchi scientifici come strumento di politica industriale*, Milan: Angeli.

Cappellin, R. and Tosi, A. (eds) (1993) *Politiche innovative nel Mezzogiorno e parchi tecnologici*, Milan: Angeli.

Centrocot (1997) *La potenzialità di Centrocot a vantaggio del sistema tessile abbigliamento: ipotesi e proposte*, Busto Arsizio: Centrocot (mimeo).

Cespri-Università Bocconi, (1997) *Cambiamenti nella struttura industriale lombarda e politiche regionali per l'innovazione*, rapporto intermedio per la regione Lombardia, Milan: Cespri (mimeo).

Cestec (1997) *Relazione sull'attività svolta nel biennio 1995–96*, Milan: Cestec (mimeo).

Ciciotti, E., Florio, R. and Perulli, P. (1994) 'Milano: competizione senza strategie?', *Quaderno dell'AIM/Associazione Interessi Metropolitani* 24 (monograph).

Cles-Fnaa/Cna (1988) *Indagine sui centri di servizi reali alle imprese artigiane: tessile-abbligliamento-calzature*, Rome: Sedart.

Del Monte, A. and Giannola, A. (1989) 'I problemi dello sviluppo industriale del Mezzogiorno ed i riflessi di questi nella determinazione del quadro di politica industriale', in Battaglia, A. and Valcamonici, R. (eds) *Nella competizione globale*, Bari: Laterza, 57–89.

—— (1997) *Istituzioni economiche e Mezzogiorno*, Rome: Nuova Italia Scientifica.

Del Monte, A. and Martinelli, F. (1988) 'Gli ostacoli alla divisione tecnica e sociale del lavoro nelle aree depresse: il caso delle piccole imprese elettroniche in Italia', *L'Industria* 3: 471–507.

Epifani, E. (1996) *L'IT a supporto di una pubblica amministrazione efficiente* (mimeo), Milan: Nomos Ricerca.

Ervet (1995) unpublished documentation, Bologna: Ervet.

European Commission (1995a) *Italia: Quadro Comunitario di Sostegno 1994–99, Obiettivo 1: sviluppo e adeguamento strutturale delle regioni il cui sviluppo è in ritardo*, Luxemburg: European Commission, Office of Official Publications.

—— (1995b) *Guida alle azioni innovatrici per lo sviluppo regionale (articolo 10 FESR) 1995–99*, Luxemburg: European Commission, Office of Official Publications.

Formez-Istituto Tagliacarne (1994) *I comportamenti innovativi delle PMI nel Mezzogiorno*, Milan: Angeli.

Gallouj, F. (1992) 'Le conseil juridique français: d'une logique professionelle à une logique d'entreprise', in Gadrey, J., Gallouj, C., Gallouj, F., Martinelli, F., Moulaert, F. and Tordoir, P. *Manager le conseil*, Paris: Ediscience Internationale, 105–56.

Giannola, A. (1982) 'Industrializzazione, dualismo e dipendenza economica del Mezzogiorno degli anni '70', *Economia Italiana* 1: 7–42.

—— (1987) 'Problemi e prospettive dello sviluppo economico nel Mezzogiorno d'Italia', in Ente Einaudi (ed.) *Oltre la crisi*, Bologna: Il Mulino, 34–82.

Giunta, A. (1992) 'Sui legami tra grande e piccola impresa nel Mezzogiorno: una verifica empirica', *Economia Marche* 3: 253–77.

Giunta, A. and Martinelli, F. (1995) 'The impact of post-Fordist corporate re-structuring in a peripheral region: the Mezzogiorno of Italy', in Amin, A. and Tomaney, J. (eds) *Behind the Myth of the European Union*, London and New York: Routledge, 221–62.

Graziani, A. (1997) *I conti senza l'oste*, Turin: Bollati Boringhieri.

Istat (1990) 'Indagine statistica sull'innovazione tecnologica nell'industria italiana 1981–85', *Collana d'Informazione* 14 (monograph).

—— (1995a) *Censimento generale dell'industria e dei servizi no. 7, 1991, fascicoli regionali (tavola 1.5)*, Rome: Istat.

—— (1995b) *Censimento generale dell'industria e del commercio no. 7, 1991*, Rome: Istat.

—— (1997) *Rapporto sull'Italia*, Bologna: Il Mulino.

—— (1998) *Rapporto sull'Italia*, Bologna: Il Mulino.

Lassini, A. (1994) 'Per un network regionale dei centri di servizio alle imprese', *Impresa e Stato* (Rivista della CCIAA di Milano) 26: 63–8.

—— (1985) *Gli interventi regionali per i servizi alle imprese*, Milan: Angeli.

Ligabue, L. (1992) 'L'industria della maglieria nell'area di Carpi e l'esperienza del Citer', paper presented at the conference on Ricostruire l'economia industriale: lezioni dagli Stati Uniti ed esperienze italiane con nuove tecnologie e strategie (Rebuilding the Industrial Economy: Lessons from the US and Italian Experiences with New Technologies and Strategies), Massachusetts Institute of Technology, Cambridge, MA, 21–22 September 1992.

Lizzeri, G. (ed.) (1983) *Mezzogiorno possibile*, Milan: Angeli.

Marinazzo, M. (1995) 'L'esperienza dei parchi scientifici e tecnologici nel contesto italiano ed europeo: stato dell'arte', paper presented at the meeting on RTD Potential in the Mezzogiorno of Italy: The Role of Science Parks in a European Perspective, promoted by the Consorzio Mario Nagri Sud, S. Maria Imbaro, 21–22 September 1995.

Martinelli, F. (1984) 'Servizi alla produzione e sviluppo economico regionale: il caso del Mezzogiorno d'Italia', *Rassegna economica* 1: 187–219.

—— (1985) 'Public policy and industrial development in southern Italy: anatomy of a dependent industry', *International Journal of Urban and Regional Research* 9(1): 47–81.

—— (1989) 'Struttura industriale e servizi alla produzione nel Mezzogiorno', *Politica economica* 5(1): 129–87.

—— (1998) 'The governance of post-war regional policy in Italy', paper presented at the Second European Conference on Urban and Regional Studies, Durham, 17–19 September 1998.

Martinelli, F. and Gadrey, J. (2000) *L'economia dei servizi*, Bologna: Il Mulino.

Mazzonis, D. (1996–7) *Il nuovo ruolo di Ervet in Emilia Romagna*, Bologna: Ervet (mimeo).

Moulaert, F., Martinelli, F. and Djellal, F. (1990) *The Role of Information Technology Consultancy in the Transfer of Information Technology to Production and Service Organizations*, Working Document 10, The Hague: NOTA (Netherlands Office of Technology Assessment).

Nomisma (Laboratorio di Politica Industriale) (1988) *I centri di servizio reale alle imprese: stato dell'arte e repertorio delle esperienze italiane*, Rapporto Nomisma No. 4, Bologna: Nomisma.

Nomos (1997) unpublished documentation, Milan: Nomos.

Papagni, E. (1995) *Sviluppo duale e progresso tecnico nell'economia italiana*, Milan: Angeli.

Pontarollo, E. (1982) *Tendenze della nuova imprenditoria nel Mezzogiorno degli anni '70*, Milan: Angeli.

Putnam, R. (1993) *Making Democracy Work: Civic Traditions in Modern Italy*, Princeton: Princeton University Press.

Silvani, A. and Prisco, M.R. (1993) 'Mezzogiorno, ricerca scientifica e innovazione', in Cappellin, R. and Tosi, A. (eds) *Politiche innovative nel Mezzogiorno e parchi tecnologici*, Milan: Angeli, 65–98.

Svimez (1997) *Rapporto 1997 sull'economia del Mezzogiorno*, Bologna: Il Mulino.

—— (1999) *Rapporto 1999 sull'economia del Mezzogiorno*, Bologna: Il Mulino.

Trigilia, C. (1992) *Sviluppo senza autonomia: effetti perversi delle politiche nel Mezzogiorno*, Bologna: Il Mulino.

Wolleb, E. (1991) 'I parchi scientifici e tecnologici nello sviluppo economico urbano', paper presented at the Conferenza del Consiglio dei Comuni e Regioni d'Europa (Conference of the Council of Municipalities and Regions of Europe), Bologna, 30 May–1 June 1991.

Wolleb, E. and Wolleb, G. (1990) *Divari regionali e dualismo economico*, Bologna: Il Mulino.

9 Greece

Knowledge-intensive services and economic development

*Pavlos-Marinos Delladetsima and
Alekos G. Kotsambopoulos*[1]

Introduction

There has been much rhetoric about services in Greece. Sentiments suggesting that Greece has little chance of being competitive in industry, and that the only way forward is through services and tourism, have become popular. Pursuit of the implications of this position, however, has been neither systematic nor continuous. A research and policy vacuum has affected assessment of the role and importance not only of knowledge-intensive services (KIS), but also the service sector as a whole. This situation, of course, is not peculiar to Greece (Daniels and Moulaert 1991), but Greece is certainly a latecomer to service analysis.

In Greece, service analysis has mostly been approached indirectly through studies of aggregate economic and industrial development. Within these studies, services have been most closely examined in relation to various aspects of industrial policy (grants and subsidies, financial mechanisms, technology development, problem firms), and the search for an effective national industrial strategy (Giannitsis 1993). The importance of services has also been revealed in the context of research and development (R&D) and technology transfer studies (Giannitsis and Mavri 1993), as developed in specific regional contexts such as central Macedonia (Tsoulouvis *et al.* 1995). These have emphasized the cultural and educational dimensions of technological change. Other commentaries touching upon the service sector have been related to tourism (Tsartas 1998), tertiary sector development (Kafkalas 1999) and management (Makrydakis *et al.* 1996). Some studies partly deal with service location in the neo-classical tradition, identifying regional/urban hierarchies in the national territory (Konsolas *et al.* 1988). The underlying basis of such approaches is the 'proximity assumption' between industry and tertiary sector firms (Koutroumanides and Loukakis 1993). Similarly, shift-share analyses of regional compared with national trends seek to identify typologies for the distribution and employment growth of services (Karakos and Koutroumanides 1998). Much of this work offers a broad contribution with little insight into the role of services in regional economies and employment growth, and even less in relation to innovation.

The growth in services and producer services is undoubtedly attracting increasing interest in Greece as a research and policy field (Sidiropoulos 1997). This is reflected in press discussion and specialized publications, as well as in a number of direct studies of the issues (Interaction/Ministry of National Economy 1991; MIET 1995). This increasing interest is a consequence of (1) the actual significance and growth of service and producer service firms; (2) the activities of professional institutions, chambers of commerce and industry and related organizations; and (3) the impact of EU policy regulations and directives. This chapter remains constrained, however, by a disjointed and piecemeal analytical and policy legacy and limited availability of data. Little is known about the operations of the service economy in Greece as a whole, or in various regional contexts, quite apart from the role of the producer services and KIS.

In addressing the theme of this chapter, it is important to take account of the meaning of 'innovation' in the Greek context. In official discussions, at least, innovation processes are still approached in a conventional manner, with overwhelming emphasis placed on production technology rather than on organizational and social processes. Innovation patterns are also influenced by the particularities of the Greek economy, characterized by the coexistence of advanced and underdeveloped sectoral and geographical segments. As a consequence, innovative performance cannot be assessed in the same manner as in other European settings, especially when the regional dimension is introduced. Innovation in certain sectoral and local contexts certainly has little to do with the influence of specific types of advanced services or KIS. This may be explained by two factors. First, the concentrated urban structure of the country and its geographical size may cause the major KIS firms to develop operations throughout Greek territory from central headquarters without regional affiliates. Second, such KIS activities focus on major public sector clients, major industrial firms, banking and insurance institutes. Access to KIS is not therefore generally spread throughout the full range of sectors and firms that compose regional economies. As a result, innovative action in specific regional contexts is primarily a highly localized phenomenon, deriving from conventional expertise and knowledge, with limited influence from national or international KIS support.

The national context and service sector development

The demand for KIS in Greece needs first to be placed in the context of recent trends and changes in the economy, including the growth of the service sector. A glance at the available census information reveals the extended role of services in the Greek economic structure. According to census information from the National Statistical Service of Greece (NSSG), in 1991 the tertiary sector constituted approximately 60 per cent of gross domestic product (GDP) and 55 per cent of total employment. The average annual rate of growth in services had been 2.85 per cent between 1981 and 1991, and continued at 1.45 per cent to 1994. These figures, however, do not demonstrate the role played by services in economic development, including the relationship of their growth to developments in other sectors such as manufacturing. The service sector in Greece has shown a more autonomous path of development compared with the service sector in other countries, related to the overwhelming growth of the public sector and the polarized structure of Greek industry. As elsewhere, services have low measured productivity. Between 1981 and 1991 annual productivity changes were 4.09 per cent in the primary sector, 2.35 per cent in manufacturing and only 0.98 per cent in tertiary activities (Katochianou *et al.* 1997: 67). Productivity in the sector remains low, reflecting the poor performance of the economy as a whole in the 1990s, in spite of some recent improvements.

Services are involved in a highly complex situation. They cover a wide and diverse spectrum of public and private activities, and vary immensely in the skills, capital and technology employed. Added to this is the extensive role of self-employment which cuts across most service categories (Vaitsos and Giannitsis 1987; Giannitsis 1993; Vaiou *et al.* 1996). These categories comprise:

1   Services employing highly skilled labour with intensive capital and technology use (telecommunications, information technology and telematics, advanced hospital services, transport).
2   Services related to tourism and commerce, with highly diverse and dualistic occupational and corporate structures. This is the dominant category in terms of employment.
3   Commercial producer services (accountancy offices, consultancy firms, training, insurance, banking).
4   Personal services operating in the social and private sphere, employing salaried labour (private education, health care, goods delivery services).
5   At the lower end of the income and formal organizational range, services at the fringe of the informal sector, gravitating to underemployed, migrant labour (cleaning and domestic services).

It is difficult to estimate the share of each of these categories in the economy. Tables 9.1 and 9.2 nevertheless give a broad idea of the shares of GDP and employment occupied by various service branches, measured by census categories. These indicate the economic importance of wholesale, retail and transport activities and the public sector, as well as the importance of tourism-related activities for employment.

A historical perspective is required to explain the significance of this complexity for the growth and role of services in the current phase of Greek development. The 1960s determined the major characteristics of the post-war Greek economy and society. At that time the country experienced high growth

Table 9.1 Greece: GDP by sector, 1994 (at current prices)

Sector	Dr 1m	%
Agriculture, animal breeding or farming, fishing	2,387,103	14.9
Mining and quarrying	186,238	1.2
Manufacturing	2,399,528	15.0
Electricity, gas and water supply	405,221	2.5
Construction	1,008,850	6.3
Transportation and communication	1,142,953	7.2
Wholesale and retail trade	2,162,625	13.5
Banking, insurance and real estate	477,150	3.0
Dwelling services	1,258,010	7.9
Public administration, defence	1,763,479	11.0
Health and education	1,252,507	7.8
Miscellaneous services	1,535,408	9.6
Total	15,979,072	100.0

Source: Elaboration of data from the NSSG.

rates in investment, production and productivity. The developmental outlook was based on a policy of concentrated investment in large industrial installations and major infrastructure works. More specifically, an industrialization boom was initiated after 1962–63 by massive investment in the metallurgical, alumina-aluminium, chemical, petrochemical and shipbuilding industries. There was also a marked shift of investment from consumer to capital goods and durables. In parallel, a restructuring of factors of production took place, with unprecedented internal and external population movement. Between 1951 and 1971 more than 1.5 million people migrated to the cities from rural areas. By the late 1960s, 29 per cent of the population was concentrated in Athens, with Salonica (or Thessaloniki) established as the second most important urban centre. Since then Athens and Salonica have carried the burden of development, followed by the agglomerations developed along the main highway corridor: Patras–Korinth–Athens–Volos–Larissa–Salonica–Kavala. The structure of Greek industry, divided between heavy industrial complexes and a mass of small firms, was thus shaped in the 1960s and has been perpetuated since. This has fostered the assumption of a subsidiary role by the service sector in subsequent decades. Under a highly protectionist environment, major firms became largely self-sufficient and tended to avoid the use of external expertise. At the same time small firms adopted a

Table 9.2 Greece: employment by sector, 1991

Sector	Dr × million	%
Agriculture, animal breeding or farming, fishing	668,766	18.7
Mining and quarrying	15,284	0.4
Manufacturing	523,120	14.7
Electricity, gas and water supply	33,357	0.9
Construction	281,186	7.9
Trade, hotels and restaurants	651,370	18.2
Transportation, warehousing and communication	243,536	6.8
Banking, insurance and real estate	73,829	2.1
Dwelling services	131,522	3.7
Public administration, defence	293,254	8.2
Education	185,509	5.2
Health	138,841	3.9
Provision of services, domestic personnel	123,763	3.5
Miscellaneous services	1,989	0.1
Did not state sector of economic activity	206,631	5.8
Total	3,571,957	100.0

Source: Elaboration of data from the NSSG.

development pattern lacking a collaborative networking culture which, in turn, made demand for external support superfluous.

From 1960 to 1974 the annual GDP growth rate of 6.7 per cent was impressive, by the standards both of the European Community (EC) and the Organization for Economic Cooperation and Development (OECD) (Lyberaki 1996: 2). These rates were primarily led by industry. The service sector participated, but its relative weight increased only in later years. In the 1970s, under the impacts of two oil crises, growth rates began to decline, initiating gradual de-industrialization. Nevertheless, the overthrow of the dictatorial regime in 1974 created exceptional political conditions which encouraged a significant recovery. Between 1974 and 1979, Greek GDP continued to grow faster than the EC average (3.3 per cent), though at a decelerating pace (Lyberaki 1996: 2). Investment showed steady growth after 1975, reaching 1973 levels again by 1979. By the late 1970s, despite political and economic crises, Greek economic growth had to some extent been revived.

The 1980s saw a reversal of these positive trends. The main characteristics of the Greek economy at this time were a decrease in the GDP annual rate of growth (to 1.5 per cent); a declining trend in fixed capital investment; a decline in the growth rate of production to 1 per cent per annum; an increase of inflation to an average of 20 per cent per annum; an increase in the public deficit fed by expansionist policies in the early 1980s; and an increase in unemployment (Sakellari 1995: 57). Such recession conditions particularly affected traditional industrial centres such as Athens, Piraeus, Patras, Volos and, in part, Salonica. In parallel, however, other areas of the country such as the northern–central region and eastern Macedonia underwent significant economic restructuring, exhibiting a dynamism which counteracted the negative trends elsewhere (Vaiou and Hajimichalis 1997: 2). The same applied to the agglomerations of Heraklion in Crete, Larissa in the region of Thessaly in central Greece, and partly to Ioannina, the region of Hepiros in north-west Greece.

In the 1980s, for the first time in post-war development, the service sector increased its share of GDP. The share of the industrial sector in GDP fell from 20.8 per cent in 1974 to 19.6 per cent in 1985 while the service sector share increased from 51.2 per cent to 56.7 per cent (Vaitsos and Giannitsis 1987: 21). Little is known in detail about the role and behaviour of services in the economic and spatial setting of the 1980s. One study, however, based on aggregate data from the 1978–88 censuses, has shown that over-concentration in Athens and Salonica was perpetuated, and moderate growth took place in some peripheral urban areas (Koutroumanides and Loukakis 1993). The latter trend was partly influenced by relative industrial growth, but mostly by the growth of consumer services and transport.

Economic development trends in the 1990s have occurred in two phases. In the early years the economy experienced a continuing crisis, based on an accumulation of all the earlier negative trends (Labour Institute of the General Trade Union of Greece 1997). This involved dramatic increases in the public debt and deficit; falls in manufacturing labour productivity and investment; increases in unemployment, to 10.1 per cent in 1998, a very high rate by Greek standards; and widespread de-industrialization, in its most acute forms in Lavrion, Euvia, Kozani, Patras and Volos. In the late 1990s, however, in spite of the crisis environment, a clear restructuring and modernization counter-trend made progress, including a strong push towards mergers and acquisitions. These processes are encouraged by, and associated with, the integration of the country into the Single European Market and monetary union. Within this process some segments of all sectors of the economy have made some recovery. This has created a demand for KIS consultancy services in a wide spectrum of private and public sector activities. In other words, behind the crisis there is an underlying mood for change, which in turn is defining new areas of demand for producer services and KIS (Kyriazis 1997a,

Table 9.3a Greece: principal types of consultancy

1 Administration and management services
2 Producer services
3 Research services
4 Human resources development
5 Information and communication
6 Marketing services

Source: MIET 1995.

Table 9.3b Greece: categories of producer service

1	Leasing
2	Fleet leasing
3	Chartered surveyors
4	Venture capital
5	Public relations
6	Factoring
7	Credit risk
8	Private banking
9	Insurance
10	Private pension plan
11	Security services
12	Auditors
13	Consultants
14	Corporate lawyers
15	Merchant banking

Source: Industrial Review 1996.

1997b). The 1990s appeared to be a period in which the KIS phenomenon dynamically emerged and thus can now be more systematically studied.

Producer services and KIS development patterns: the supply structure

Types of KIS supply

The supply structure of producer services is strongly influenced by discontinuities in demand patterns, to be examined in the next section, as well as the dynamics of the sector itself. An impression of the supply structure of Greek producer services and KIS is provided by preliminary classification attempts, for example the sixfold classification of consultancy employed by the Ministry of Industry, Energy and Technology, and the fifteen categories of producer services employed by the periodical, *Industrial Review*, shown in Tables 9.3a and 9.3b. Such classifications vary in their content and objectives, and only hint at the nature and activities of KIS firms. They have little to offer in the understanding of their economic significance, especially regionally. Clearly, producer service firms tend to diversify into a variety of activities, with the possible exceptions of marketing, advertising, auditing and shipping which remain more specialized (ICAP 1995; Logothetis 1997). Accountancy firms have diversified their activities over the years from basic auditing to tax consultancy and business planning.

Moreover, the composition of supply broadly includes a number of relatively large firms partly represented, for example, by the Greek Union of Management Consultants (Syndesmos Etaireion Symboulon Management Ellados: SESMA), and an array of small firms. According to SESMA estimates, total turnover had reached Dr 20 billion by the early 1990s, of which their members provided 68 per cent (MIET 1995). A peak year in terms of turnover for management consultants in Greece was 1999, when the sector scored a 21 per cent increase, while the average rate in Europe was 15.9 per cent (FEACO 1999). The patterns of services supplied, represented by the activities of SESMA members, was 58 per cent in management consultancy, 4 per cent in human resources advice, 8 per cent in IT consultancy, and 30 per cent in other consultancy services (SESMA 2000). The scale of activity is, however, difficult to assess because of problems of definition. A good deal of consultancy is done outside management consultancy firms. The operations of the large firms also vary according to market characteristics and may involve other large firms, national and regional networks, franchising and on-the-spot collaboration with small–medium enterprises (SMEs) through subcontracting. Without taking account of performance, work quality, turnover, profitability and efficiency, simple distinctions between large and small firms may be misleading. This becomes even more complex when examining areas of innovation.

Patterns of KIS supply

Our main concern here is to analyse KIS supply structure and also to examine its possible regional significance. Since service development in Greece is an urban phenomenon, our account focuses on the Athens metropolitan area and on the three other major urban centres of the country: Salonica, Patras and Heraklion.

Salonica constitutes the second major urban/industrial agglomeration in the country. Its strategic position in relation to the Balkan hinterland consolidates its importance for KIS. There are indications of a potentially expanding role for the city as a service centre in the Balkan region (Lambrianidis 2000).

Patras, the capital of the administrative periphery of western Greece and a traditional industrial centre, is currently experiencing an acute de-industrialization crisis. At the same time, the agglomeration includes some innovative institutions developed around the university including the technology park, and services related to the port economy.

Heraklion is the capital of the administrative periphery, in Crete. Successive periods of public investment have led to the creation of a university and a series of innovative institutions such as a business information centre, the Institute of Molecular Biology, the Institute of Technological Applications, the Institute of Marine Biology and the Technological Information Foundation of Crete.

KIS developments must be understood in association with the geography of the wider productive system. The analysis here is based on two principal sources of information: (1) census data, based on industrial establishments for 1978–88; and (2) data from the data bank and business information company, ICAP Hellas, which enables the structure of activity in 1995 to be examined. Beyond the usual information difficulties faced in the study of any service economy in Europe (Gaebe *et al.* 1993), Greece presents special problems. There are different data sources, adopting different classification and sampling systems, pursuing different objectives, over different time periods. To achieve some degree of comparability, emphasis is given here to KIS sectors possessing a maximum definitional convergence in both sources. This at least allows the analysis of long-term developmental trends that neither source could provide independently.

Trends in the 1980s: census evidence

The census data have important weaknesses. The last census year was 1988 and no data are available for more recent years. It is impossible therefore, to elaborate on a phenomenon that, at least in Greece, has been shaped primarily since then. The census is based on a national version of ISIC (International Standard Industrial Classification) and ISCO (International Standard Classification of Occupations). It does not include certain important service categories such as public administration, educational institutions and research institutions. In 1993, the classification system was changed with the adoption of NACE (Nomenclature des Activités Communauté Européen 1991, Revision 1).

The service sector, according to the available census data, is classified into four principal categories (6–9). Category 8 encompasses most of the business services, and it is here that our analysis concentrates (see Table 9.4). In aggregate, between 1978 and 1988, total employment increased by only 13 per cent, while the whole service sector (including commerce, transport and communications, finance, etc. and miscellaneous services) experienced a 20 per cent increase, and 23 per cent in the number of establishments. The average establishment size in the service sector remained relatively small (2.7

Table 9.4 Greece: NSSG census classification of the service sector, 1993

---

6   Commerce – Hotels – Restaurants

7   Transport – Communications – Warehousing

8   Banking – Finance – Real estate – Producer services, etc:

    1   Banking

    2   Stock exchange and related services

    3   Leasing

    4   Insurance

    5   Real Estate

    6   Accounting and fiscal consulting

    7   Electronic data processing and related services

    8   Advertising

    9   Business consulting

    10   Miscellaneous services (typing, photocopying, rental of production equipment, other)

9   Miscellaneous services (health provision, leisure, domestic services)

Source: Elaboration of data from the NSSG.

employees). Employment in category 6 (commerce, hotels and restaurants) in 1988 covered almost 68 per cent of total service employment (see Table 9.5b).

Between 1978 and 1988, business services experienced an increase in employment of over 45 per cent (see Table 9.6b) and there was a particularly impressive increase of over 80 per cent in the number of establishments (Table 9.6a). The category is dominated by banking and insurance, with over 55 per cent of total employment. There were rapid rates of employment growth of 220 per cent for business consultancy, advertising, electronic data processing and related services.

The most common establishment size for category 8 is four employees, although this varies between sectors and geographical areas. Nationally, very small firms dominated employment in many miscellaneous business services, as well as accountancy and real estate, which remain dominated by the structure of family or sole-practitioner businesses. Larger establishments, suggesting a relatively secure organizational basis for enterprise, were more dominant in banking, insurance, advertising and business consultancy, and establishments of over fifty employees accounted for more than one-quarter of jobs (see Table 9.7). In Greater Athens (Attica), the average establishment size is six employees, while for Patras and Heraklion it is 3.2. If sub-sectors such as

Table 9.5a Greece and major cities: number of establishments, 1978–88

Major sectors	Greece		Athens		Salonica		Patras		Heraklion	
	1988 Estab ×1,000	1978–88 %ch	1988 Estab ×1,000	1978–88 %ch	1988 Estab ×1,000	1978–88 %ch	1988 Estab ×1,000	1978–88 %ch	1988 Estab ×1,000	1978–88 %ch
Mining	1.2	–5	0.1	–41	*	*	*	*	*	*
Industry	144.7	12	48.3	3	14.7	36	2.3	10	2.1	16
Electricity	4.6	190	0.1	–2	*	*	*	*	*	*
Commerce	294.0	21	83.8	15	23.4	31	4.9	24	4.5	29
Transport/comm.	10.7	18	4.3	9	0.8	53	0.2	43	0.2	49
Finance/real est./producer services	26.0	82	9.2	32	2.6	110	0.6	110	0.6	50
Misc. services	28.4	16	11.2	13	2.3	24	0.5	90	0.4	30
Total	509.6	44	157.0	19	43.8	51	8.5	23	7.8	27

Source: Elaboration of data from the NSSG.

Note
* less than 50 establishments/employees.

Table 9.5b Greece and major cities: employment, 1978–88

Major sectors	Greece		Athens		Salonika		Patras		Heraklion	
	1988 Empl ×1,000	1978–88 %ch	1988 Empl ×1,000	1978–88 %ch	1988 Empl ×1,000	1978–88 %ch	1988 Empl ×1,000	1978–88 %ch	1988 Empl ×1,000	1978–88 %ch
Mining	20.8	–4	1.1	–13	*	*	*	*	*	*
Industry	795.8	5	246.4	–13	71.3	27	14.1	–10	6.5	15
Electricity	32.1	23	10.1	–19	1.2	90	0.4	49	0.2	22
Commerce	646.2	24	230.1	17	54.7	30	11.0	18	10.5	41
Transport/comm.	139.2	–9	79.4	–21	11.9	12	2.2	2	2.9	45
Finance/real est./producer services	104.2	46	55.1	25	9.5	92	1.8	54	1.8	50
Misc. services	64.0	31	32.9	38	4.7	25	1.1	5	0.8	27
Total	1,802.3	13	655.1	–1	153.3	38	30.6	5	22.7	33

Source: Elaboration of data from the NSSG.

Note
* less than 50 establishments/employees.

Table 9.6a Greece and major cities: finance, real estate and producer services, number of establishments, 1978–88

	Greece		Athens		Salonika		Patras		Heraklion	
	1988 Estab ×1,000	1978–88 %ch	1988 Estab ×1,000	1978–88 %ch	1988 Estab ×1,000	1978–88 %ch	1988 Estab ×1,000	1978–88 %ch	1988 Estab. ×1000	1978–88 %ch
Banking	2.2	67	0.7	59	0.2	131	*	*	*	*
Stock exchange, etc.	0.1	-10	0.1	6	*	*	*	*	*	*
Insurance	2.2	38	0.9	23	0.2	81	*	*	0.1	13
Real estate	0.9	-28	0.4	-40	0.1	-21	*	*	*	*
Accounting & fiscal consulting	3.1	169	0.9	73	0.4	155	0.1	314	0.1	245
Electr. data processing, etc.	0.3	1,583	0.2	1,307	*	*	*	*	*	*
Advertising	0.4	5	0.3	-8	0.1	71	*	*	*	*
Business consulting	0.4	75	0.3	60	0.1	181	*	*	*	*
Miscellaneous	16.4	98	5.5	35	1.6	122	0.4	113	0.4	41
Total	26.0	82	9.3	32	2.7	110	0.5	54	0.6	42

Source: Elaboration of data from the NSSG.

Note
* less than 50 establishments/employees.

Table 9.6b Greece and major cities: finance, real estate and producer services, employment, 1978–88

	Greece		Athens		Salonika		Patras		Heraklion	
	1988 Empl ×1,000	1978–88 %ch	1988 Empl ×1,000	1978–88 %ch	1988 Empl ×1,000	1978–88 %ch	1988 Empl ×1,000	1978–88 %ch	1988 Empl ×1,000	1978–88 %ch
Banking	44.7	31	23.0	15	3.82	74	0.9	52	0.7	53
Stock exchange, etc.	0.3	37	0.3	51	*	*	*	*	*	*
Leasing	0.1	—	0.1	—	*	*	*	*	*	*
Insurance	13.2	56	9.4	42	1.0	70	0.2	59	0.2	154
Real estate	1.4	-24	0.7	-33	0.1	-15	*	*	*	*
Accounting & fiscal consulting	5.7	175	2.0	88	0.8	210	0.1	338	0.1	230
Electr. data processing, etc.	1.5	1,366	1.2	1,140	0.1	4,600	*	*	*	*
Advertising	2.3	44	2.1	40	0.2	122	*	*	*	*
Business consulting	3.0	123	2.2	88	0.6	454	*	*	*	*
Miscellaneous	31.9	47	14.0	15	2.9	83	0.6	47	0.8	21
Total	104.1	46	55.0	25	9.5	92	1.8	54	1.8	80

Source: Elaboration of data from the NSSG.

Note
* less than 50 establishments/employees.

Table 9.7 Greece: finance, real estate and producer services, establishment size groups, 1988

	Percentage of establishments and employment by establishment site groups							
	0–4		*5–9*		*10–49*		*>49*	
	Estab	*Empl*	*Estab*	*Empl*	*Estab*	*Empl*	*Estab*	*Empl*
Banking	19.6	2.5	25.3	8.8	48.9	49.4	6.2	39.6
Stock exchange, etc.	68.9	28.5	24.3	37.4	5.4	14.9	1.4	24.1
Leasing	42.9	8.3	28.6	21.2	28.6	70.5	—	—
Insurance	79.5	23.4	8.7	9.4	10.8	36.2	1.1	20.1
Real estate	97.9	85.4	1.4	5.2	0.7	9.4	—	—
Accounting & fiscal consulting	97.1	85.8	2.3	7.6	0.6	6.7	—	—
Electr. data processing, etc.	72.5	34.3	18.2	24.4	8.3	28.0	1.0	13.3
Advertising	71.0	24.1	15.3	15.5	10.9	32.4	2.7	28.1
Business consulting	69.1	18.1	14.7	11.8	13.3	30.6	2.9	39.5
Miscellaneous	95.3	71.5	3.3	10.6	1.3	11.3	0.1	6.7
Total	86.7	33.4	6.0	9.9	6.5	31.9	0.8	24.9

Source: Elaboration of data from the NSSG.

Note
Establishment size: number of employees.

banking are excluded the average size drops to 2.1 employees, indicating the very small size of the majority of enterprises.

Athens is the major centre of producer services, with 53 per cent of national business service employment in 1988 and 35 per cent of establishments, although these shares had fallen since 1978, from 61 per cent and 49 per cent respectively (see Tables 9.6a and 9.6b). The role of Athens is nevertheless reinforced by an overwhelming concentration of headquarters, for example in banking, and of the more advanced sectors. In 1988 Athens contained 76 per cent of business consultancy employment, 94 per cent of stock exchange and related services, 98 per cent of leasing, 79 per cent of computer services and 88 per cent of advertising. Athens similarly dominates the share of larger establishments, with 65 per cent of those with more than ten employees. Location quotient (LQ) data, comparing local and national shares of activities, confirm this picture (Table 9.8). For the sector as whole, Athens exhibits an LQ of 1.4, indicating a share 40 per cent above the national average. Excluding banking, the LQ exceeds 1.6. The LQs of the more advanced producer services, such as business consultancy, financial services, advertising, and information and computer services, generally exceed 1.5.

The only other significant centre of service concentration is Salonica. The sector as whole contained 9.1 per cent of employment and 9.8 per cent of establishments in 1988. The city has a significant role in accounting, computer services and business consultancy. Location quotients are around 1.0, with the exception of stock exchange and related services and leasing, which are highly concentrated in Athens. Salonica has an impressively high LQ of over 2.2 in business consultancy, indicating a distinctive strategic role in northern Greece. Patras and Heraklion have LQs of around 1.0 for more routine banking, insurance, real estate, accounting and miscellaneous services. They have low LQs for business consultancy, computer services and advertising, reflecting their status as local centres.

Polarization has also occurred within regions. Almost 65 per cent of sector employment in central Macedonia in 1988 was in Salonica, and the comparable regional shares for Patras and Heraklion were 45 per cent and 40 per cent respectively. For Salonica, polarization is more intense for key sub-sectors such as computer services, advertising and business consultancy. On average for these the regional employment concentration in Salonica exceeded 90 per cent.

In summary, the pattern of advanced producer service development in Greece in the 1980s was overwhelmingly dominated by the Athens metropolitan area. This is probably a greater concentration of such services than in any other European country. Salonica nevertheless began to exhibit an emerging role, presumably strengthened by its (potential) position in the Balkan region and Eastern Europe. The other urban centres are small, although they show some potential for service and KIS development at the regional level.

Developments during the 1990s

The more recent ICAP data are highly detailed even at the firm level. They are not comparable with the census figures, however, and they are available only for 1995. The statistical unit measured is the firm rather than the establishment, and only for companies with published accounts (mostly limited liability companies). Therefore they generally exclude private companies, and SMEs are probably under-represented. There also appears to be an Athenian registration bias compared to peripheral regions.

The analysis is based on a sectoral classification comparable to that for the census data (see Table 9.9). The first outcome of this analysis is that employment appears to be much higher than that indicated from the census in 1988, for both the total and individual sectors (see Table 9.10). Banking accounted for 59,100 employees instead of 47,700 measured in 1988; business services

Table 9.8 Greece: location quotients in urban centres, 1988

	Greater Athens	Salonica	Greater Patras	Greater Heraklion
Finance, real estate, business services	1.27	1.08	1.18	1.07
Banking	0.98	0.94	1.14	0.90
Stock exchange and related services	1.77	0.28	0.00	0.64
Leasing	1.85	0.26	0.00	0.00
Insurance	1.34	0.82	0.81	0.97
Real estate	0.97	0.95	0.99	0.70
Accountancy and fiscal consultancy	0.67	1.44	0.91	1.07
Electronic data processing, etc.	1.50	1.06	0.74	0.35
Advertising	1.67	0.84	0.27	0.26
Business consultancy	1.43	2.23	0.00	0.06
Miscellaneous	0.83	0.99	1.07	1.33

Source: Elaboration of data from the NSSG.

Note
LQs for category totals have been estimated assuming employment in the sector as a whole as denominator. LQs for sub-categories have been estimated assuming employment in the relevant category as denominators.

Table 9.9 Greece: ICAP Hellas service sector classification

1	Banking
2	Stock exchange and related services
3	Services auxiliary to finance (leasing and mutual funds)
4	Insurance
5	Computer services
6	Advertising
7	Business services
8	Miscellaneous

Source: Service sector classification adopted by ICAP Hellas 1995.

Table 9.10 Greece: financial and business services, 1995

	No. of firms		Employment ×1,000		Turnover ×Dr 100 billion	
	Greece	*Attiki*	*Greece*	*Attiki*	*Greece*	*Attiki*
Banking	63	55	59.1 (62)	57.0 (55)	26.1 (29)	25.6 (26)
Insurance	416	345	11.7 (264)	10.0 (236)	6.1 (179)	6.0 (162)
Computer services	753	544	4.4 (357)	3.8 (271)	0.4 (198)	0.4 (167)
Advertising	626	522	4.1 (323)	3.7 (277)	3.2 (192)	3.2 (169)
Business services	1,653	1,307	14.4 (753)	12.4 (615)	1.3 (430)	1.1 (352)
Miscellaneous services	1,089	792	90.3 (437)	86.7 (347)	15.6 (255)	15.4 (193)
Stock exchange, etc.	77	75	1.2 (71)	1.3 (70)	0.2 (58)	0.2 (57)
Auxiliary to finance	218	193	1.2 (105)	1.1 (99)	1.4 (83)	1.4 (76)
Total	4,895	3,833	186.4 (2,372)	176.0 (1,970)	54.3 (1,424)	53.3 (1,202)

Source: KISINN survey and elaboration of data from ICAP Hellas.

Notes
Figures in brackets indicate number of valid responses.
Attiki is the region around Athens, but is not the same as Greater Athens.
Figures affected by rounding.

for 14,400 instead of 3,000; computer services employed 4,400 instead of 1,400; advertising 4,100 instead of 2,300; stock exchange and related services 1,200 instead of 300; and services auxiliary to finance services 1,200 instead of 100. Only insurance showed a lower figure in the 1995 data compared with the census. These figures obviously reflect a combination of real growth during the period with the effects of classification changes. The comparison is likely to be most accurate for the sectors that closely match those adopted by the census, including banking, insurance, computer services, advertising, stock exchange and related services, and services auxiliary to finance. Business services should also be broadly comparable, even though ICAP data include auditing services.

Banking, insurance and miscellaneous services constitute large proportions of employment and turnover. Advertising, business services and services auxiliary to finance nevertheless account for Dr 600 billion, almost 5 per cent of gross national product (GNP). Advertising alone accounts for more than 5 per cent of the industrial product. The concentration of service activities in Greater Athens is again pronounced. The agglomeration contained the headquarters of companies employing 97 per cent of the total national employment and 99 per cent of turnover in banking; 85 per cent of employment and 98 per cent turnover in insurance; 87 per cent of employment and 90 per cent of turnover in business services; 90 per cent of the employment and 98 per cent of turnover in advertising; 85 per cent of both employment and turnover in computer services; and almost 100 per cent of employment and turnover in stock exchange and related services and in services auxiliary to finance (including leasing and mutual funds). The data also confirm the relative role of Salonica as the second service centre in the country, with 7.5 per cent of employment and 10.5 per cent of turnover in business service companies and 10.5 per cent of both employment and turnover in computer service companies, and the negligible significance of the other urban areas.

Table 9.11 demonstrates the high concentration of larger firms in two key sectors, computer and business services, as well as their huge concentration in Athens, compared even with Salonica. Overall, the twenty largest firms in computer services account for 43 per cent of sector employment and 53 per cent of turnover, while the top twenty business service firms account for 36 per cent of employment and 43 per cent of turnover. The competitive structure of the sector thus means that 400 firms compete for an 80 per cent share of the market and 2,000 firms for the remaining 20 per cent.

The internationalization process

The supply structure is also dominated by multinational consultancy firms. Until the early 1990s their presence was moderate, and their operations highly selective, related to privatization, business planning for public, banking and other institutions, and auditing. Especially in privatization projects most

Table 9.11 Greece: computer and business services, larger firms, 1995

	Firms		Employment		Turnover	
	No.	%	No.	%	Dr 100 billion	%
Greece						
Computer services	157	—	3,672	—	36.99 (135)	—
Business services	339	—	12,673	—	104.89 (188)	—
Greater Athens						
Computer services	136	87	3,206	87	32.04 (88)	86
Business services	290	86	11,082	87	94.88 (164)	90
Salonica						
Computer services	16	10	371	10	4.00 (12)	12
Business services	20	6	972	8	7.52 (10)	8

Source: KISINN survey and elaboration of data from ICAP Hellas.

Notes
Figures in brackets indicate numbers of respondents included in turnover estimates.
Larger firms: more than eight employees.

multinational firms have used Greek SMEs on a subcontracting basis. Nowa-days nearly all the major multinational consultancy firms, including Arthur Andersen, Ernst & Young, PricewaterhouseCoopers and Deloitte & Touche, are represented in Greece, the main exception being McKinsey & Co and Booz Allen & Hamilton. Recently two other major multinational firms, Grant Thornton and BDO, have established offices. The main operators in Greece have been Coopers & Lybrand (now merged with Price Waterhouse), KPMG and Arthur Andersen, along with two major Greek firms, Kantor and Planet, and corporate financing dominated by public and private bank subsidiaries.

On the whole, multinational firms have appeared reluctant to be fully present and involved in the Greek market. One reason may be that 85 per cent of client firms prefer Greek consultancy firms to multinationals (MIET 1995). Also, multinational firms may find difficulties in the complexities of the legislative framework, and the particularities of demand and work patterns, especially at the regional level. There may thus be a 'disguised resis-tance' to internationalization, assuming different proportions in different sectors. There is also a marked divergence in the penetration of multinational capital in various service sectors. Only one multinational firm is represented in the list of the larger firms in leasing. In mutual funds, only two are repre-sented; in the computer services only four. The representation of multina-tionals increases relatively in the management consultancy sector (see Table

Table 9.12a   Greece: major management consultancy firms in Greater Athens, 1995

	Employment	*Turnover Dr billion*
Ethiniki Management SA	500	7.60
ICAP Hellas SA	330	3.65
Alpha Finance SA	33	2.49
Ethiniki Kefaleou SA	15	2.39
E.E.T.A.A.S.A	110	1.39
Planet SA	52	1.28
Candor SA	42	1.09
C & C International SA	15	0.95
B.C.C SA	24	0.92
Rapp Colins Hellas SA	20	0.83
A.G.B Hellas SA	20	0.76
Erasmos Horizon Ltd	9	0.65
Andersen Consultants SA	53	0.64
LDK Consultants Ltd	75	0.61
Infogroup SA	25	0.54

Table 9.12b   Greece: major auditing firms in Greater Athens, 1995

	Employment	*Turnover Dr billion*
S.O.L Greek Cooperation SA	410	4.37
KPMG Kryiakou SA	130	2.05
Arthur Andersen SA	95	1.77
Coopers & Lybrand SA	120	1.17
Ernst & Young SA	105	1.13
Deloite & Touche SA	100	1.11
Price Waterhouse SA	60	1.02
Union of Auditors SA	17	0.37
Kinopraxia Auditors SA	33	0.36
Diethnis Elegtiki SA	20	0.28

Source: KISINN survey and elaboration of data from ICAP Hellas.

9.12a) and is particularly strong in auditing (Table 9.12b). These supply characteristics may well reflect the transitional stage of the country's current development. It is difficult to assess where this will lead. It may become impossible,
for example, to separate the activities of Greek and foreign firms, with all
tending to cooperate with multinationals in various ways, through subcontracting, royalties, affiliations and franchising. A major recent development,
indicative of this 'unification' trend is the merger of the Greek firm, Planet SA,
with the management consultancy sector of Ernst & Young in Greece,
Cyprus, Romania and Bulgaria, and Fyrom in Yugoslavia, Albania and
Moldova, creating Planet Ernst & Young.

The structure and development of producer service demand: key sectors affecting KIS demand

The demand for producer services and KIS relates, in principle, to the operations
of the economic institutions and sectors that promote innovative change. For
Greece, these include the information technology sector, the European,
national, regional and local states, advanced segments of conventional sectors
and newly emerging sectors. It is difficult to assess demand potential at the
regional level without detailed empirical research. Demand arising from the
dominant growth sectors seems to develop predominantly in the major
agglomerations, most evidently, as we have seen, in Athens and Salonica.

The growth and influence of information technology in the Greek market

The expansion of information and telematics technology defines an emerging
set of activities which go hand in hand with demands for producer services.
Since the early 1980s, the Greek information technology industry has exhibited steady growth, although mostly through widely used products, such as
PCs and servers, and conventional software. This growth has been associated
with the modernization of the market, supported by the activities of universities and research institutions. A further boost has been given by the enforcement of certain laws. For example, TAXIS is an EU/Ministry of Economics
programme for the modernization and computerization of the tax system.
There have also been reforms in public administration and local government,
and the implementation of various EU directives and programmes. In addition, developments in telecommunications have shaped a fairly dynamic infor-
matics sector, producing in turn an escalating demand for KIS.

Anticipated improvements in the economy include further potential growth
in the demand for services related to information technology and telematics.
The information technology sector is currently going through a phase of
maturity in the Greek market (Haikalis 1994; Moumouris 1997), with a
marked shift of firm strategy. This primarily involves a push towards the
territorial expansion of the operations of IT firms. Information technology,

software and hardware dealers are found everywhere and there is potential for further growth of demand in various regions of the country. These developments also involve a diversification of IT business activities, placing emphasis on consultancy and the provision of services more attuned to local demand.

The sector is also currently characterized by systematic attempts by multinational firms to expand their share in the conventional IT market. One of the particularities of the Greek market is that it is dominated by SMEs offering low-cost, locally assembled hardware components. To counteract this, multinationals are introducing new software products and shifting their strategic objectives to the provision of support services. Thus, the current consolidation of IT multinationals in the market is accompanied by a move towards new products such as networks, multi-user systems and, above all, services. This move, in combination with the fusion of information and telematics in administration and production, is shaping new patterns of KIS demand. A recent study in two peripheral areas of Greece has shown the effects of the need to comply with new norms and regulations derived, for example, from the taxation system or EU regulations, and to adjust to the changing competitive environment. This requires new information and (more recently) telematics technology. The process is particularly evident among large firms and other leading regional sectors, occurring first through local dealers backed by major Athens-based consultancies (Delladetsima and Moulaert 1998). These key local institutions or firms may also seek support directly from Athens-based KIS firms.

EU policy and regulations

The impact of the EU is strongly felt in Greece, both directly through the Union's Structural Funds and the promotion of investment initiatives, especially those supported by the Community Support Framework, and indirectly, through the introduction of new common regulations and policy guidelines at various levels. In this context EU institutions and policies constantly define new needs for specialized knowledge. On the whole, 'take-off' of the KIS consultancy business in Greece took place in the early 1980s when the European Commission required compulsory business plans and feasibility studies by independent consultancy firms as a prerequisite for the financing of programmes and projects.

The most direct and probably most widely felt EU impact in Greece has related to European Regional Development Fund (ERDF) investment. The availability of ERDF funding determines, at a meso-level, a wide spectrum of activities, such as infrastructure works in transport and telecommunications, which employ high-ranking and specialized consultancy firms. These have included major firms such as R. Parsons (Athens Airport), Brown and Root (Egnatia Highway), GIBB (Attica Free Highway), Atkins (Evinos River Deviation) Ramboll-Precho (the Aktion–Prevesa connection) and many others.

The Structural Funds impact can also be seen in a multiplicity of fields, such as the improvement of production sectors and the financing of local development strategies. Each of these fields has generated significant demand for consultancy. Moreover, the pursuit and implementation of programmes predominantly backed by the European Social Fund (ESF) tend increasingly to be the responsibility of local agents, such as local authorities and chambers of commerce and industry. These require new fields of local action based on the expansion and systematization of appropriate knowledge-based processes. Finally, in addition to the ERDF, the ESF and the related financial instruments, an array of European initiatives (STAR, Valoren, Resider and Renaval) and EU initiatives have been adopted since the reform of the Structural Funds in the early 1990s (including Rechar, Envireg, Interreg, Regis, Regen, Now, Horizon, Stride, Leader). Particular reference must also be made to the series of regional development initiatives promoted by the EU concerning research and technological innovation. A specific regional dimension to innovation has been given by the Sprint programme (Strategic Programme for Innovation and Technology Transfer 1989–93) and since 1994 through the regional technological programmes such as the Regional Technology Plan (RTP), the Regional Innovation Strategy (RIS) and the Regional Innovation and Technology Transfer Infrastructure and Strategy (RITTS). These each define potential fields of new regional action, posing further challenges to local organizational capabilities and negotiating powers between the public and private sectors. EU policy initiatives thus in principle import the need to resort to specialized knowledge.

The indirect impacts of EU policy arise first through the introduction of new common regulatory norms and standards. This requires the replacement of national standards by agreed supranational standards and norms, designed to remove existing barriers to competition. This process confronts firms and institutions with a new environment, delineating other areas of KIS need. In addition, policy conceptions such as 'integrated programming' or the 'principles of subsidiarity' prescribe novel frameworks with which localities and firms have to comply to gain access to funding. This makes new organizational demands on the spatial domains and implementation capabilities of established institutions and companies. Outside support thus may again be of critical importance.

The role of the state

Since the 1980s, especially after the entry of the country into the European Union, the Greek state has embarked, even though in a disjointed and piecemeal manner, on a modernization strategy involving many facets of the institutional, economic and social environment. This has affected all basic functions of the public sector, at central, regional and local levels, including planning, administrative and regulatory functions, infrastructure networks, collective consumption goods, the provision of inputs to various sectors of the economy (for instance agriculture) and major productive public investment

(Caloghirou 1993). Within these programmes there have been specific fields of state action, especially in planning, regulatory functions and public works, which have encouraged the growth of existing, and the emergence of new, consultancy markets. In particular, developmental incentives legislation (Law 1262/1982, subsequently modified and amended) has sponsored demand for the formulation of feasibility studies and technical reports. Environmental and related legislation (Law 1650/1986) has given impetus to the formulation and implementation of environmental impact studies. Law 1947/1991 and the complementary Law 2052/1992 (subsequently modified and amended), allowed for the privatization of public land, which promoted demand in real estate development projects. The liberalization of the auditing market in 1991, until then under the monopolistic control of the Corporation of Chartered Accountants, has allowed the entry of independent firms and multinationals into the consultancy market. Consequently, demand has expanded and been subdivided into a multiplicity of specialized areas, including accounting, internal auditing, information technology, and management, strategy and business development (Industrial Review 1996).

The state, and in particular the central state, has also been the agent *par excellence* for promoting innovation policy. The most systematic attempts to formulate and implement such policy date from the mid-1980s and early 1990s, carried out by the following institutions and organizations.

1   The Ministry of Development (formerly Industry, Energy and Technology) and its supporting institution the General Secretariat of Research and Technology (GRST), with affiliate regional centres. The GSRT, in particular, has been involved in initiatives for the promotion of predominately technological innovation, such as EPET I & II (Epixeirisiako Programma Erevnas kai Texnologias – Operational Programmes for Research and Technology), Stride Hellas (an EC initiative directed towards regional capacities for research, technology and innovation). It has also recently embarked on the implementation of EPAN (Epixeirisiako Programma Anatgonistikotitas – Operational Programme for Competitiveness).

2   The Hellenic Organization of Small and Medium Size Enterprises and Handicraft (Ellinikos Organismos Mikromesseon Epixeiriseon kai Xeirotexnias – EOMMEX) and its sectoral institutions (Leather, Clothing, Silver, Marble), innovation centres, and jointly shared services, including the European Information Centre. The organization is currently undergoing major restructuring, resulting in the closure of its peripheral centres.

3   Chambers of commerce and industry and local authorities, and the institutions which they have co-established, such as technology parks, business information centres and limited liability companies (Hassoula and Kirtsoudi 1997).

Under these policies, 'innovation' has been narrowly defined, associated exclusively with manufacturing products, processes and machinery, but this has generated demand for consultancy firms. Unfortunately, as we will see later, consultancy related to innovation has been confined to the preparation of technical reports primarily to gain access to EU financing. However, the emphasis placed on innovation has promoted other areas of KIS demand, often involving the broader, more qualitative evaluations of the business needs arising from change.

Developments in conventional and new sectors

Within Greek economic modernization, demand for KIS has been generated by the development of both so-called 'conventional' and newly emerging sectors. One conventional sector which has experienced remarkably high rates of growth is advertising, as a spin-off from the development of related businesses such as public relations, and the private media following the abolition of the state TV and radio monopoly in the early 1990s. The tourist sector and especially its most advanced segments, such as major hotels and hotel chains, face international competition and also rely increasingly on the provision of specialist support services. The same applies to advanced areas of the agricultural sector, and other growing specialist services, which include private health care, security services, express package deliveries, mail-order businesses and real estate (Alogoskoufi-Agapitidou 1994). Statutory changes have also encouraged some newly emerging activities. These include leasing, introduced for the first time under Law 1665/1986 (amended in 1991, 1995 and 1996); venture capital, affected by legislation enacted in 1988 (amended in 1993 and subsequently systematized with Law 2367/1995); and factoring, introduced through Law 1905/1990. Finally, the growth of e-business has recently had a catalysing impact on KIS demand, embracing all sectors.

The client basis for KIS

There is limited and scattered information on the actual client basis for producer service demand in Greece. The only study attempting to deal with the issue was undertaken in 1995 by the Ministry of Industry, Energy and Technology (MIET 1995). This study, *Consultancy Service Networks in Attica*, defined eight client categories: the EU and local authorities; industrial firms; the public sector; public welfare institutions; banking institutions; commercial firms; academic institutions; and tourist firms. Although restricted to a sample of management consultancy firms, this study indicated the producer service client base. Authors' interviews with major consultancies in the mid-1990s, for the KISINN study which formed the origins of this book, identified the following producer services and KIS client areas: the public sector; industrial–commercial firms; banking and insurance companies; shipping and maritime activities; and foreign investment by Greek capital. These will be examined in turn below.

The public sector

In Greece, the public sector constitutes the core client basis for producer services and KIS. The MIET study estimated that it absorbed approximately 60 per cent of the total demand for consultancy. This is now probably significantly higher because of growing demand in the 1990s generated by central government privatization and restructuring policies. The privatization process was still in progress, and even intensified, in the late 1990s, expanding into such sectors as the defence industries, publicly controlled shipbuilding, telecommunications and the Post Office, energy, transport (for example aviation, duty free, Piraeus and Salonica port authorities), mass media (TV, radio), the Athens and Salonica water authorities, capital markets, stock exchanges and major public works, along with the market liberalization of public monopolies. Privatization takes various forms, for example through international tenders, stock market flotation, and the proposed adoption of investment management strategies in 2006. Recent information from SESMA (2000) stresses that in 1999, 14 per cent of the client basis of management consultancy was in public sector-controlled industries and institutions, and 28 per cent in public administration itself, amounting in total to 42 per cent of the total demand.

Some indicative examples of how the restructuring and other activities of the Greek state have generated KIS demand include: a project carried out by a consortium of companies, with the participation of Coopers & Lybrand Hellas, developing information technology and management restructuring for IKA, the major national health care and pension fund institution (Akavalos 1997); a reorganization plan prepared by McKinsey for the state-owned Olympic Airways; and the auditing and financial management services provided by Coopers & Lybrand for the same company. Also of relevance here is the appointment of PricewaterhouseCoopers Business Advisers as financial–economic consultants to the Olympic Games Organization Committee (Athens 2004 SA) for the contracts and promotion of special works, including the Olympic Village Redevelopment Project.

Industrial–commercial firms

KIS clients in the industrial–commercial service sector are concentrated in the largest and most advanced firms. The continuing processes of merger and acquisition also involve most facets of economic activity. A recent study has shown that, although in manufacturing mergers mainly consist of leading corporations acquiring smaller firms, the great majority involve service firms. Such activities involve the presence of KIS firms. In manufacturing, Lyberaki has noted, 'A steady process of mergers and acquisitions is progressively changing the management structure in a number of spheres of industrial activity. As a result the creation of a network for producer services is gaining momentum' (1996: 4–5). A number of studies, with diverse analytical objectives, illustrate the development of the producer service and KIS client basis in Greek manufacturing (ICAP 1995; MIET 1995; Lyberaki and Mouriki 1997). The study by ICAP, based on a random sample of 250 industrial–commercial firms, demonstrated that, between 1990 and 1995, one in four firms used a management consultancy at least once. The proportion was higher for industrial firms. A positive correlation between the use and number of services offered by consultants has also been identified (Sakellari 1995: 72). The MIET (1995) study, based on a sample of 144 firms, showed that 90 per cent had made use of consultancies, and 34 per cent did so regularly. Moreover, one in three firms declared their intention to use such services regularly. A study undertaken between 1995 and 1996 of seventy-six firms in Greater Athens showed a trend towards the use of external support (Lyberaki and Mouriki 1997). Fifty-four per cent often used such support, 16 per cent used it occasionally, 22 per cent never did so, and 8 per cent did not respond. Broadly, all studies indicate that a significant percentage of industrial firms tend to resort to consultancies and that demand is concentrated in conventional areas of industrial activity. Regularity of use of external expertise and knowledge is related to the size of the enterprise, the expertise offered and the entrepreneurial culture embodied in the operation of the client firm. Currently, SESMA (2000) estimates that the private sector amounts to 27 per cent of the client basis for the larger consultancy firms that make up its membership.

In the retail and commercial sectors the move to employ external expertise is undoubtedly more systematic (though less documented and studied) and this relates to the overwhelming growth of chain stores, new superstores and out-of-centre retail units throughout Greece. Retail firms have been expanding their activities reciprocally with the growing presence of multinational capital in the sector. These include BHS, Marks & Spencer, Carrefour-Promodes/Continent-Lectrec, Delhaize Le Lion and Makro Cash and Carry, operating either independently or through mergers and acquisitions. Demand for KIS from retail firms relates especially to administration/management, modern marketing systems, automation, new processes and new products (Sakellari 1995: 72). Moreover, retailing has now become a strategic sector nationally for the application of advanced IT services on behalf of multinational IT firms (Mourmouris 1997). Another indication of KIS presence is the support provided by KPMG in the design and implementation and expansion of the McDonald's network in Greece. The retail sector has been a core actor in stimulating new specialities in the KIS market, such as development and location studies (Travlos 1997).

Banking and insurance companies

Banking and insurance have a long-standing tradition in Greece and have exhibited relatively rapid signs of modernization and development during the last decade. Both are leading client sectors for producer services and KIS. The processes of modernization and development in the banking sector began with its liberalization in the mid-1980s. This led first to the appearance of new banking institutions, including affiliate companies and/or foreign branches, aiming to acquire a growing share of the market based on specialized products such as mortgages, mutual funds, leasing, venture capital and factoring. This modernization trend was accompanied by a gradual push towards mergers and acquisitions, which has intensified in recent years, and the systematic internal reorganization of institutions. This has generated considerable demand for external KIS expertise (Union of Greek Insurance Companies 1996). For example, a project by Booz Allen for the state-controlled National Bank of Greece introduced advanced computer technology and operation-research systems (Stratigaki 1996). A major business re-engineering project has been undertaken by Andersen Consulting, for the Commercial Bank of Greece. Moreover, other major firms, such as KPMG, have also been widely involved in the banking sector. Consultancy activities have extended to the insurance sector. Insurance has for years been highly regulated but has recently been liberalized under EU directives. It is thus also undergoing rapid restructuring, including mergers with other activities such as banking and private health care. This reflects constant efforts to introduce new market products, such as group life insurance, private pension plans and financial planning services. Demand for KIS is therefore currently escalating and relates both to insurance company reorganization and to market expansion strategies.

Shipping and maritime activities

In spite of fluctuations in its overall performance, shipping remains a stronghold of the Greek economy. It is probably the most distinctive component of the Greek service market but its impacts on demand have been poorly studied. Shipping is supported by a multiplicity of other services, listed in Table 9.13, which are inherently internationalized. Shipping support services also show distinctive geographical trends, with a historical concentration in the city/port of Piraeus, although this is now losing its growth momentum.

Significant developments within the maritime sector, related to the growth of new products, are undoubtedly reshaping the KIS client basis. A newly developing financial product, for example, is 'freight swaps', used in the dry cargo market and currently expanding in the oil-tanker market. Above all, however, the shipping client basis for KIS received a strong boost after the early 1990s, when the market became more regulated (Logothetis 1997). Escalating insurance costs related to expanding accident rates have led to the

Table 9.13 Greece: shipping sector support services

Management/agencies
Salvage and offshore services
Ship management services
Shipbuilding/ship repair
Banking/finance
Insurance
Maritime legal support
Marine consultant surveying
General maritime organization

Source: Logothetis 1997.

imposition of compulsory regulation. The first stage of International Safety Management (ISM) began in June 1999 and will be augmented in a second stage in 2001 and 2002. This is a regulation with which all shipping firms will have to comply and which requires the use of further knowledge-intensive expertise.

Foreign investment by Greek capital

In recent years Greek firms have tended to expand their operations abroad, especially in the Balkans and the ex-socialist countries. For some of the Balkan countries, Greece represents the most important market in the Balkan area. Although limited information is available, it seems that Greek capital operates both jointly with multinational firms, for example in the telecommunications and computer information industries, and/or independently, as in the extractive industries: food, milk and dairy products; private health care; banking and insurance; and education and training. In many respects this process is highly dependent on the support of the EU. Programmes designed to assist former socialist states include TACIS (Technical Assistance to Confederate Independent States) and PHARE (Poland–Hungary Assistance for Economic Reconstruction; this programme later provided assistance to other countries of Central and Eastern Europe). Support for Mediterranean countries comes through the MEDA programme (Measures to Accompany the Reform of Economic and Social Structures within the Euro-Mediterranean Partnership), as a component of the broader Euro-Med Partnership with the European Union. There are cross-border cooperation programmes such as Interreg II (European Regional Fund Community Initiatives for Border Regions). Support also comes from the major European financial institutions and other international institutions, such as the OECD and the World Bank. The European Bank for Reconstruction and Development (EBRD), created in 1991, has sponsored the establishment of investment funds such as the Danube Fund, operating primarily in Romania, and the Euromerchant Balkan Fund, in Bulgaria (survey carried out by the *Industrial Review*, 1997).

On the whole the performance of Greek firms in their investment activities abroad, either as members of wider consortia or autonomously, seems to have been successful (MIET 1995). Taking into account also the fact that Greece's exports to the Balkan countries increased from US$0.3 billion to US$0.8 billion, 165 per cent between 1990 and 1994, an expansion of the client basis for Greek producer service and KIS firms for such regional markets might be anticipated.

Obstacles to demand

So far, we have highlighted inherently positive trends and an expansion in the client basis for producer services and KIS in Greece. There are, however, many factors that operate negatively and act as obstacles to the development of such demand, producing severe distortions of the service market. These relate to the role of state policy, EU financing and the dominant entrepreneurial culture.

State policy

It is arguable that for Greece the main factor constraining service demand is the inefficiency of state policy. This generates an atmosphere of uncertainty and mistrust, affecting relationships both between the state and enterprise and between enterprises themselves. Uneven relationships, based on clientism and favouritism between central, regional or local state institutions and particular large or small private consultancy firms, critically distort the rules of competition. Another factor is the lack of policy continuity. Many areas of the economy where state action has led to the formation of new service markets have been stifled by the provisional character and lack of continuity in policy actions. For example, the implementation of environmental legislation has been seriously delayed, especially the enforcement of the necessary legislative decrees. Banking liberalization has been slow because established operating rules are too restrictive and reforming legislation had not been sufficiently flexible (survey carried out by the *Industrial Review*, 1997). Leasing exhibited a strong growth trend following the enactment of legislation but is now at a standstill because subsequent amendments have not provided for immovable assets, a most critical aspect of the law for market potential (Alogoskoufi-Agapitidou and Dimitriou 1995). Similar drawbacks affect developing activities such as factoring or even joint-venture capital management. There is also a typically trivial situation with regard to auditing. The law requires firms to use members of the previous Corporation of Chartered Accountants (now represented by the private company SOL), but for their international operations they are bound to rely on independent consultancy firms. This undoubtedly creates severe functional and especially cost problems.

EU financing

In contrast to earlier arguments, it is posited here that EU financing may in some ways negatively affect the demand for producer services and KIS. A good deal of consultancy work does not come from direct client needs, but as a condition imposed by the EU as a third party. The EU, in taking over a share of development costs, has apparently encouraged the development of the producer service market, through programmes such as Mendor, to promote competitiveness through the use of consultants, and Retex (Reconversion Programme for Former Textile Areas), to assist industrial SMEs in quality control and assurance. Unfortunately, they may also undermine the role of consultancies as 'project providers', since they can do no more than enable access to EU funding. For instance, in the context of innovation policy programmes, consultancies act merely to produce 'appropriate reports', which allow clients to obtain subsidies. The actual needs and problems to be tackled may remain untouched. In the Greek market an array of consultancy firms operate as intermediaries in this way, absorbing a considerable share of the spending by Greek firms on consultancy (Sakellari 1995: 74). Critical innovative aspects of the consultancy process, such as training and the transfer of skills and knowledge, virtually disappear. There is thus a real danger, in the generation of fictitious demand, that there could a contraction or collapse in the consultancy market if EU regulations or financing change.

The entrepreneurial culture

A third set of factors negatively influencing KIS markets relates to the prevailing entrepreneurial culture in Greece. This is characterized by short-term views of business development, conservative and autarchic modes of business management, a tendency to preserve family structures in the administration of firms, underestimation of the importance of product quality and, in many cases, attempts to run the business at the fringe of legality. To these could be added the lack of a collaborative networking spirit among SMEs and an overall inability, with a few exceptions, of local chambers of commerce and industry to act as cohesive forces and centres of innovation at a regional level. This affects the demand for knowledge-based services, especially in peripheral regional economies, or at least confines such demand to the largest and most advanced firms operating there.

The interplay of supply and demand

Clearly, demand for producer services and KIS is embracing all sectors of the economy, and is associated with a marked growth in the presence of consultancy firms. The disjointed and possibly fictitious character of demand has repercussions on the supply side, complicating the study of demand–supply interactions, especially in identifying areas of innovative action. A multiplicity of relationships governs KIS firms and their client basis, with no clear link easily identifiable between large KIS firms and client firms or small KIS firms and SMEs. A large client may easily resort to small KIS firms for certain consultancy services. There is undoubtedly a clear correlation between the size of the client firm and the use of consultancy, but there is little correlation with the type of demand and even less with turnover. Four focal issues emerge from investigation of KIS demand–supply interplay in Greece, which pose questions related to the developmental potential of the sector.

First, there is a need to look beyond the 'innovative façade' and identify the real incidence of innovation through KIS action. For example, client firms are increasingly using consultancies to address very specific problems, in hardware–software development, legal matters, auditing or ISO 9000 quality certification, rather than for the long-term purposes of strategic planning and management, total quality promotion, sectoral analysis or environmental policy. This may imply an inherent limit to the innovative development of KIS or, alternatively under a more 'evolutionist' view, that they are now simply in an early stage of market development.

Second, the regional dimension is difficult to evaluate because of the concentrated spatial structure. Athens/Salonica KIS obviously serve most other parts of the country. Thus the association of demand and service concentration does not fully apply in Greek circumstances. Location quotients in Athens and Salonica for certain advanced producer services, such as business consultancy and computer services, reach up to 2.0, compared with around 1.0 for industry and commerce. This could mean that a major segment of KIS demand is not dependent on the local regional context. Such an inter-regional, export orientation of supply firms is clearly a critical issue, especially in understanding the Greek situation, requiring further research at national and regional levels.

Third, the concentration of advanced producer services in major cities is undoubtedly supported by supply factors, including agglomeration economies and economies of scale, the availability of human resources and the supply of high-productivity labour, and the existence of supporting services such as data banks and R&D institutions. But the type of demand also encourages such concentration. This comes not just from decision-making centres and financing institutions serving Athens/Salonica, but also from international agencies such as the EU, state functions, the headquarters of multinational and major Greek firms, and other sources of knowledge and innovation within the economy at large. For KIS-related innovation, spatial concentration is important. Cities are the centres of all sorts of innovative interaction between supply and demand, within which KIS are playing an increasing role, both within Greece, and linking Greece to international developments.

Finally, demand–supply interaction in the near future will be critically influenced by the impacts, whether positive or negative, of new investment waves.

The first is obviously the European Union Community Support Framework (CSF) that for the next decade will constitute an economic development cornerstone. The CSF contains an inherent potential for increasing demand for KIS, and thus further changes in patterns of supply. The administration and implementation of the framework may thus determine, with other influences, the role, development and composition of KIS firms in the near future. A further investment wave will arise from Athens' status as the host city for the 2004 Olympic Games. This event will clearly have a major impact, especially in the next few years, on both the Athens agglomeration and the Greek economy and society as a whole. There are already clear indications of KIS demand related to the games developing in a multiplicity of areas, including financial management, public works, telecommunications and real estate.

Conclusion

Recent research has revealed the indisputable presence and growth of producer service and KIS activity in Greece, embodying a critical potential in relation to turnover and the generation of employment. It has also revealed the relatively innovative impact of KIS, although this is constrained by the small size of the market and the inherent complexity of identifying innovative action in a number of sectors. Further specific empirical research is obviously required at different spatial and sectoral levels.

Policy needs to address forms of intervention that may remove what we have described as 'obstacles to demand'. These include the adverse effects of state policy concerning competition, lack of continuity and relationships between the state and the private consultancy sector. There is also a need to develop strategies that augment conventional perceptions of innovation, placing additional emphasis on organizational and social processes and long-term development.

In relation to EU financing, emphasis needs to be shifted to innovative issues such as the transfer of skills and knowledge, and training, rather than 'project logic' in the search for access to financing. Control over the quality of consultancy work, and of the firms themselves, is also important. Emphasis must further be placed on counteracting the prevailing entrepreneurial culture, possibly through training and related actions, the consolidation of relationships with the appropriate policy institutions and the promotion of inter-firm cooperation.

With respect to positive policy initiatives, priority must be given to expanding the territorial limits of KIS innovative action to other regions of the country. New intermediate institutions could assume a leading role, acting as liaisons between local firms and consultancies. Another priority is the expansion of innovative action in peripheral areas, including the establishment of networking relationships between KIS and established innovative institutions in the periphery of the country such as the university research institutions, science parks and innovation centres in Heraklion, Patras, and the northern

Aegean Islands. In addition, since one of the problems identified has been the weakness of KIS firms in Greece, emphasis should be given to consolidating current specialization in key fields, such as shipping, public works, environmental impact assessment or retail studies.

Finally the 'externally' orientated character of the KIS consultancy business in Greece should also be considered more systematically and possibly strengthened, since the country appears to have developed significant comparative advantages owing to its links and proximity especially with the Balkan states.

Note

1 This is a summary of the research programme carried out for the Knowledge-Intensive Services and Innovation Thematic Network (KISINN) Framework IV. The programme was carried out at the University of the Aegean (UA) by Dr P.M. Delladetsima (Associate Professor at the University of the Aegean, BArch, DIP, MPhil, PhD, MRTPI) and by A. Kotsambopoulos (economist at the University of Athens, MA Kent); with the collaboration of P. Stavrou (economist) as research assistant.

References

Akavalos, A. (1997) 'Learning process in management' (in Greek), paper presented at the KISINN workshop, Athens, 11 June 1997.

Alogoskoufi-Agapitidou, D. (1994) 'Real estate as an investment means', *Success* 1: 110–11.

Alogoskoufi-Agapitidou, D. and Dimitriou, C. (1995) *The Extension of Leasing in Real Estate* (in Greek), Athens: Commercial Bank of Greece.

Caloghirou, J. (1993) 'State markets, industrial structures and public policies in the Greek territory', in Giannitsis, T. (ed.) *Industrial and Technological Policy in Greece* (in Greek), Athens: Themelio, 95–129.

Daniels, P. and Moulaert, F. (1991) (eds) *The Changing Geography of Advanced Producer Services*, London: Belhaven Press.

Delladetsima, P. and Moulaert, F. (1998) *Producer Services and Regional Development in the Aegean: Lesvos and Chios*, research report, Brussels: European Commission, D-G XII-F5.

FEACO (Federation of European Management Consulting Associations) (1999) *Annual Report*, Brussels: FEACO.

Gaebe, W., Strambach, S., Wood, P. and Moulaert, F. (1993) *Employment in Business-Related Services: An Inter-Country Comparison of Germany, the United Kingdom and France*, report to D-G V, Brussels: European Commission.

Giannitsis, T. (1993) (ed.) *Industrial and Technological Policy in Greece* (in Greek), Athens: Themelio.

Giannitsis, T. and Mavri, D. (1993) *Technological Structures and Technology Transfer in the Greek Industry* (in Greek), Athens: Gutenberg.

Haikalis, S. (1994) 'High-tech losses for information technology firms', *To Vima* 6(3), 59.

Hassoula, P. and Kirtsoudi, M. (1997) 'Innovation policy in Greece', paper presented at the KISINN workshop, Athens, 11 June 1997.

ICAP (1995) 'Consulting business market research: main findings', *I Epilogi* 4.

Industrial Review (1996) 'Oiling the wheels: a survey of Greek business services', *Industrial Review* 22 (December): 5–41.

Interaction/Ministry of National Economy (1991) *Greek Business Consulting Firms Market*, Athens: Ministry of National Economy.

Kafkalas, G. (1999) *Thessaloniki: Reduction of the Mono-Central Performance of the Urban Agglomeration of Thessaloniki and the role of the Tertiary Sector*, Thessaloniki: Diti.

Karakos, A. and Koutroumanides, T. (1998) 'Regional structure of the service sector in northern Greece', *Topos* (Athens) 14: 3–14.

Katochianou, D., Todikidou, P. and Kavadias, P. (1997) *Basic Data of Regional Socioeconomic Development in Greece*, Athens: Centre of Planning and Economic Research.

Konsolas, N., Sidiropoulos, H. and Papadasklopoulos, A. (1988) *Structural and Interregnal Changes in Industrial and Service Sectors in Greece 1971–1981*, Athens: IPA (Instituto Periferiakis Anaptyxis).

Koutroumanides, T. and Loukakis, P. (1993) 'Occupational analysis of space distribution in Greece during the period 1978–88', *Technical Chronica* 13(4): 261–83.

Kyriazis, H. (1997a) 'Country article: Greece', *The Euro*, 2–7.

—— (1997b) 'Knowledge as a factor of economic development' (in Greek), *O Ikonomikos Tachidromos*, 8 May, 75–6.

Labour Institute of the General Trade Union of Greece (1997) *Working Paper 10–11* (April), Athens: Labour Institute of the General Trade Union of Greece.

Lambrianidis, L. (2000) 'Greek–Balkan economic relations: the city of Thessaloniki in search of a new role', paper presented at the international conference: Economic History of Thessaloniki, Thessaloniki, 20–21 March 2000.

Logothetis, S. (1997) 'Shipping and consulting needs' (in Greek), paper presented at the KISINN workshop, Athens, 11 June 1997.

Lyberaki, A. (1996)'Adjustment and resistance to change virtues and vices of the Greek economy', *Journal of Modern Hellenism* 12: 1–19.

Lyberaki, A. and Mouriki, A. (1997) *The Soundless Revolution* (in Greek), Athens: Gutenberg.

Makrydakis, S., Papagiannakis, E. and Calogirou, J. (1996) *Greek Management* (in Greek), Athens: Association of Chief Executive Officers.

MIET (Ministry of Industry, Energy and Technology) (1995) *Consultancy Service Networks in Attica* (in Greek), Athens: MIET.

Moumouris, N. (1997) 'Computer from your country even if it is ... assembled' (in Greek), *I Elftherotypia* 5(2).

Sakellari, E. (1995) 'Business consultants: their biggest client is the state' (in Greek), *I Epilogi* 3: 68–76.

SESMA (2000) *Greek Union of Management Consultants: 2000 Report* (in Greek), Athens: SESMA.

Sidiropoulos, H. (1997) 'Producer services and the urban system' (in Greek), paper presented at the KISINN workshop, Athens, 11 June 1997.

Stratigaki, M. (1996) *Gender-Labour and Technology* (in Greek), Athens: O Politis.

Travlos, S. (1997) 'Possibilities of survival of Greek consultancy SMEs in the European market' (in Greek), paper presented at the KISINN workshop, Athens, 11 June 1997.

Tsartas, P. (1998) *La Grèce: du tourisme de masse au tourisme alternatif*, Paris: L'Harmattan.

Tsoulouvis, L., Besta, H., Bakouros, I., Deniozos, D. and Komninos, N. (1995) *Analysis of Best Practices and Schemes of Technological Development in the European Union*, research report by the Regional Technological Project of Central Macedonia, Aristotle University of Thessaloniki.

Union of Greek Insurance Companies (1992–97) *Economic Results of Greek Insurance Companies*, Annual Reports, Athens: Union of Greek Insurance Companies.

Vaiou, N. and Hajimichalis, C. (1997) *With the Sewing Machine in the Kitchen and the Poles in the Fields: Cities, Peripheries and Informal Labour* (in Greek), Athens: Exandas.

Vaiou, N., Golemis, H., Lambrianidis, L., Hajimichalis, C. and Chronaki, Z. (1996) *Informal Forms of Industrial Production: Labour and Urban Space in Greater Athens* (in Greek), Athens and Thessaloniki: Aristotle University of Thessaloniki.

Vaitsos, K. and Giannitsis, T. (1987) *Technological Transformation and Economic Development* (in Greek), Athens: Gutenberg.

10 Portugal

Knowledge-intensive services and modernization

Isabel André, Paulo Areosa Feio and
João Ferrão

Introduction

This chapter offers a systematic perspective on knowledge-intensive services (KIS) in Portugal. The research was carried out by members of a small consultancy firm, where academic and technical staff combine to act as innovation carriers. This experience has helped us to understand the complex of subjective and objective factors which acts as a powerful barrier to change and innovation among clients. To observe and to be observed is thus the singularity of the research underlying this chapter. It is divided into four parts. The first summarizes the main characteristics and recent trends of the Portuguese economy, with special emphasis on employment and business structure. The second section outlines the supply structure of KIS, both in terms of employment and within the broader business structure of Portugal, taking account of its key sectors and regions. The third section presents some ideas on the development of KIS, including their evolution and territorial foundations, emphasizing the institutional background to their growth, in particular the current European Union Community Support Framework (CSF). Its inherent policy of industrial modernization especially favours the role of consultancy firms in stimulating innovation. The final section analyses the dynamics of KIS use in seven sectoral situations (two in agriculture, three in industry and two in services). In each case an attempt is made to establish how far consultancy–client relationships have led to the adoption of truly innovatory procedures. The conclusion summarizes the opportunities and threats offered by consultancies as innovation carriers in a changing regulatory environment and a highly segmented market place.

Distinctive characteristics of the Portuguese economy

Profound and rapid changes have taken place in the Portuguese economy over the last three decades. By the beginning of the 1970s some of the large financial and economic conglomerates had consolidated their position. The origin and development of these companies can in many cases be traced to colonial markets and production, and from the 1960s onwards they coexisted with foreign investment originating in Europe and North America. By the time of the 1974 revolution the country was divided in economic terms. On the one hand there was an archaic and very traditional productive segment made up mainly of small family businesses, not only in agriculture, but also in many areas of industrial production and retailing. The strategy of these firms was largely guided by the need to keep down costs of production. In contrast, there was a more modern part of the economy, heavily concentrated in the Lisbon region and dominated by large economic conglomerates and foreign investment. This segment contained the strategic industries and the financial sector. This economic dichotomy had developed against the backdrop of a highly repressive and protectionist state apparatus, and a state budget directed towards covering the expenses of the colonial war.

Under the new socialist political regime, the economic conglomerates were nationalized and the state came to control strategic productive sectors, in particular the whole of the financial sector. Public sector involvement in the social sphere was also considerable. There was a complete restructuring of the education system, which was now extended to the whole population. State intervention began to decline from the beginning of the 1980s, however, as more liberal economic policies were adopted. Following Portugal's entry into the European Community in 1986, the Portuguese economy began to develop into a market economy within the European framework. From the mid-1980s, retailing and other service activities took off. Staff qualifications started to be of greater concern to employers than mere maintenance of very low salaries, particularly in the service sector. With these developments came very substantial public investment, concentrated into infrastructure and professional training, financed largely by European Union Structural Funds and large-scale corporate foreign investment.

Employment

In Portugal, a relatively high percentage of the population is in active employment compared to the situation in other EU countries. In recent years this percentage has in fact increased (see Table 10.1), with unemployment declining, generally early entry into the labour market and high numbers of women in active employment. This is accounted for by the low salaries prevailing in most areas of the economy, especially in traditional industry and in retailing. However, the employment rates of young people (aged 14–24) fell significantly, from 56 per cent in 1989 to 47 per cent in 1999, reflecting a trend towards a better-qualified labour force with incentives for people to remain in the education system for a longer period, and the increasing availability of professional training courses.

Table 10.2 shows the growing importance of the tertiary sector in the economy. Between 1989 and 1999, the share of employment in the tertiary sector rose slightly from 46 per cent to 49 per cent. There were, however, significant increases in the retail, hotel and catering sectors (by almost 30 per cent) and

Table 10.1 Portugal: employment and unemployment for total and female populations, 1989–99 (%)

	1989	*1999*	*1989–99* *% change*
Total population ×1,000	9,804.2	9,987.8	1.87
Active employment	47.2	50.5	7.0
Women	39.0	44.2	13.4
Active employment 14–24 years old	56.2	47.3	−15.8
Women	51.8	40.8	−21.3
Unemployment rate	5.3	4.4	−17.2
Women	7.8	5.1	−34.6
Unemployed, first job	28.0	15.5	−44.6
Unemployed, new job	72.0	84.5	17.3

Source: INE various dates.

particularly in banking and insurance, following the privatization and liberalization of the financial sector, and professional business services (60 per cent). The growth of employment in the tertiary sector has been widespread throughout the country, although there are regional differences. In the Lisbon region the impact of growth was relatively greatest through its domination of the business services and financial sub-sectors. In other regions retailing, hotels and catering have seen the main increases.

The difference between Lisbon and other regions is accompanied by a difference in levels of qualification. Almost 30 per cent of personnel in the Lisbon region had intermediate or university-level technical or scientific qualifications in 1999. This compared with only around 16 per cent in the north region, 17 per cent in the central and Alentejo regions and 18 per cent in the Algarve. In spite of these generally low levels, increases in qualifications have been marked in recent years. In 1987 only 14 per cent of the employed population had more than nine years' basic schooling. This had risen to 22.5 per cent by 1997, with slight higher values for the female working population.

The structure of business

There is a large proportion of sole traders and small companies in Portugal. In 1993, 17 per cent of companies had no employees and 47 per cent had between one and four employees. Table 10.3 shows the small average size of companies in the agricultural sector, in retailing and the hotel trades, the business services, and personal and social services, contrasted with generally larger firms in manufacturing, banking and insurance. However, levels of sales per

Table 10.2 Portugal: employment by activity, 1989–99 (%)

	Portugal			North	Central	LTV	Alentejo	Algarve
	1989	1999	1989–99 % change	1999	1999	1999	1999	1999
Total (employers and employees ×1,000)	4,400.5	4,825.2	9.7	1,746.6	933.2	1,564.1	209.2	159.3
Agriculture and fishing	17.6	12.6	–28.5	12.6	26.1	4.3	14.5	10.6
Mining industries	0.5	1.0	93.0	0.9	0.7	1.2	1.1	1.1
Manufacturing	26.1	22.7	–13.0	33.3	20.3	17.1	11.4	5.7
Electricity, gas and water	0.9	1.0	7.0	0.8	0.7	1.2	1.2	1.1
Construction	8.8	11.1	26.2	11.5	10.7	9.8	14.5	12.6
Retail, restaurants, hotels	14.9	19.3	29.8	16.5	16.3	22.3	21.1	35.2
Transp./communications	3.8	3.4	–9.4	2.7	2.3	5.1	2.8	3.6
Banking and business services	3.7	5.9	59.6	4.1	3.3	10.3	2.9	6.3
Public administration	7.1	6.1	–14.6	3.1	4.5	8.6	11.6	8.2
Social and personal services	16.6	16.9	1.8	14.5	15.2	20.2	18.9	15.6

Source: INE various dates.

Note
LTV Lisbon and Tagus Valley region.

Table 10.3 Portugal: average company employment and productivity, 1990–98

	Employment/company %			Total sales/company ×Es 1m			Total sales/employee ×Es 1m		
	1990	1993	1998	1990	1993	1998	1990	1993	1998
Total (all sectors)	12.5	11.1	9.1	116.0	128.7	171.9	9.3	11.6	18.9
Agriculture	n.d.	n.d.	5.9	n.d.	n.d.	56.5	n.d.	n.d.	9.6
Mining industries	20.0	17.3	14.8	151.5	145.5	186.8	7.6	8.4	12.6
Manufacturing	29.7	26.3	21.8	221.9	231.6	295.0	7.5	8.8	13.5
Construction	11.8	11.2	8.9	73.8	95.1	129.8	6.3	8.5	14.6
Retail, restaurants, hotels	6.3	6.2	5.7	102.1	122.1	154.2	16.2	19.7	26.9
Transp./communications	19.7	14.6	13.0	152.1	156.0	218.1	7.7	10.7	16.7
Banking and insurance	n.d.	n.d.	43.4	n.d.	n.d.	2,416.6	n.d.	n.d.	55.6
Real estate, business services	5.0	4.7	5.4	41.4	48.2	62.2	8.3	10.1	11.5
Social, personal services	7.5	7.9	5.2	25.7	34.1	39.5	3.4	4.3	7.5

Source: INE various dates.

employee are higher in the tertiary sector, especially in retailing and the hotel trade. In recent years there has been a strong growth in the numbers of companies, particularly in business services. Those sub-sectors which saw significant employment growth also had the greatest increase in turnover, showing effective demand for business, as well as for personal services. Growth in these sectors reflects the modernization of the economy through increasing demand for specialized services from companies and public institutions, as well as for social and personal consumption, including education, health and social security, and welfare protection systems.

Manufacturing companies are concentrated in the North (see Table 10.4), where there is a predominance of small and medium-sized companies involved in labour-intensive activities. Service companies are quite heavily concentrated in the Lisbon area, not only because a large number of offices are located there but also because the region houses most of the head offices of multi-office companies. Proximity to the nation's capital, reflected in better road access, larger market size and better availability of support services, together with the better-qualified workforce in the Lisbon area, are decisive factors for service and retail companies, especially those targeting their business at a national or international level.

In summary, the Portuguese economy in recent years has shown the following characteristics.

1 A rapid increase in the level of qualifications, particularly through improvement in levels of education of the younger population. This trend is visible above all in the Lisbon and Tagus Valley region.
2 Related employment growth in the tertiary sector, through strong increases in the retailing, financial and business service sectors.
3 An increasing percentage of women in the labour market.
4 Increased instability in some industrial sectors, especially small and medium-sized businesses in traditional sectors, which face difficulty in meeting the challenge of the internationalization of the economy.

The supply structure of knowledge-intensive services

The statistical analysis of KIS corporate structure and employment focuses on the category, 'business services' (Economic Classification (CAE) 832), covering a broader variety than formally defined KIS. Data from the Ministério para a Qualificação e o Emprego (Ministry of Employment and Training) do not allow a more precise definition. The analysis focuses on sectoral and regional changes in these business services, including:

• legal services;
• accountancy services;
• data processing;
• technical services (such as engineering, architecture, topography, geology);

Table 10.4 Portugal: regional distribution of companies, 1998 (%)

	North	Centre	LTV	Alentejo	Algarve	Portugal
Total (all companies)	29.9	14.1	44.6	3.4	4.4	100.0
Agriculture	17.0	16.5	38.3	19.0	6.3	100.0
Mining industries	34.7	24.2	25.9	9.1	3.0	100.0
Manufacturing	45.7	18.0	29.7	2.8	1.9	100.0
Construction	29.1	16.2	44.1	2.5	5.5	100.0
Commerce and retail	29.8	14.2	45.4	3.4	4.0	100.0
Restaurants and hotels	24.3	11.8	49.1	3.5	7.9	100.0
Transp./communications	24.9	15.4	48.6	2.4	2.9	100.0
Banking and insurance	23.1	10.0	55.5	2.3	2.9	100.0
Real estate, business services	24.4	9.9	51.7	2.0	5.0	100.0
Social, personal services	26.5	1.3	51.0	3.1	4.1	100.0

Source: INE Statistical Yearbook, 1999.

Note
LTV   Lisbon and Tagus Valley region.

- advertising and marketing;
- other business services.

In the regional analysis, we have divided the country into nine regions as listed in Table 10.5. Overall, business services accounted for 4.6 per cent of total companies and 3.2 per cent of total employment in 1994. Although these percentages are tiny in the European Union context, the actual rates of growth in this sector were very high, with an 86.7 per cent rise in the number of companies and a 58.9 per cent growth in employment during 1988–94.

Corporate structure

The significance of business services within the overall fabric of corporate activity varies considerably from region to region. The Lisbon metropolitan area, and to a lesser extent the Porto metropolitan area and the Algarve, stand out clearly as having the highest representations. By contrast, the importance of these services in the northern coastal region, the inland North and Centre, the West and Ribatejo, and Alentejo is not very great, representing a considerable constraint on the progress of economic modernization of those regions.

Since the late 1980s, the imbalance between the regions has become less marked. The West and Ribatejo and the northern coastal regions had the highest growth rates. These regions are adjacent to the two metropolitan areas and are undoubtedly benefiting from the growth which has taken place there. Nevertheless the geographical concentration of this sector is still very strong. In 1994 the Lisbon metropolitan area accounted for 47.3 per cent of companies and 63.3 per cent of employment in business services.

There are significant variations in the relative sizes of the different segments of business services. Accountancy firms represent about one-third of the sector. Technical and legal services occupy 14.6 per cent and 11.4 per cent of the number of companies respectively. A third group includes advertising and marketing firms with 7.8 per cent and data-processing companies with 3.4 per cent. The remaining services, including such diverse activities as financial information agencies and photocopying services, make up 30 per cent of companies in the sector.

There are marked regional differences in the dominant types of business service activities. In the metropolitan areas, services involving higher levels of qualifications are more significant (Table 10.5). In the remaining regions, there is a clear preponderance of the more common and routine types of services, including non-qualified lawyers and accountants acting as local intermediaries. The degree of geographical concentration is also related to type of demand. Technical services show a fairly uniform pattern of distribution reflecting growth especially in public sector demand, supporting many large infrastructure projects, as well as regional and urban planning and facilities.

The average size of companies in business services is small, and is tending to decrease. In 1989, 58.4 per cent of companies employed fewer than five people and 7.7 per cent employed more than twenty. In 1994 the corresponding figures were 66.8 per cent and 5.2 per cent. Companies are small in all regions of the country, but slightly larger in the metropolitan areas, particularly Lisbon, where 8 per cent of companies have over twenty employees. Sole trader companies and individuals accounted for 22.3 per cent of all companies in 1994. There are some differences between sectors in the predominant types of legal entity. Individuals dominate in legal services, but private limited companies are most common in other sectors. Public companies are quite significant in data processing, technical services, advertising and marketing, and 'other services'. These are nearly all (90.9 per cent) located in the metropolitan areas, with 78.4 per cent in Lisbon and 12.5 per cent in Porto.

Table 10.6 shows 1998 service turnover and salaries information for the more broadly defined regions for which data are available. It confirms the dominant position of Lisbon and the Tagus Valley (LTV) in terms of share of turnover, levels of salaries and staff productivity, and the generally unfavourable position of the central, Alentejo and Algarve regions. While the relatively slower growth of business-related services in the inland regions can be explained by low economic growth, the position of the Algarve is less obvious, since an apparently sophisticated demand from the tourism sector for specialized services is not reflected in the pattern of business service growth, as seems to have happened in the Madeira region.

Table 10.5 Portugal: regional characteristics of business-related service companies, 1994

	Share of sector in regional economy (%)	Share of region in national sector (% of companies)	Growth 1989– 99[a]	Over- represented segments[b]	Over- represented types of organization[b]
Coastal North	2.5	10.0	+ + +	Legal Accounting	Sole trader Individuals
Porto metropolitan area	5.0	14.5	+	Advertising Other services	Private ltd co. Sole trader
Inland North and Centre	2.7	5.5	+ +	Legal Accounting	Sole trader Individuals
Central coastal	3.8	7.0	+ +	Accounting	Private ltd co. Sole trader Individuals
West and Ribatejo	3.2	5.4	+ + +	Legal Accounting	Sole trader Individuals
Lisbon metropolitan area	7.6	47.3	+	Advertising Data processing Technical services Other services	Public cos. Private ltd co. Foreign cos.
Alentejo	2.9	3.1	+ +	Legal Accounting	Sole trader Individuals Foreign cos.
Algarve	4.4	4.5	+ +	Legal Accounting	Individuals
Autonomous regions (Madeira/ Azores)	3.4	2.9	+ +	Legal Accounting Advertising	Private ltd co. Individuals
Portugal (total)	4.6	100.0	—	—	—

Source: INE various dates.

Notes
a growth pattern 1989–99:

+	between 60 and 80%
+ +	between 90 and 120%
+ + +	over 120%

b values higher than the national average.

Table 10.6 Portugal: economic indicators for business-related services, 1998

	Portugal	North	Centre	LVT	Alentejo	Algarve	Azores	Madeira
Turnover/company (10⁶ Es)	116.6	99.9	31.4	158.9	22.2	36.5	32.6	221.2
Turnover/employee (10⁶ Es)	20.1	20.7	10.7	20.5	8.2	10.7	8.1	47.4
Average annual salary (10³ Es)	2,221	1,815	1,365	2,505	1,382	1,574	1,564	1,262
Staff costs/turnover (%)	11.0	8.8	12.8	12.2	16.8	14.8	19.2	2.7

Source: INE 1998.

Employment

There are very high numbers of young people and women in employment in business services. In 1994, 48.9 per cent of employees were under the age of 30 and 37.6 per cent were women. These features are mainly attributable to strong growth in the sector, but marked differences between segments and regions reveal different approaches to employment (see Table 10.7). The youth of the workforce is particularly marked in accountancy and data processing, where many young people start their professional careers gaining initial training.

Regional patterns in turn show the impact of rapid growth in recent years and the types of qualifications required in the various activities. Higher than average numbers of women in employment are clearly associated with routine tasks requiring lower qualifications, although this does not necessarily mean that the work is less specialized. Examples are accountancy, in which women make up 67.2 per cent of the workforce, and especially legal services, with 74.8 per cent women employees, concentrated in clerical and administrative support grades. In most cases these two segments are less demanding in terms of time commitment, geographical mobility and ongoing training than, for example, technical services. The regional patterns of women in employment also reflect higher rates of growth in recent years, especially where more routine jobs predominate. The proportion of women in employment is due to the effects of improved levels of women's education, their lower wage costs and other contractual requirements, as well as lower geographical mobility (André 1996).

In the European context, the average education level in business services in Portugal is low, but it is much higher than the national average. There was a

Table 10.7 Portugal: sectoral and regional employment patterns in business-related services, 1994

	*Youth**	*Proportion of women**	*Levels of education**
Sectoral patterns	Accountancy Data processing	Accounting Legal services	Technical services Data processing
Regional patterns	Northern coastal area West and Ribatejo	Northern coastal area West and Ribatejo Algarve	Lisbon metropolitan area

Source: INE various dates.

Note
* higher than national average figures for business-related services.

significant improvement here between 1989 and 1994. The percentage of workers who had secondary, intermediate or higher educational qualifications rose from 31.4 per cent to 42.7 per cent. This trend goes hand in hand with the increasing recruitment of young people. The metropolitan areas of Lisbon and Porto show a significant advantage related to the easier access to higher education. In contrast, Alentejo has very low levels of educational attainment.

In summary, business service employment has shown a strong growth in the employment of young people and women, associated with an enhanced level of qualifications in the workforce. These characteristics are important for the more dynamic segments of the economy and for raising general educational competence as the basis for professional training.

KIS development and the institutional framework

The development of consultancy activity in Portugal

Consultancy activities are narrower than those encompassed in the definition of business services employed in the above statistical analysis and closer to the world of KIS providers. Until the 1960s they had hardly developed in Portugal. This reflected the country's backwardness and its poorly developed regulation through legislation and technical specification, which created little demand for this type of service. In spite of this, in the 1940s and 1950s large public infrastructure projects and a few multinational companies were instrumental in developing some consultancy activity, mainly in engineering and financial auditing.

The Portuguese economy gradually opened up from 1959 onwards following Portugal's membership of the European Free Trade Association

(EFTA), and more generally during the politically liberalizing phase after 1968. This led to a significant increase in management consultancy work through the 1960s. Economic conglomerates were formed and new foreign-owned companies appeared, while new and innovatory market sectors also grew to cater for the growth of consumption. This created demands for investment and market research studies, and for expertise in corporate organization. The establishment in Portugal of companies such as AC Nielsen dates from this period.

The 1974 revolution put a brake on the economic expansion of the preceding decade, although political and monetary instability encouraged growth of legal and financial consultancy firms. However, the activities of these firms had more to do with specific problems, such as inflation and the legal ownership of companies, than with the promotion of innovation and competitiveness of client companies. By the early 1980s, with relative political and economic stability, new trends came to the fore. There was a concern with energy prices and production quality under increasing international competition. The influence of new information technologies was also growing, although they still focused on data-processing software designed to reduce the cost of routine administrative tasks, including accounting, stock control and sales, and the outsourcing of certain former in-house functions.

The first half of the 1980s was a period of expansion for both Portuguese and international consultancy firms. Price Waterhouse, for example, set up in Portugal during this period. Some of the currently leading KIS providers also set up at this time, although engineering consultancy was already well established. It was after Portugal joined the European Community (EC) in 1986, however, that a significant growth in the number of KIS providers took place. There was a marked increase in foreign investment, including access to EC funds. In more practical terms, the PEDIP programme (Programa Estratégico de Dinamização e Modernização da Indústria Portuguesa: Strategic Programme for the Improvement and Modernization of Portuguese Industry) was in force after 1988. These developments created a favourable climate for consultancy activities. The supply of these services widened and became more sophisticated. Portugal's new position as a member of the European Community, with its strong impact on fiscal, financial and technical regulation, tended to favour the growth of companies specializing in training, quality and process improvement and, not least, on the preparation of applications for Community and national grants. The period between 1986 and 1990 undoubtedly saw the highest growth of KIS companies.

In the 1990s these earlier tendencies were strengthened and a number of new areas began to emerge, including strategic consultancy, re-engineering, systems and environmental consultancy. At the same time, integrated and global solutions became increasingly important, and contributed to a partial blurring of the traditional distinctions between management, process and even product innovation. Three main factors underlie these developments.

First, the progressive liberalization and deregulation of the Portuguese economy, which included re-privatizaion of companies nationalized in the 1970s, and mergers and acquisitions. Second, the increasing role of new information technologies, especially in the widening context of internal and external corporate communications. Finally, a new generation of economic incentives were developed which encouraged the use of consultancies. These will be analysed in the next two sections.

A study of management consultancy firms published in the *Diário de Notícias* special supplement on 24 February 1997, based on the IF4 Company database, emphasized the following features of the sector:

1   It contains about 400 companies, employing close to 4,000 people and billing Es 60 billion (over ECU 300 million) annually.
2   The top ten companies in the sector account for half the total billings.
3   Total billings represent 0.3 per cent of gross domestic product (GDP) (the European average is 0.8 per cent).

The 400 companies identified are essentially the larger companies in the sector, but the results of this study confirm that its significance is more strategic than quantitative. There is a dual corporate structure, with a very unequal distribution by segments. Auditing and information systems stand out in terms of billings and staff numbers, while recruitment and training are more significant for the numbers of companies involved. Also, the multinationals occupy a dominant position in key sectors. Table 10.8 shows the list of the top twenty-five consultancy companies according to this study, their main areas of activity, levels of turnover and numbers of staff.

KIS, territorial models of economic development and internationalization

In view of the regional variability of KIS already noted, the Portuguese development of KIS must be understood in the context of different territorial models of economic development (Ferrão 1997). Table 10.9 shows the relative importance, for Portugal, of four basic territorial models and, for each model, the relative importance of KIS in terms of employment and the types of KIS represented in comparison with the national average.

The scope for the internationalization of Portuguese KIS providers is also limited by their territorial context. Table 10.10 outlines the characteristics of three different territorial contexts which have an impact on the internationalization of KIS: core urban regions, including the Lisbon metropolitan area and, to a lesser degree, the city of Porto; specialized regional markets, including the northern coastal region, the Algarve and local pockets of industrialization, such as the central coastal region; and peripheral regions, in most rural areas and small towns.

Table 10.8 Portugal: top 25 consultancy companies, 1995

Companies	Main activity[a]	Turnover $Es \times 10^6$	Total employees
Edinfor	6	8.020	314
Andersen Consulting	2, 6	5.518	276
Ernst & Young	2	3.800	n.d.
Arthur Andersen	2	3.000	224
Coopers & Lybrand	2	2.800	n.d.
Price Waterhouse	2	2.700	300
Fernave	5	2.421	211
Eurociber	6	1.658	88
EDS	6	1.617	71
Unisys	6	1.568	165
Manpower	5	1.469	43[b]
KPMG Peat Marwick	2	1.458	n.d.
Deloitte & Touche	2	1.381	170
Geslógica	6	1.260	73
ES Data Informática	6	1.230	56
SSF-Soc Software Financeiro	6	1.200	55
SAP Espanha e Portugal	6	770	11
BDO-Binder	2	750	80
Cegoc-Tea	5	700	45
Efacec GI	6	622	63
Consiste	6	599	n.d.
Quatro-Sistemas de Informação	6	579	61
Marktest	4	550	70
Development Systems	6	550	35
Prisma	6	528	45

Source: *Diário de Notícias Especial*, 24 February 1997.

Notes
a The numbers listed in 'Main activity' are the following:
   2 Economic and accounting consultancy and asset valuation
   4 Personnel, training and recruitment
   5 Information systems, software and services
   6 Investment, planning and environmental consulting investment
     Advertising/marketing companies are not included.
b Total permanent employees only.

Table 10.9 Portugal: territorial models of economic development and presence of KIS

Regions	Territorial models of economic development	Employment (%) (continental Portugal = 100%) 1993		KIS over-represented (LQ > 1)
		Total employment	Employment in KIS	
Lisbon metropolitan area	Urban agglomeration – 'eurocity' type	35.4	55.7	R&D Advertising and marketing Information technology
North coastal region	Industrial districts specializing in traditional and labour-intensive activities (textiles, clothing, shoes, etc.)	37.5	31.7	Technical services
Algarve	Mass tourism	3.5	2.8	Real estate
Others	Rural areas	23.6	9.8	—

Source: INE various dates.

Consultancy and innovation: the context of the Community Support Framework, 1988–92 and 1994–99

Investigations carried out in the context of the 1994–99 EU CSF, and the many studies of business innovation in Portugal (GEP-MIE 1994; Simões 1997), point to the common characteristics of Portuguese innovation:

1 Efforts towards innovation have been concentrated on the purchase of new equipment (technical and technological modernization).
2 Per capita expenditure on research and development (R&D) is very low: one-tenth of that in Germany and half that in Spain. Within this expenditure, the proportion representing in-house R&D, 26 per cent, is substantially below the Community (EUR12) average of 65 per cent.
3 The potential for scientific and technological investigation based, for example, on past infrastructure and human resources investment, and qualification levels, are low.
4 Responsiveness to innovation, especially in small–medium enterprises (SMEs), is rare. When it does occur, it is more likely to happen as a result of initiatives taken by suppliers and clients rather than through strategic options chosen by company managers.

Table 10.10 Portugal: territorial contexts for internationalization of KIS firms, 1993

Aspects of internationaliz-ation	Metropolitan area of Lisbon and city of Porto	Specialized regional markets	Peripheral regional markets
Degree	Differentiated, but potentially relevant	Reduced, with possible isolated exceptions	Nil or very small
Type of KIS segments	Engineering/technical services	Technical support services	—
	Management consultancy	Possibly others, related to specialization and development of other local systems of production	
	Information technology		
	Market research		
	Training		
	Transport/logistics		
Comparative advantages	Value for money specialization	Value for money specialization	—
	Local knowledge*		
Main target markets	Portuguese-speaking countries	Portuguese-speaking countries	—
	Spain	Spain (especially frontier regions)	
	Eastern Europe		
	Mercosul		
	Others, depending on location of international tenders		
Prevailing method of market penetration	Export	Export	—
	Direct investment		
	Networks, partnership		
	Cooperation		

Source: Ferrão 1997.

Note
* Local knowledge is considered here to have four main dimensions:
  1   knowledge of the regulatory environment (from the culture of corporate and public administration to current legislation)
  2   knowledge of the local terrain and of sources of statistical and qualitative information
  3   network of personal contacts (inter-personal relational capital)
  4   identification/linguistic similarity.

Given these structural deficiencies, and the prevailing view of technology as the privileged innovation carrier, policies for economic modernization and business innovation have focused on three main elements: technological equipment and infrastructure, human resources, and R&D activities. The attitude of most businessmen to innovation and the structure of the various support programmes have placed a low premium on the role of consultancy firms. In addition, in those programmes where express provision is made for business service support, such as in regional programmes, the results have been very disappointing in relation to the number of opportunities created.

Although the fundamental role of services as a factor in modernizing the economic fabric may therefore have been recognized in the 1980s, there was a prevailing policy view that the market would grow on the basis of the growth of demand, with the state filling any gaps in the supply of relevant services by means of public agencies or by public–private partnerships, particularly in relation to innovative technology. With the exception of technological infrastructure initiatives, and to a lesser extent the provision of information technology, no specific measures were introduced to encourage the service sector until a new generation of policies emerged with the second European Union Community Support Framework (1994–99).

Nevertheless, the earlier programme for the development of Portuguese industry, PEDIP I, between 1988 and1992, showed how the institutional framework might encourage the expansion of KIS by providing new market opportunities (see Table 10.11). At the sectoral level, the expansion of KIS was uneven. Companies providing economic or management consultancy and training services developed faster than those providing production-related services such as marketing support, design, quality control or, in general, technological services. This reflected the incentives available for training schemes and for reinforcing companies' general management capabilities. Along with the economic and financial consultancy work inherent in the application phase of all such programmes, it was these schemes which essentially made up the core of policies that had an impact on demand for KIS.

Under the 1994–99 PEDIP II, a new support programme for industry has been implemented (Table 10.12). This most significantly adopts a broader concept of innovation and, as a consequence, of the processes and carriers which stimulate it. In this context, the role of producer service firms is now seen as potentially significant: 'Innovation is not only product and process technology: it is also the management of information and of technology, it is the qualification of managers and workers, it is design, it is marketing, it is suitable financing'. Also, 'Innovation assumes a climate which is conducive to the appearance, development and exploitation of new ideas: if the underlying culture is not favourable, innovation cannot flourish' (PEDIP 1995: 5).

These statements make clear the fundamental differences between the guidelines of PEDIP II and those which applied in the earlier programme, which are:

1   A greater concern to act on the external environment of innovation, ranging from communication infrastructures to the encouragement of corporate associations.

Table 10.11 Portugal: PEDIP I (1988–92) and the development of the market for KIS

PEDIP I and sub-programmes (incentives in Es ×10⁹)	Link with industrial policy priority objectives	Potential impact on market for KIS		
		Supply	Demand	Environ-mental
1.1 Basic infrastructure (n.d.)	General	–	–	–
1.2 Technological infrastructure (41.1)	Technology policy Industrial quality Human resources and information	+ +[a]	+	+
2 Training (37.6)	Human resources and information Flexibility	–	+ +[b]	–
3.1.1 Purchase and development of technology (7.7)	Technology policy Industrial quality Human resources and information Flexibility	–	+	–
3.1.2 Innovation and modernization (94.3)	Technology policy Industrial quality Flexibility	–	+ +[c]	+
3.1.3 Quality and environmental management (14.7)	Industrial quality	–	+	–
3.1.4 One-off investment in equipment (3.8)	—	–	–	–
3.2 Rational use of energy (3.3)	Energy policy Technology policy Industrial quality Flexibility	–	+	–
3.3 Sectoral restructuring and modernization (13.4)	Technology policy Industrial quality Marketing/sales strategy Flexibility	–	+	–
3.4.1 Service and information industry support, capital goods and vehicles (3.9)	Technology policy Industrial quality Flexibility	+	–	–
4 Financial restructuring (n.d.)	Financial aspects Flexibility	–	+ +[d]	–

Table 10.11 (cont.)

PEDIP I and sub-programmes (incentives in Es ×10⁹)	Link with industrial policy priority objectives	Potential impact on market for KIS		
		Supply	*Demand*	*Environ-mental*
5 Productivity missions (24.5)	Technology policy Industrial quality Marketing and sales policy Flexibility	–	+ +[c]	+
6 Quality and industrial design missions (11.3)	Technology policy Industrial quality Marketing and sales policy Flexibility	–	+ +[c]	+

Source: Based on 'O PEDIP e a evolução da indústria portuguesa: perspectivas de uma avaliação 1988–1992' (PEDIP and Portuguese industry: outline for an assessment 1988–1992), unpublished study drawn up for the Ministério da Indústria e Energia by the Universidade Católica, 1995.

Notes
– insignificant.
+ moderate.
+ + considerable.
a Support for public agencies; public–private partnerships; technology centres, institutes and transfer centres; demonstration units, incubation centres.
b Incentives for training and retraining workforce; improvement of management.
c Incentives for technology, organization and management consulting.
d Incentives for financial consulting.
e Incentives for technology, organization and management consulting, economic studies.

2   A greater emphasis on investment not directly related to production, but to dynamic competitiveness factors, including studies and audits, design, technical assistance, quality, market studies, information, professional training, environment, and health and safety at work.

At the same time there has been a change in the philosophy of state intervention, with fewer grants, less direct support to companies, and greater selectivity in its attribution; more proactive measures to increase awareness and stimulate change; and a greater emphasis on demand-side involvement, to encourage industrial companies to be more selective in their approach to, and use of, specialized service providers.

This approach to innovation and to the philosophy of state involvement in economic modernization has opened up a new role for consultancy firms. Indeed, consultancies are named along with six other types of privileged

Table 10.12 Portugal: PEDIP II (1994–99) and the market for KIS

System of incentives	Priority action areas	Eligible KIS-related activities	Main predictable effects on the market for KIS	Implementation 1994–96
1 Sindepedip: industrial enterprise strategy incentives	Company evaluation	Strategic analysis Investment analysis Partial economic studies Audits	On demand for management and financial consulting services, auditing, and economic studies	Applications: 3,404 Approved projects: 2,076 Investment (Es 10^9): 543.1 Incentives (Es 10^9): 173.6
	Research and development	R&D preliminary studies R&D projects	On demand for technology consulting services	
	Integrated company strategy	Diagnostic work Innovation and internationalization projects Strategic project contracts	Demand for management consultancy, economic/market studies, marketing, advertising	
	Smaller SMEs	Promotion of growth Support for modernization	On demand, as above	
	Promoting industrial quality	Certification and calibration Support for total quality management implementation	Supply of production support services, quality control; certification Demand for management consultancy services	
	Assistance with access to capital	Assistance with access to capital	Demand for financial and management services	

	Productivity and company presentation; Promotion through case-study success; Advanced technology demonstrations; Several; Intercompany cooperation	Supply of technological services; Demand for economic/market studies; General demand	
2 Sinfrapedip: support for technological/quality infrastructure	Consolidation of technological infrastructure: technology centres; technology institutes, transfer centres, parks; and industrial quality: companies, associations, universities, etc.	Supply of technology and quality management services	Applications: 206; Approvals: 149; Investment (Es 10^9): 23.1; Incentives (Es 10^9): 8.7
3 Sinaipedip: incentives for industrial support services	Consolidation of capabilities of associations; Reinforcing technical capability of consultancy firms, support and technical assistance to industry	Supply of various types of KIS	Applications: 178; Approvals: 101; Investment (Es 10^9): 12.5; Incentives (Es 10^9): 6.4
4 Sinetpedip: incentives to strengthen technological colleges	Consolidation of capacity of technological colleges	Supply of training services	Applications: 7; Approvals: 7; Investment (Es 10^9): 12.7; Incentives (Es 10^9): 9.9
5 Sinfepedip: incentives for financial restructuring	Corporate financial restructuring plans	Demand for financial and management services	Applications: 25; Approvals: 19; Investment (Es 10^9): 41.1; Incentives (Es 10^9): 16.6
Other measures	New industrial support infrastructures; Innovation and transfer of technology; Human resources improvement	Demand for technological and training services	Applications: 84; Approvals: 50; Investment (Es 10^9): 22.6; Incentives (Es 10^9): 15.0

Source: Data from the PEDIP Management Office for September 1996.

innovation carriers, including companies; owners and managers; R&D institutions and companies; institutions responsible for the transfer of technology and the interface between knowledge creation centres and companies; educational and professional training establishments; and financing entities. One of the incentives systems created under PEDIP II, SINAIPEDIP (Sistema de Incentivos de Apoio à Indústria PEDIP: Incentive System for Industrial Support Services PEDIP) has as its objective precisely the encouragement of the provision and utilization of this type of service: 'Some of the industry support service providers will be encouraged to improve their competence in various different technical areas and to obtain relevant qualifications, in order to provide services to industrial companies, especially SMEs' (PEDIP 1994: 5).

Support for these consultancy activities is to be further encouraged through the assistance which the programme can provide in creating a market by subsidizing up to 75 per cent of the costs to industrial enterprises of acquiring these services.

The SINAIPEDIP sub-programme finances three of the measures contained in the PEDIP II programme:

1   *Measure 1.4:* Support for cooperative associations, directed at consolidating their ability to take part in the processes of modernization undertaken in the context of the programme.
2   *Measure 1.6:* Directed towards 'other industry support services', aimed at strengthening the technical ability of service providers whose main activity is the provision of support and technical assistance to industry, with an impact on the capacity for innovation.
3   *Measure 5.1:* Promotion of professional training courses for auditors with the capacity for promoting innovation.

This group of measures and support initiatives represents formal recognition of the role which consultancy firms can and should play as innovation carriers.

Taking into account the amounts involved, in the form of direct public investment or financial incentives, and the impact on the market for KIS, the PEDIP programme is likely to continue making a major contribution to the further development of KIS in Portugal.

The involvement of the state, however, is not limited to PEDIP II. Other areas of government regulation, many operating within the EU Community Support Framework, are also influential for the development of KIS (see Table 10.13), especially their role in maintaining or helping to expand specific KIS markets. Training and industrial retraining schemes, as well as regional policies, are prominent here. These schemes combine investment incentives for SMEs in economically less-favoured regions with direct action on the part of the state administration, and they undoubtedly stimulate demand for KIS.

Table 10.13 Portugal: EU Community Support Framework, 1994–99: outline of policies impacting development of KIS

Regional development plan guidelines	Operational involvement	Areas (estimated public investment 1994–99 in ECU×1 million)	Impact on the market for KIS		
			Supply	Demand	Environmental
1 Qualification of human resources and employment	Innovation and knowledge base	Science and technology (524,969)	+		+
	Training and employment	Starting qualifications and job-finding (787,326)		+ +	
		Improvement in level and quality of employment (759,989)		+	
		Training and management of HR (288,000)		+ +	
		Training for public administration (68,000)		+ +	
2 Reinforcing the competitive factors of the economy	Modernization of economic fabric	Agriculture (4,873,245)		+	
		Industry (4,460,350)	+ +	+ + +	+ +
		Tourism and heritage (1,159,872)		+	
		Retail and services (838,000)		+	
		Community programmes Retex and Textile (541,867)		+	
3 Promoting quality of life and social cohesion	Environment and urban renewal	Environment (346,666)		+	
		Urban renewal (487,053)		+	

(continued on next page)

Table 10.13 (cont.)

Regional development plan guidelines	Operational involvement	Areas (estimated public investment 1994–99 in ECU×1 million)	Impact on the market for KIS		
			Supply	Demand	Environmental
4 Strengthen the regional economic base	Promoting potential for regional development	Rural and local development (272,596)		+	
		Regional incentives (637,200)	+	+ +	
	Action at regional level	Regional operating programmes (3,603,964)	+	+	+
		Technical assistance to the whole of the Regional Development Programme (216,000)		+ +	

Source: Based on SEPDR 1993.

Note
+ low   + + moderate   + + + high.

The dynamics of KIS use

The market context

The Portuguese market for KIS is still relatively weak (Ferrão and Domingues 1995). Nevertheless, there has been significant development in recent years, to a great extent as a result of Portugal's integration into the framework of the European market, and, in more general terms, as a result of the internationalization of the Portuguese economy. These factors have encouraged new players in the KIS market and attracted branches of the multinational service companies; furthermore, significant changes have occurred in patterns of demand, resulting in particular from:

1   the development of demand itself, as new players stimulate new types of demand and at higher quality;
2   entry into the market of economic players who are more demanding, particularly those who are more highly qualified;
3   increased outsourcing by public administration;
4   incentives towards the modernization of Portuguese producers, especially at the levels of management, technology and corporate organization;
5   in general terms, the new competitive environment both nationally and internationally.

Demand patterns pose the greatest barriers to the further development of KIS, but their level is very different according to the type of KIS. There are also great variations between sectors and within sectors as a result of the differences between companies. Overall, the relatively widespread services are most in demand. A recent study, based on research into industrial and service companies, showed that only accountancy and taxation, and software and personnel training services could be classed as having more than 'low usage' (Godinho 1996). There was a low level of demand for specialized services, but also a direct relationship between KIS demand and the degree of technological and organizational development of client companies, often related to their size (Delgado 1996, 1997; Ferrão 1992).

Demand for KIS: case studies of agriculture, manufacturing and services

Seven case studies, relating to different areas of economic activity, offer a comparative overview of demand for KIS. In agriculture, two regions were chosen: the Algarve in the extreme south of the country, and the central border region in the middle of the country. Although these two are structurally different, they are both relatively modernized systems in the context of Portuguese agriculture as a whole, and have a clear commercial orientation. Three industrial case studies reflect different situations, at both regional and sectoral levels:

1 The industrial system of Castelo Branco, in the inland centre of the country. It is a region traditionally dominated by the textile industry where there has been substantial recent development, essentially as a result of the establishment of branches of Portuguese and international companies in the refrigeration and cabling sectors.
2 The ceramics sector, with characteristics very similar to most of the traditional sectors of Portuguese industry, including a predominance of SMEs, highly specialized areas of production, location along the western coastal strip, a low level of technological sophistication and production directed essentially towards export.
3 The automotive components industry, including companies in such diverse sectors as textiles, glass, plastics, electronics and metalwork, which has grown considerably in recent years as a result of the establishment of large assembly plants, especially the mega-investment of Auto-Europa (Ford-VW) which has been earmarked as one of the new areas of industrial specialization for Portugal.

Finally, the two case studies in the services sector – international transport and management consultancy – represent tertiary sector activities that show strong expansion and ever-increasing internationalization, but which are clearly differentiated in the qualifications offered and the amount of knowledge which they embody.

Table 10.14 Portugal: utilization of KIS in seven case studies

Type of KIS	Central border region	Algarve	Castelo Branco	West and Lower Vouga region	All Portugal	Greater Porto region	Greater Porto
	Agriculture	Horticulture and fruit farming	Industry	Ceramics	Automotive components	Transport	Management consulting
Management/administration	+	+	+	+	++	+	+
Production	+	+	+	++	++	−	−
Research	+	−	−	+	+	−	−
Human resources	+	−	+	+	+++	++	+++
Information/communication	−	−	++	++	+++	++	+++
Marketing	−	−	++	+	++	++	++

Source: Table devised by authors.

Note
Strength of demand:
+ low;
+ + moderate;
+ + + high.

The general characteristics of KIS demand in the different cases, including its relatively low level and the significant variations in pattern and strength of demand as between different sectors, are summarized in Table 10.14.

Demand for KIS in agriculture

The two cases examined here show the general weakness of KIS in this sector (Moreno 1992; Pereira 1994). Consultancy in agricultural management is quite an important field in the context of demand for KIS in Portugal. On the whole, the more common services are employed, based on local demand, such as accounting organization and elementary management training. Utilization of services requiring higher qualifications, such as economic and market studies, management consultancy or investment advice, is restricted to a small number of large agricultural companies. Demand for technical services is also of some significance. Traditionally, services such as soil sampling, crop selection, or the introduction of new crops and new cultivation methods, were provided by the state. Over time, private companies, generally at the local level, have taken them on. Again, services requiring more highly qualified personnel, such as the development of new crop types, are provided by national or foreign companies, especially from Spain, and their use is restricted to a small number of larger agricultural companies.

The use of other services, such as information and marketing support, is also significant. However, in both the Algarve and the central border region, as in Portuguese agriculture generally, this demand is met essentially by cooperative or state entities.

Demand for KIS in manufacturing

KIS utilization in manufacturing is strongly affected by sectoral characteristics, in particular by the degree of technological development, and by the prevailing features of company structure (Costa 1992; Feio 1997; Vale 1997). In the traditional sectors and among SMEs, which are generally poorly organized, the more common locally available services are predominantly used, which include accountancy, taxation and legal consultancy, or repair and maintenance of equipment. In larger companies these functions are provided in-house. Demand for more specialized management support services, such as consultancy and research studies, is restricted to larger companies or to companies which are part of a Portuguese or foreign economic group. For technical services, SMEs join this group, where they have aggressive market penetration strategies, developed either in isolation or cooperatively, as do most of the components industry firms. These are encouraged by their clients who contractually require them to maintain a high standard. Both of these types of companies are more demanding of specialized technical services, including automation and technology consultancy, quality control and certification, and the concept and design skills required of product innovation.

This differentiation is also visible in the demand for services relating to human resources, information and communication, and marketing. Only the more common services, including basic training and computerization, show moderate demand, sometimes complemented by companies' own resources, state or cooperative agencies. The more advanced services are used more selectively, and include recruitment, specialized professional training, complex IT and communications solutions, advertising, market research, and services to export.

Demand for KIS in services

The two sample types of services generally show a high level of demand for KIS (Domingues 1993), since they are themselves highly specialized areas of activity. In general terms, management consulting services show higher and more selective levels of demand for KIS, as is to be expected of a field which is itself knowledge intensive, and where the pace is set by large corporations, many of them multinationals, with well-established organizational structures. KIS utilization is particularly high in the areas of human resources, information and communication, and marketing.

The pattern of demand for KIS in international transport services is similar, though at a lower level of qualifications. Both for management and administration, and for training and information, the more common services are more visible, reflecting a greater diversity of company types in which large Portuguese or multinational companies coexist with small operators whose organizations are more fragile. It should also be noted that the internationalization of these two fields, both through the presence of multinationals and with Portuguese companies being active abroad, is also reflected in their use of KIS outside Portugal.

Conclusion

Although demand for KIS in Portugal is still weak compared to demand in the more developed countries of the European Union, it has grown strongly in the last few years. In general terms, and even though there are considerable differences between types of companies, sectors and regions, the following factors, ranked in decreasing order of importance, have been instrumental in achieving this growth:

1   A stricter regulatory environment.
2   The introduction and expansion of information technology and information systems.
3   Pressure from increasingly knowledgeable and highly qualified clients.
4   Proactive differentiation and market repositioning strategies.

However, these developments have taken place more as a result of externally imposed factors, especially the first three of these, than by reason of voluntary strategic choices made by the companies themselves.

Figure 10.1 attempts to show, in a necessarily simplified form, the relation-

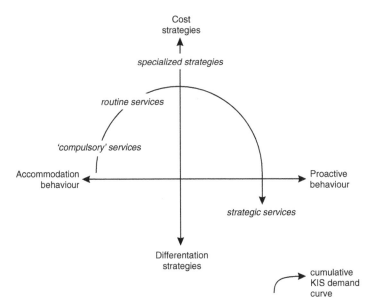

Figure 10.1 KIS demand and strategic attitudes

ship which appears to exist between the pattern of service use by companies and their prevailing strategic and philosophical approach.

In most cases in Portugal (the upper left quadrant) there is an attitude of resigned acceptance which is reflected in an infrequent use of KIS, essentially limited to 'compulsory' and relatively routine tasks. For companies whose strategic vision is more proactive, but predominantly based on cost-control measures, specialized services are used in addition to the 'compulsory' and routine services. It is only for those companies with a proactive strategic vision based on differentiation that the prevailing pattern is one which seeks out a range of different types of services, from the 'compulsory' to those of a more strategic nature. An increasingly strict regulatory framework, especially after Portugal joined the EC, was therefore a first stimulus for many companies to cease pursuing a merely defensive policy of accommodation with the status quo, and to start appreciating the advantages of using KIS.

The regulatory role played by the state, combining the implementation of certain requirements, such as strategic studies, audits, quality control, environmentally friendly solutions, with a subsidy policy which has assisted companies in gaining access to KIS providers, has contributed decisively to the consolidation of the consultancy market. The question which remains is whether this has in fact led to the adoption of new attitudes to the use of KIS or whether it merely generated reactions to a specific set of policy circumstances. This question is important because current subsidies will tend to be reduced and will eventually disappear. If this context helps to consolidate positive attitudes to demand for KIS, then a sustainable market for these types of services will have been developed, including an appetite for innovation and recognition of its needs. If, on the other hand, current demand is viewed merely as a means to an end by KIS clients, and not really a strategic option, then the decline in public support could lead to a significant reduction in the market for KIS.

There is a general consensus that the favourable context for the development of consultancy activities in recent years has led to the proliferation of service companies with poor quality standards. These companies took advantage of an opportunity based on guaranteed markets and a not too demanding clientele. The future of KIS-related activities and the role they will effectively play as innovation carriers for client companies therefore requires several measures to ensure the overall quality of KIS providers, whether directly or indirectly. Examples of such measures include:

1   Administrative measures developed by the state, such as formal accreditation schemes.
2   Publicity for 'best practice', including promotional seminars describing successful case studies, the publication of results to show which KIS providers are best capable of carrying out certain tasks, or publicizing KIS databases together with the names of their clients.
3   A greater role for consultancy firms' associations, particularly in defining professional ethics and guidelines.

Regardless of the precise measures to be taken, it seems obvious that the state should continue to promote attitudes and practices which contribute to the consolidation of quality in the KIS market. This requires moving away from the individual project-based approach, which currently prevails, to a philosophy of ongoing specialized assistance to clients and towards client retention. In fact, much of the easily identifiable demand for KIS continues to be associated with preparing applications for subsidy schemes or with similar one-off events. This view of KIS as a mere means to an end will have to give way to a more lasting relationship between KIS firms and their clients. International experience suggests that this is an essential prerequisite for consultancy activities to become effective innovation carriers. The availability of better-quality services and a concern to develop more interactive corporate relationships will affect the way in which the consultancy market develops over the next few years.

In general terms, the supply of KIS tends to be based on four types of entity. These in turn link in with different segments of clients' activity.

Client companies with higher strategic capability and resources will essentially use:

1   *Global consultancies*   Branches of multinationals or companies which are part of an international network, able to sell to their clients integrated and global solutions (strategy, process improvement and implementation

capability), to develop their own research and to stay alongside their clients for considerable periods of time (up to ten years).

2 *Niche consultancies* Portuguese or international companies which provide highly specialized services, particularly in the process area and, for that reason, have a strong technological base. Consultancies of this type are often subcontracted by the global consultancies. Client companies with lesser strategic capability and fewer resources will tend to favour the third type of service provider.

3 *Proximity consultancies* Generalists or consultancies with some degree of specialist technical knowledge. They have a regional base and intimate knowledge of the corporate environments in which they operate.

4 *Public, quasi-public and associative bodies* These can provide a full service directly, or may simply carry out first-stage diagnostic work and give preliminary advice, with a view to identifying needs and options. They thus transform hidden demand into effective demand which may subsequently be channelled to the private sector. Institutions of this type play a crucial role in regions or sectors, including SMEs which are economically fragile.

These types of situation seem to point to strong segmentation in the KIS market, with four typical potential client sectors in respect of which national and European Union policy should adopt different approaches:

1 Companies which are completely outside the scope of KIS and which it would be extremely difficult to bring into that scope for reasons connected with the personalities of the owner and/or the company. In these cases, a careful assessment needs to be made, on the basis of value-for-money criteria, as to whether any support mechanisms at all should be developed.

2 Companies which have already used KIS, or may do so in the future, but only on a one-off basis and mainly for 'compulsory' or relatively routine services. To consolidate this sector the state should positively promote deeper interaction between KIS firms and their clients. If this does not happen, there is a risk that any impact in terms of innovation will tend to dissipate over time, becoming eventually marginal or even completely irrelevant.

3 Companies which use a varied range of services provided by qualified consultants, but may cease doing so when subsidies have disappeared. This situation has to be addressed by means of strategies and actions aimed at providing stability in those cases where there is potential but the client remains economically vulnerable.

4 Companies with a track record of sustained use of specialized and strategic services and for this reason are able to take advantage of structural events in terms of product and process innovation in a framework of competitive-enhancing change. In the future neither these companies nor the

consultancies they use should receive direct state support other than in exceptional circumstances.

Any wider role for KIS as innovation carriers thus depends heavily on the capacity of the state to stimulate an entrepreneurial culture among SMEs, sensitive to the requirements of modern competitiveness. KIS will then become accepted as an investment, rather than a cost to be avoided.

References

André, I. (1996) 'At the centre on the periphery? Women in the Portuguese labour market', in Garcia-Ramon, M.D. and Monk, J. (eds) *Women of the European Union, the Politics of Work and Daily Life*, London: Routledge, 138–55.

Costa, M.E. (1992) 'Reestruturação económica e desenvolvimento local: o caso de Castelo Branco', master's thesis, University of Lisbon.

Delgado, A.P. (1996) 'Serviços à produção e indústria: o exemplo do noroeste de Portugal', *Sociedade e Território* 23: 81–94.

—— (1997) 'Les PME locales et les flux externes de services: quelles opportunités pour le développement des centres urbains de dimension moyenne?', *RERU* 1: 23–48.

Domingues, A. (1993) 'Serviços às empresas – concentração metropolitana e desconcentração periférica', PhD thesis, University of Porto.

Feio, P.A. (1997) *Território e competitividade: uma perspectiva geográfica do processo de internacionalização do sector cerâmico*, Lisbon: Edições Colibri.

Ferrão, J. (1992) *Serviços e inovação: novos caminhos para o desenvolvimento regional*, Oeiras: Celta Editora.

—— (1997) 'Empresas prestadoras de serviços intermédios e internacionalização', *Economia e Prospectiva* 1: 125–34.

Ferrão, J. and Domingues, A. (1995) 'Portugal: the territorial foundations of a vulnerable tertiarization process', *Progress in Planning* 43(2–3): 241–60.

GEP-MIE (Grupo de Estudos e Planeamento, Ministério da Indústria e da Energia – Department of Studies and Planning, Ministry of Industry and Energy) (1994) *O PEDIP e a evolução da indústria portuguesa: perspectivas de uma avaliação 1988–1992*, Lisbon: GEP-MIE.

Godinho, M.M. (1996) 'Mercado de serviços técnicos e de informação para negócios: inquérito revela procura incipiente', paper presented at the conference Dez Anos de Integração Europeia: Portugal, a Indústria e o Papel das PME (Ten Years of European Integration: Portugal, Industry and the Role of SMEs), Lisbon (Centro de Documentação Europeu Jacques Delors), 25–26 January 1996.

INE (Instituto Nacional de Estatística) (various dates) *Regional Statistical Yearbooks*, Lisbon: INE.

—— (1998) *Estatísticas das Empresas* (Business Statistics), Lisbon: INE.

Moreno, L. (1992) 'Informação na agricultura algarvia: os anos oitenta', master's thesis, University of Lisbon.

PEDIP (1994) 'PEDIP II: um instrumento ao serviço da competitividade da indústria portuguesa', PEDIP Working Paper 1, Lisbon: Ministério da Indústria e da Energia.

—— (1995) 'Inovação na indústria portuguesa, imperativo vital para a competitividade', PEDIP Working Paper 3, Lisbon: Ministério da Indústria e da Energia.

Pereira, T.A. (1994) 'Serviços e reestruturação produtiva: utilização de serviços pelas explorações agrícolas da raia central e desenvolvimento regional', PhD thesis, University of Lisbon.

SEPDR (Secretaria de Estado do Planeamento e do Desenvolvimento Regional) (1993) 'Preparar Portugal para ol século XXI', Regional Development Plan 1994–99, Lisbon: Ministério do Planeamento e da Administração do Território.

Simões, V.C. (1997) *Inovação e gestão em PME*, Lisbon: GEPE, Ministério da Educação.

Vale, M. (1997) 'Reconfiguração espacial da indústria automóvel em Portugal', paper presented at the conference Novos Modelos de Produção na Indústria Automóvel (New Production Models in the Portuguese Automotive Industry), Lisbon (Faculdade de Ciências e Tecnologia da Universidade Nova de Lisboa), 7–8 February 1997.

11 Spain

Knowledge-intensive services: a paradigm of economic change

Luis Rubalcaba-Bermejo and Juan R. Cuadrado-Roura

The national context

The current situation of the Spanish economy arises from the significant changes experienced by the country over recent decades. Forty years ago, a semi-autarchic system still persisted, based on post-war isolation from the rest of the world under the Franco regime. After 1959, a period of liberalization and opening-up to the outside world was initiated, later incorporating four decisive processes: the transition towards democracy between 1975 and 1978; the entry of Spain into the EC in 1986; the economic convergence required by the Treaty of Maastricht in 1992; and the successful fulfilment of the conditions established for the entry to European Economic and Monetary Union in 1998. The impacts of all these positive steps seemed to come together in the 1995–2000 period. Economic growth rates were higher than in most European countries, private sector dynamism was growing, unemployment rates were significantly reduced, the public sector was restructured and regulation reduced.

In just over twenty years Spain has moved from a social, economic and political system which was very different from the countries surrounding it, to one more clearly in line with the other EU economies. This change, which some have called the 'Spanish miracle', has its roots in a model democratic transition as well as in the opening-up to European-orientated aspirations shared by the Spanish people and its main political leaders. The marked improvements in the Spanish economy have nevertheless not occurred in a linear, uniform or automatic way. They were not linear because there were serious crises in the second half of the 1970s and early 1980s, echoed again in the short but intense recession of the early 1990s. Improvement has not been uniform because integration with more advanced economies saw increasing unemployment and the persistence of serious structural problems (Fuentes and Barea 1997). Finally, developments have not been automatic because economic policy has not always been successful, especially in exploiting the expansion period at the end of the 1980s (Velarde 1996; Tamames 1995; Myro 1996).

The impacts of Spanish integration into the EU have undoubtedly been positive, establishing favourable conditions for economic growth. However, major negative features remain. Income is still generally below 80 per cent of the EU gross domestic product (GDP) per capita and the unemployment rate is still over 15 per cent of the population capable of work. The situation in Spain seems to confirm the theory that there are limits to the degree of convergence within the EU (Raymond 1995). Further structural reforms are required to make some factor and product markets more dynamic and to stimulate growth through investment in infrastructure, innovation and research (Parejo Gamir *et al.* 1995; Koedijk and Kremers 1997).

This diagnosis is reinforced by an analysis of regional differences in Spain. The strong economies of the Mediterranean regions and the Ebro Valley, including Catalonia, Valencia, Aragón, Navarre and La Rioja, contrast with the industrial problems of the Cantabrian regions, the Basque Country, Asturias and Cantabria, and the persistence of serious structural and unemployment problems in Andalusia and Estremadura. Moreover, these differences do not seem to be diminishing. Recent research (Cuadrado *et al.* 1998) indicates that, despite all the efforts to reduce economic disparities, there has been no significant process of regional convergence. The regional roles of knowledge-intensive services (KIS) are therefore likely to be very different, depending on relative regional economic strengths, processes of industrial reconversion, and the centrality to them of information and innovation systems, for example in Madrid.

The innovation context

Several indicators may be used to study the context of innovation in Spain. Research and development (R&D) spending is the most common, although this excludes many of the most important innovation processes, including the inputs of KIS. Spending percentages on R&D in Spain have traditionally been among the lowest of developed countries, at 0.9 per cent of GDP in 1999 compared with the EU average of 1.81 per cent (OECD 2000a). Spanish R&D spending still does not reflect the economic advances of the country in recent years, or its current economic growth. This reduces its competitive capacity, also hindering the assimilation of the advanced KIS based on knowledge and innovation. The comparative backwardness of Spain is compensated for to some degree by the rapid assimilation of foreign innovation and technology and by export earnings from some services, based either on low to medium technology, as in tourism, or intensive creative capacity, for example in advertising design.

It has often been emphasized that the positive effects of R&D on Spanish economic growth are reduced by a series of conditioning factors (Martín 1988; Molero and Buesa 1995). As far as business innovation is concerned, few companies carry out R&D to a high level. The majority resort to importing technology, leaving responsibility for indigenous R&D to the state. There have been low spending levels in the public domain, however, with no scientific or innovation policies until the 1980s. Finally, the problems of developing innovation in Spain are particularly severe among small–medium enterprises (SMEs). These lack financial support for innovation, information about its potential and advantages, and cooperation among themselves. They also have limited ability to acquire and benefit from technology. These various limitations are made worse by the regional concentration of innovation activities. For instance, Madrid alone takes up 40 per cent of the national R&D spending (Rubalcaba *et al.* 1998).

From the entrepreneurial point of view, the research carried out by the Círculo de Empresarios (Businessmen's Circle) in 1995 is illustrative, based on a survey of 500 major companies. Innovatory firms are normally among the largest (more than 1,000 employees), although some employ fewer than 500 workers. Most are integrated into multinational groups, based in Madrid or Catalonia. Fifty-four per cent carry out product innovation, 19 per cent pure process innovation and 27 per cent combine both types. In general, innovation seems orientated more towards traditional technology and communications than towards the development of information and knowledge processes.

Demand for KIS

Within this context, KIS have a special significance. The economic growth experienced by Spain and its opening-up to the outside world have coincided with the consolidation of the service economy (Cuadrado and del Río 1993; Rubalcaba 1997). Advanced business services are paradigmatic activities, reflecting the organizational changes faced by Spanish industry. In the new competitive context, Spanish companies need to be more flexible, to transfer fixed into variable costs, emphasize quality, develop economies of scale through processes of specialization, and enter new international markets. To achieve all this they need the help of specialist services, and particularly of the most advanced, innovatory and knowledge-intensive services.

On the other hand, KIS in Spain are still underdeveloped. At first, advanced service growth took place internally within the larger manufacturing and service corporations. Since the late 1980s, however, independent KIS functions have become more important, associated with processes of outsourcing and externalization. Nevertheless, surveys show that these processes have been orientated more towards routine services than towards those of high-technological value or knowledge (Cuadrado and del Río 1993; Mañas 1992). In spite of the increasing need for such knowledge, there is a dominant culture, especially among SMEs, in which companies tend to subcontract more to reduce costs than to improve processes and quality.

The pilot survey on services to manufacturing carried out by the Ministry of Industry and Energy (INE 1995), including 401 companies from diverse manufacturing sectors, provides an analysis of business service demand complementing previous research (see Table 11.1). Most companies provide

Table 11.1 Spain: use and degree of outsourcing of services by sample of 401 manufacturing firms, 1995

Services	% used	% outsourced
Management	41.8	65.2
Production	31.5	42.8
Information and communications	48.4	61.3
Personnel	35.6	50.4
Financial	51.8	74.9
Sales-related	23.1	54.1
Research and development	21.4	39.4
Operative services	50.3	82.3
Total		60.1

Source: MINER 1995.

specialist services for themselves, but many are externalized. The highest
degree of externalization was found in smaller companies, in highly developed
industrial regions, and in sectors such as paper, publishing or electronics. The
services that are most regularly externalized (82 per cent of cases) are 'opera-
tive services' (including cleaning, surveillance and security, business catering,
transport), followed by financial services (75 per cent). Lower levels of outside contracting characterize some of the medium-knowledge services associated with production, personnel and sales. A low level of in-house R&D is also
poorly compensated for by outside support. KIS thus still have considerable
potential for further externalization. This conclusion has recently been confirmed from input–output analysis. Business services in general, and KIS in
particular, are strategic activities for both manufacturing and service industries, rather like energy, electronics or commerce provision. Outsourcing is
important but by no means the key basis for KIS demand (Cuadrado and
Rubalcaba 2000).

Three statements thus summarize the current role of KIS in Spain.

1   They have only recently developed, mainly in relation to the needs of
     major companies and a limited public sector, although demand from both
     of these is growing rapidly.
2   Demand is also highly polarized with a focus among most SMEs on rou-
     tine, cost-cutting uses of KIS.
3   There is little use of KIS support in innovation/SME policies, even
     though theoretically there is wide scope for their use to support such
     policies.

The rest of this chapter reviews the recent rapid growth of KIS supply capacity in Spain, and the potential for KIS support in relation to economic development and innovation policies at European, national and regional levels. The actual and potential scope for interaction of KIS supply and demand in support of regional policies will then be examined in three case studies, in Madrid, Valencia and the Basque Country.

KIS supply structure and trends

KIS supply

The classification of KIS used here is based on a 1993 survey by the Instituto Nacional de Estadística (INE: National Institute of Statistics) of consultancy services, the annual reports of the Asociación Española de Consultoría (AEC: Spanish Consultancy Association), and an INE 1995 survey of computer services. The INE survey of consultancy services included four groups of activity: management consultancy; market research; accountancy, tax advice and auditing; and labour recruitment. The AEC reports concentrate on various activities related to management consultancy: consultancy; management training, selection and recruitment; information technology; the development and installation of information systems; and market research. These classifications do not cover all KIS in the Spanish market. For example there is insufficient statistical or comparative information on electronic communication, R&D, trade fairs and exhibitions, and public relations.

Table 11.2 shows the main results of the INE studies. The combined surveys examined over 40,500 companies, representing a total turnover of more than Pta 1,200 billion (around ECU 7.3 billion; 2 per cent of GDP). Approximately half of this was in computer services. Most of the companies were based on individual entrepreneurs (55.6 per cent), in accounting and tax consulting (60.7 per cent) and to a lesser extent in computer services (55 per cent). On the other hand, limited liability companies are more common in market research (82.7 per cent), management consulting (60.2 per cent) and labour recruitment (67.1 per cent). This dominance is even greater when measured by financial turnover, about 94–98 per cent. A similar share of turnover is held by private companies in the computer services, although these include a much higher proportion of individually based enterprises.

Almost 85 per cent of the measured KIS companies have a turnover less than Pta 25 million (ECU 0.15 million) and only 3.2 per cent generate more than Pta 100 million (ECU 0.61 million) per year. The latter, however, represent 63.3 per cent of the total market turnover, while the former only about 18 per cent. The market thus tends to be dominated by large companies. Similar conclusions can be drawn from an analysis of employment data. More than 93 per cent of companies have fewer than ten workers, although these support only 35.5 per cent of total turnover.

Table 11.2 Spain: supply structure of KIS

	Computer services	Accounting / tax experts / auditing	Market research	Management consultancy	Labour recruitment	Total
Number of enterprises	12,934	23,025	1,314	2,789	612	40,674
of which % individuals	55.0	60.7	17.3	39.8	32.9	55.6
of which % companies	45.0	39.3	82.7	60.2	67.1	44.4
of which % <Pta 25m	83.6	87.2	70	76.7	72.7	84.6
of which % >Pta 100m	5.1	1.3	10.8	6.2	8.2	3.2
of which % <10 workers	93.5	94.1	90	92.4	74.9	93.4
of which % >10 workers	6.5	5.9	10	7.6	25.1	6.6
Turnover	626,372	376,041	69,015	109,791	27,422	1,208,641
of which % individuals	4.3	28.5	2.1	6.5	4.8	11.9
of which % companies	95.7	71.4	97.9	93.5	95.1	88.1
of which % <Pta 25m	8.5	36.8	13.9	13.3	11.9	17.9
of which % >Pta 100m	81.0	31.8	70.6	67.2	66.2	63.3
of which % <10 workers	19.2	57.0	43.5	40.6	74.2	35.5
of which % >10 workers	80.8	43.0	56.5	59.4	25.8	64.5
of which % secondary activities	—	7.3	4.3	7.8	7.9	7.1
Exports	5.1	1.1	7.1	3.9	1.4	3.8
Imports	9.0	2.6	1.1	7.0	0.6	6.5
Employment	65,554	79,408	11,373	11,360	13,938	181,633
of which % employees	85.4	76.7	92.2	81.9	96.9	82.7
of which % part-time	10.0	11.3	21.7	10.4	23.7	12.4
of which % women	32.7	46.9	62.9	40.9	64.6	43.7

Sources: INE 1993, 1995.

The table also demonstrates a higher level of imports (6.5 per cent of total turnover) than exports (3.8 per cent), especially among computer services and management consultancy, which both seem to be most open to outside influence (the sum of imports and exports representing around 14 per cent and 11 per cent of market turnover, respectively). Market research is notably more export orientated than the other activities. All five consultancy activities also concentrate on their core area of expertise, with only about 7 per cent of turnover generated by secondary activities.

Finally, the employment distribution is generally balanced between men and women, except in the case of computer services, in which about 68 per cent of employees are men. Women nevertheless represent a higher proportion of salary earners, at 56 per cent, than men at 44 per cent. Women also have more than 68 per cent of the part-time jobs. Part-time jobs represent around 12.4 per cent of the total, although in market research and labour recruitment this share is above 20 per cent.

Recent trends

As indicated earlier, the KIS market has changed markedly in Spain in recent years. The consolidation of economic growth and internationalization have contributed to the development of advanced service supply. An increasing number of companies, especially the larger ones, employ these types of services as a source of organizational and technological innovation. By the end of the 1980s, the major multinationals in the consultancy sector had established themselves in Spain, promoting the use of such services and supporting a 20–30 per cent annual growth in turnover. This situation collapsed, however, during the economic crisis of the early 1990s, resulting in a notable contraction in most KIS activity. During the crisis, many companies considered that KIS were unnecessary. However, the second half of the 1990s showed a clear and strong recovery and expansion.

Management services offer an example of the impacts of such perceptions of KIS. Data from the AEC, focusing on activities linked with management consultancy, show the trend of the KIS cycle from 1991 to 1999, and estimates for 2000. Figure 11.1 indicates market growth for the service sector, based on the turnover of surveyed firms, excluding outsourcing, deflated by the Consumer Price Index (CPI: there are no price indexes for KIS services), compared with the rate of the GDP growth, in constant prices. The graph shows the pronounced sensitivity of management service activities to the phases of the economic cycle, during the recession phase of 1992–93, the recovery of 1994–96 and the new expansion between 1996 and 2000. This sensitivity was particularly notable during the recession when the market declined, with a fall in real terms of more than 16 per cent in 1992, well below a slight growth of GDP. After 1993, management service growth was consistently above the GDP rate, at 6.5–8 per cent, although more closely paralleling it than during the recession and the recovery phase. The more recent

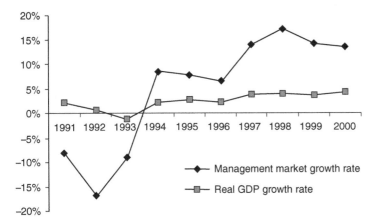

Figure 11.1 Spain: comparative growth of management market and GDP

Sources: AEC 1996–99; Ministry of Economy 1997; OECD 2000b.

Notes
2000 Estimate for 1999/2000 real growth rates.
Rate of growth of management market deflated from CPI (consumer price index) series.

period, 1996–2000, shows much higher growth in management services than in the economy as a whole: 17.1 per cent in 1997–98, 14.1 per cent in 1998–99, and 13.7 per cent estimated for 1999–2000.

The varying importance of the different management activities is illustrated in Table 11.3, which presents estimates of total turnover for different activities and service products. Forty-five per cent of the estimated market share in 1999 came from the development and installation of information systems, with a further 11.3 per cent for information technology support. Management consultancy covers 15.1 per cent of the KIS market identified by the AEC. Market research support accounted for 18.4 per cent of the total market share, while management training took 7.1 per cent. All activities were experiencing accelerated annual growth rates up to 1999. The highest growth rates from 1993 to 1996 were in market research (43 per cent) which more than doubled to 13 per cent of the estimated total. During 1996–99, however, the highest growth rates were in information technology and information systems: 28 per cent and 25.8 per cent respectively. Estimates for 2000 were still positive for these two services, but at lower rates: 16 per cent and 22 per cent respectively.

Outsourcing, estimated in 1995 for the first time, had reached a turnover of Pta 66,700 million (ECU 400 million) in 1999, equivalent to 29 per cent of the total turnover of the management service sector, supporting high annual growth rates, at 37 per cent for 1998–99 and a further 33 per cent estimated for 1999–2000. These growth rates should nevertheless not disguise the fact that the management service market in Spain is still less than 0.2 per cent of GDP, compared with the European average level of 0.8 per cent.

Table 11.3 Spain: turnover (market volume) of management services according to type of activity

Types of services	1993	1996	1999	1999	1993–96	1996–99	1999–2000
	×1 billion			Market share %	% growth	% growth	Est. % growth
Management consultancy	20	25	34.5	15.1	8	13	13
Strategy	3.8	4.9	6.5	2.9	10	11	13
Organization	7.3	8.5	11.5	5.0	5	12	12
Operations/ processes	6.3	8.2	11.2	4.9	10	12	16
Human resources	2.6	3.4	5.3	2.3	10	19	15
Management training	9	11.6	16.2	7.1	9	13	15
Selection and recruitment	3	4.3	6.5	2.9	14	18	16
Information technology	11.1	14.1	25.8	11.3	9	28	22
Development and installation of information systems	46.0	58.3	103.0	45.2	9	26	18
Market research	12.0	27.3	42.0	18.4	43	18	13
Total	101.0	140.6	228.0	100.0	13	21	17

Sources: AEC 1996, 2000.

Although process re-engineering and outsourcing continue to expand very fast, the services with the brightest future at present are those associated with staff recruitment and training, and linked with so-called 'knowledge management'. The latter is one of the newest services promoted by management consultancy in Spain. It relates mainly to the management of information, expertise and intangible assets to optimize the knowledge-based resources of the company, capitalizing on the experience and expertise of its qualified staff.

In 1999, 14.2 per cent of the turnover for these services, excluding outsourcing, was for markets in public administration and 85.8 per cent for private companies. In the three years to 1999, private companies increased their market share by 10.1 percentage points compared with the public sector.

Table 11.4 Spain: regional percentage distribution of advanced business service establishments, 1997

Regions	Total business services	Computer services	R&D	Quality control	Personnel services
	(Total sector ×1,000)	(NACE 72)	(NACE 73)	(NACE 74.3)	(NACE 74.5)
North West	6.29	6.23	7.46	8.27	6.02
North East	9.50	7.80	4.68	10.21	7.76
Centre	7.58	5.53	6.35	9.83	5.33
East	32.58	37.32	44.08	31.25	35.77
South	17.37	14.00	16.77	20.55	13.82
Madrid	26.70	29.12	20.67	19.89	31.30
Total	100.00	100.00	100.00	100.00	100.00

Source: INE 1997.

Public administration attracts high proportions of management consultancy (11 per cent) and the development and installation of information systems (22 per cent).

The location of KIS in Spain is consistent with the economic characteristics of the different regions. Multinationals have chosen to set up primarily in Madrid and Barcelona, opening head offices in other provinces later in response to local business opportunities. These arise from restructuring or privatization, production developments, especially in strategic sectors, or commercial links with local affiliates or commercial partners. Despite the recent decentralizing trends, there is a strong focus of supply and demand in the main regions and cities, with the most advanced KIS predominantly concentrated in Madrid and Barcelona.

At the regional level, Table 11.4 shows how the concentration of the most advanced business services, defined by the Nomenclature des Activités Communauté Européen (NACE) classification, is located in the eastern and Madrid regions. Madrid includes 26.7 per cent of all local business service units in Spain, but 29.1 per cent of computer services and 31.3 per cent of personnel services. The eastern region, led by Catalonia and Barcelona, has the highest percentage of all business services, at 32.6 per cent, and clearly exceeds this share in computer services (37.3 per cent), R&D (44.1 per cent) and personnel services (35.8 per cent). These figures show the KIS dynamism of this Mediterranean region.

The role of regulation and policy in KIS development

The state has many policies promoting innovative processes and the possible uses of KIS (Rubalcaba 1999). However, there are two main channels for state intervention: through direct or indirect legal regulation, or through policies supporting innovation, regions and SMEs. Some legal regulations, such as those governing financial auditing, support the use of advanced services. Others, such as privatization, create business for KIS companies. There are also regulatory or deregulatory processes affecting the KIS professions themselves. Further, recent steps taken by the Competition Court could make KIS markets more dynamic, for example by eliminating certain corporate restrictions and introducing greater supply competition. Legal regulation directly affecting KIS, however, is unimportant compared to the promotional role of the state. In fact, KIS markets are quite free, and professionals do not consider themselves to be significantly affected by either legal restrictions or stimulation.

There are no specific KIS-directed policies in Spain at local, national or European levels. KIS have developed without any substantial support from public authorities. Nevertheless, many new industrial, technological and innovation policies have in fact affected KIS directly or indirectly. Some national SME policies include the provision of publicly supported advisory services, for example on computer resources or quality control. Business service centres supported by regional development bodies also offer public or semi-public consultancy services, especially for SMEs. These affect private KIS, stimulating competition and regional demand, and increasing the need for collaboration between public and private agents.

There are many clear examples of policies that support the growth of publicly provided KIS. The influence of policy on the demand for private KIS is much more difficult to identify. This is usually only indirect, through the effects of innovation, SME or technological programmes available to any type of enterprise. The policies discussed here therefore are not primarily addressed to the role of private KIS. They nevertheless offer the best evidence available relating to the influence of public instruments on KIS markets.

The policies affecting KIS markets operate at three political levels: the European level, the national level and the regional level. The main instruments used in each case are here briefly considered. However, the development of autonomous regions has transferred to them many industrial and innovation policy roles that were formerly exercised by central government.

EU policies

Innovation promotion through European Union policies comes basically from three sources: the R&D Framework Programmes, the Structural Funds (mainly from the Fund for Regional Development (FEDER)) and the European Social Fund (ESF), which influences the diffusion of certain innovations.

The first represented 82 per cent of spending in Spain during the five-year period between 1994 and 1999, although it grew more slowly than ESF spending up to 1996. Innovation may thus arise from very different actions under these programmes, including infrastructure support, especially for telecommunications, the improvement of qualification levels of the workforce, and measures to strengthen productive capacity.

During the 1989–93 period, most of the funds were concentrated on infrastructure, mainly highways, rail transport and energy. After 1994, although infrastructure continued to be important, the emphasis moved towards strengthening the productive system, favouring innovation and the use of KIS. Among these measures, those related to technological innovation and promotion deserve special attention, including support for SMEs and the financing of advanced technology centres. These include the construction of mini-satellite platforms in Madrid and the financing of the Instituto de la Ciencia de los Materiales e Inteligencia Artificial (Institute of the Science of Materials and Artificial Intelligence) in Barcelona (an EU Objective 2 region). Such R&D programmes received the highest percentage of funds related to technology between 1994 and 1996, at 78.5 per cent, far above that for universities (2.5 per cent) and the training of researchers in the Objective 1 regions (19 per cent). However, the university-related actions have been carried out most rapidly under the EU's R&D 1994–96 Framework Programme for Support, receiving more funds than initially planned. Spending in other categories has thus been slower.

Policies to strengthen the productive system have affected the share of R&D in national gross value added, which rose from 0.72 per cent in 1989 to 0.91 per cent in 1993, and is estimated at 1.2 per cent in 1999. In addition, EU resources for research, development and innovation are estimated to rise to 3.1 per cent of the total EU Structural Funds designated for Objective 1 regions in Spain between 1994 and 1999, compared with 1.2 per cent between 1989–93 (Quasar 1998). These modest figures suggest that the effect of innovation policy is better measured through its long-term micro-economic impacts rather than through the immediate macro-economic R&D spending impacts.

Spanish participation in the EU Framework Programmes for Research has grown over the years. In Programme II, projects including Spanish involvement represented 21 per cent of the total, while in Programme IV they reached 30.4 per cent, although the number of projects led by Spanish scientists was still very low. In addition, the 'return rate', which is the percentage of total EU Framework Programme funds obtained by Spain compared with the total it contributed, increased from 5.5 per cent in Programme II to 6.2 per cent in Programme IV. Taking into account the 'additionality' of these funds, this provided financing for many new projects. Of those arising from the Framework Programmes, the Eureka and Esprit projects are of particular importance for their impacts on KIS development. While there is no specific programme for the development of KIS, the technology-orientated priorities of Eureka generally encourage advisory technical and managerial KIS inputs. Esprit is specifically directed towards R&D policies in information technology, again affecting the need for KIS, including their public provision.

National policies

In recent years there has been a significant change in the design of national government industrial policies, largely stimulated by membership of the European Union. Generally, horizontal measures, potentially affecting all sectors, have acquired greater importance than vertical, sector-based approaches. Policies also now seem to be targeted at improving industrial competitiveness rather than simply helping companies in crisis. This change of direction has stimulated actions promoting innovation and KIS. After 1995, there was also an intense process of privatization of public enterprises, with other actions designed to liberalize and deregulate some industrial activities and key services such as telecommunications, transport, energy and the professional services.

In assessing its innovation policies, the former Spanish Ministry of Industry and Energy (MINER – Ministerio de Industria y Energía) concluded that:

1   The National Plan and EU Framework Programme for Research should give priority to funding the most peripheral markets.
2   There is a better balance between vertical and horizontal measures, although horizontal programmes are growing in number owing to the increasing importance of generic technologies.
3   Technological infrastructures benefit only from MINER programmes that aim to boost the supply of technological services to companies.
4   Some deficiencies are evident in the diffusion of R&D, because the plans have only recently considered the creation of networks to facilitate access to, and the assimilation of, technologies as a priority.
5   The resources designated to support technology transfer have increased in all programmes, at both a national and regional level (MINER 1994).

The former Ministry of Industry and Energy (MINER), now the Ministry of Science and Technology, has been responsible for coordinating industrial innovation policies in Spain. The most important plans include (1) the Plan of Action for Industrial Technology (PATI); (2) the National Plan for R&D; (3) the SME Initiative; (4) the Technology, Quality and Industrial Support Initiative (ATYCA); and (5) the recently (1999) approved Technical Research Development Programme (PROFIT). These are described briefly below.

1   Plan of Action for Industrial Technology

PATI is designed to encourage Spanish companies to improve their technological development and the incorporation of advanced technology. Its first phase ended in 1993. The second phase, up to 1996, introduced some new elements, such as the strengthening and creation of technological infrastructure for SMEs, the promotion of business cooperation in technological initiatives, and greater coordination in regional and community policies. Subsequently most of the PATI measures were incorporated into the ATYCA programme (see below).

2   National Plan for R&D

The third National Plan for R&D (1996–99) had, as its main lines of action, the coordination of R&D activities and improvement of the scientific-technological environment for the production sector. Innovation is targeted through the so-called PACTI (National Programme for the Promotion of Coordination of Science, Technology, Industry). In 2000, a new programme was launched for the period after 2001. Sixty-five per cent of public researchers work on R&D projects for the National Plan.

3   SME Initiative

The SME Initiative, promoted by the Instituto Madrileño de Promoción Industrial (IMPI – Institute for Small and Medium-Sized Industrial Companies), was approved in 1994, and was effective until 1999. It included two types of action under which private KIS can be actively involved:

1   Financial support for collaborative agreements among SMEs, including private or public institutions/enterprises supplying support services for such collaboration. These may include intermediate institutions (such as associations, foundations), where KIS firms can play an important role as advisers, or through their own trade associations.
2   Financial support for spatial innovation and technological networks, to advise and support proposals for innovation programmes (such as, under the EU Framework IV Programme, the National Transfer of Technology Programme). In these cases KIS-orientated public institutions (such as technological institutes and business service centres) play the lead role, but private KIS may also participate.

4   ATYCA

The Technology, Quality and Industrial Support Initiative (ATYCA) has been the latest of the Ministry of Industry and Energy's new tools designed to increase the competitiveness of Spanish companies in the new century. MINER's previous support programmes were gathered together under one umbrella, which was broader in scope and adapted to the recommendations of the European Commission's *Green Paper on Innovation*. In 1996 the initiative also supported a 42.1 per cent increase in MINER's initial budget.

ATYCA directs particular attention to creating a propitious environment for innovation in fields barely covered by private initiatives, especially related to the needs of SMEs. These include the creation and consolidation of collective technological infrastructures; staff training; cooperation between companies; cooperation between R&D centres and companies; networking, in which various companies and one or more technological centres in different autonomous regions participate to promote the spread of technology at an inter-regional level; and the development and promotion of horizontal technology.

5   PROFIT

The Technical Research Development Programme (PROFIT) was approved in November 1999 to reinforce the centralization of all scientific and technological innovative initiatives into a single programme. The programme includes direct support for enterprises and other institutions that promote technological R&D, design, sustainable growth, and human capabilities and skills. The two specific objectives are to support (1) the application of new ideas and knowledge to productive processes; and (2) the conditions that favour the absorption of new technologies, fast-growth sectors, and high-technology-orientated enterprises.

This is currently the most powerful state instrument promoting the use of KIS. KIS SMEs are as eligible to receive support from the programme as any others, and the implementation of many PROFIT projects requires the engagement of KIS services (such as human resources, computer, R&D expertise).

Regional policies

An important element in industrial policy is the responsibility of regional governments. The regional promotion of KIS depends on the varying regional environments of business, the concentration of innovative resources, and the existence of bodies such as regional development agencies. KIS-related policies, such as business service centres or technology parks, are targeted to compensate for market failures in non-central regions that lack effective KIS. Regional policies are thus not necessarily coordinated with R&D policies. These favour Madrid, Catalonia and the Basque Country where 80 per cent of total R&D policy spending is directed, nearly twice their share of value added.

KIS-based policies may succeed in other regions. For example, technological institutes and diffusion innovation policies have been successful in Valencia where there has been relatively low R&D effort. In contrast, programmes such as laser diffusion in Madrid, which has the largest concentration of both R&D spending and KIS supply, have been less successful because of political mistakes and lack of enterprise participation.

The three case studies presented in Box 11.1 are for three sample regions: Madrid, Valencia and the Basque Country. Madrid is selected as the area of

Box 11.1 Spain: KIS and regional technology-related policies

Madrid

Technological innovation policy in the Madrid region is articulated around the following actions:

1 Those carried out by the regional development agency Instituto Madrileño para el Desarro llo Empresarial (IMADE) are fundamentally of two types: support for the provision of equipment and technology to companies (for example, construction of technology parks, logistic and technological centres, and the creation of telecommunication networks); and measures for business promotion (such as internalization, sector promotion and the Regional Plan for Industrial Innovation).

2 Scientific policy, carried out through the Second Regional Research Programme (PRI), initiated in 1994 with a dual objective. First, 'horizontal' measures, through the provision of research infrastructure as required and the promotion of researchers' training. Second, the establishment of priorities, taking into account different strategic areas in coordination with 'industrial R&D' and the Regional Plan for Industrial Innovation.

3 Other actions include the Agreement for Industry and Employment, signed in 1993, the Directive Plan for Innovation and Employment, and actions aimed at the creation of technology parks (for instance, Tres Cantos is fully functioning and that in Alcalá de Henares is in its initial phase).

All three forms of innovation policy include important measures related to the development of services to companies. The actions carried out by IMADE regarding its programmes for technology centres are notable. These have become true service centres for companies, offering technical advice and consultancy, supporting R&D actions and ensuring coordination with the other institutional agencies. Similarly, the measures carried out within the Regional Research Programme centre on helping SMEs which, in many cases, deal with business service enterprises contracted out by industrial corporations.

Valencia

Regional policy orientated towards SMEs is developed by IMPIVA (the Institute of Valencia for Small and Medium-sized Enterprises), the regional development agency, founded in 1984 as a result of the transfer of responsibilities from the national to the regional government.

IMPIVA is a public company governed by private law in its relations with third parties, whose actions are developed at two levels: the building of service infrastructure to support SMEs, through centres and institutes promoted by IMPIVA (e.g. technological institutes, European centres for enterprise and innovation, the Technology Park of Valencia); and the provision of funds to intermediate and autonomous private institutions, such as chambers of commerce and the Quality Promotion Centre, for the development of specific activities in the fields of training, information, technology, international cooperation, and diversification of the industry.

The IMPIVA program support SMEs in the following two areas:

1   Enterprise modernization programmes, carried out by means of the fifteen technological institutes in the region, mostly sector based. These programmes focus on the application of new technologies in traditional sectors, and on product diversification.
2   Industry diversification programmes, developed by the centres for enterprise and innovation. These offer infrastructure for entrepreneurial projects in non-traditional sectors or new product lines, as well as locations for new enterprises and information and advice services, including those related to financing. This kind of project is also supported by the IMPIVA Annual Technological Plan, through direct funds at different stages of R&D (creation, dissemination and application), as well as through other actions promoted by the Technology Park of Valencia and the technological institutes.

Enterprise modernization programmes, focused on the application of new technologies in traditional sectors, and on product diversification, are carried out through fifteen technological institutes, mostly sector orientated, based in the region. These institutes, with important basic and applied research activities, offer support through advanced services related to their sectoral specialization (such as footwear, furniture), These include specialized training, technological knowledge promotion, information management, and the promotion of technological Cupertino among companies.

The Basque Country

Many of the technological and KIS-related policies in the Basque Country are led by SPRI (Society for Industrial Promotion and Restructuring), the Basque Regional Development Agency founded in 1981, through which the Basque government manages the policies orientated towards enterprise competitiveness. The integrated support offered by SPRI to Basque businesses is aimed at three levels:

1   Business promotion, where the main programmes embrace support to new business projects involving the creation of between five and thirty jobs, participation in investment through a public risk capital company, and the provision of services by mean of SPRI technological infrastructure, mainly focused on technology parks and business centres.
2   Business innovation, with programmes orientated towards SMEs, including qualitative and quantitative support in the recognition, planning and definition of plans of action; economic aid for external consultancy work in the application of new organizational structures, operational techniques, innovation diagnostics, and total quality management; the provision of specialized human and material resources for training in management; and economic support to certain stages of joint ventures or inter-companies cooperation processes.
3   An internalization programme including loans for promoting commercial and industrial activity abroad, channelling and brokerage for access to aid from state bodies (ICEX[a], ICO[b], COFIDES), and networks of consultancy agencies in other countries to facilitate cooperation between Basque and foreign companies.

As well as these SPRI actions, there are other programmes of the Basque Department of Industry, Agriculture and Fisheries to support business, including grants for projects above a threshold level of investment and the generation of jobs; and providing interest rate advantages in loans and leasing for investment in industrial projects.

Notes
a   Instituto Espagñol de Comercio Exterio.
b   Instituto de Crédíto Oficial.

strongest concentration of Spanish KIS and R&D activities in a service-based region, where services make up 79 per cent of employment and 77 per cent of GDP (FIES 1997). As it is the national capital, many national and multinational headquarters are concentrated there. Valencia and the Basque Country are much more decentralized and have a stronger manufacturing presence. Valencia has key manufacturing sectors such as textiles, footwear, toys and furniture, with many small–medium and export-orientated enterprises, together with the traditional horticulture and citrus goods. The Basque Country is a restructuring industrialized area, with a high representation of traditional heavy manufacturing industries such as iron and steel, mining, shipbuilding and engineering. However, it is becoming a much more diverse region, where the service sector already accounts for close to 60 per cent of employment and 55 per cent of GDP (FIES 1997).

The regional agencies supporting innovation are usually public and private organizations that assist small and medium-sized companies. Their main objective is to provide companies with technical information, assessment and support, as well as to promote innovation and support R&D projects. With some notable exceptions, most were created or strengthened in the 1980s and work in specific autonomous regions.

Generally speaking, the groups of organizations and associations dedicated to the promotion of business innovation and KIS-related policies at a regional level are (1) organizations coordinating R&D, including managing and coordinating regional policies to promote fundamental R&D; (2) technological institutes to undertake basic and applied research, and technological development services; (3) technological service centres disseminating business innovation; (4) strategically located and developed technology parks to attract new businesses; and (5) business promotion centres to support the creation of businesses. Many of these actions are promoted by regional governments and their regional development agencies in relation to specific regional needs.

The interaction of supply and demand for innovative expertise

Having outlined the different aspects of the Spanish market that have a bearing on KIS provision, including demand, private supply and the impacts of public policy, this final section addresses a number of questions affecting the way KIS markets function. First, we consider the often neglected interactive nature of innovation. Second, we analyse the organization of KIS markets and the justification for public intervention. Finally, some examples of KIS in the case study regions will be presented.

Innovation as an interactive process in Spanish KIS markets

If innovation arises from KIS intervention this is because there is a high degree of mutual understanding between KIS supply and demand. Although innovation may be a two-way street, however, the principal onus is on the client to convert innovation potential into reality. Spanish regional development agency staff emphasize that the innovative force must be in client companies. KIS companies are facilitators of change. Even where they are indispensable, the innovative behaviour of the client is even more so. Many Spanish SMEs do not innovate, not because there is inadequate KIS supply, but because they do not have an innovative outlook and are unaware of the potential advantages of innovation. Many of the services they employ thus relate to the day-to-day running of the company rather than to significant change. In practice, when innovation occurs, it is usually incremental rather than radical. Suppliers of

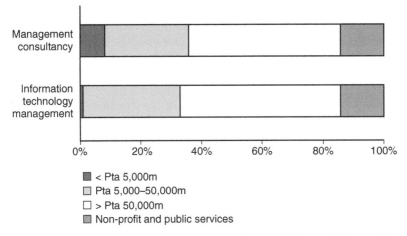

Figure 11.2 Spain: use of consultancy services by size of client, 1996

Source: Provisional data from the main consultancy firms in Spain.

services, for example in Madrid, have greater know-how than many SMEs, which are not sufficiently willing or qualified to make the best use of the services on offer. In fact, client executives who know what help they need and are open to outside influence require the most innovative KIS services.

Innovation increasingly requires training and learning. For this reason many consultants train themselves by maintaining links with universities or academic centres allowing them access to up-to-date information and to a more global perspective. Although demand is not directly implicated in these processes, the learning potential of KIS in Spain is also determined, *inter alia*, by client demand and openness to innovative processes. One of the problems raised by the Spanish Ministry of Industry and Energy's study (MINER 1994) is the relatively impermeable nature of company innovation processes, traditionally focused high in the Spanish business hierarchy. The transfer of technology and knowledge down the line towards the SMEs that constitute the majority of companies is very limited.

The Spanish KIS market is bipolar. A few large companies, making up the majority of the market, coexist with many SMEs who lack market power. The main clients of the large KIS companies are large manufacturing firms, banks and a number of other large companies, as well as the public sector. Many SMEs supply KIS, usually dealing with other SMEs. Figure 11.2 illustrates this situation. Very few of the clients of the large consultancy firms have a turnover of less than Pta 5,000 million (ECU 30 million). They represent 8 per cent of clients for management consultancy and 1 per cent for information technology (AEC 1996). The dominant clients of large consultancy firms are large companies, especially those with a turnover of over Pta 50,000 million (ECU 300 million).

There are several reasons for the duality of KIS markets in Spain. The size of the market, while growing, does not force large and small to medium-sized KIS companies to compete strongly with each other. Markets are specific to each sector, so that there are well-defined entry and exit barriers between them. Of course, competition is much greater within the large corporate segment, and within the SME segment. The large KIS companies compete on brand name, while SMEs compete on the specialization of their founding partners. As a consequence, there is a low rate of client turnover, and low specialization in small medium-sized KIS firms.

Small and large KIS companies have nevertheless recently grown closer together in several ways. Some of the large consultancy firms have set up departments for SMEs, for example Andersen Consulting. Others tend to collaborate with smaller consultancy firms to settle or resolve work peaks or to externalize secondary services. Moreover, a number of partners of large consultancy firms have left their companies to establish their own consultancy firms, encouraging greater integration between large and small companies. Small companies have become more specialized and can thus compete with the larger consultancies, especially where the reputation of the main partners inspires confidence. The duality of the markets is reproduced in part at a regional level. More outlying areas with few large multinational companies maintain a fundamentally local SME business service supply. In contrast, the more developed core regions have an evolving pattern of KIS supply in which large companies coexist with SMEs.

The SME clients of KIS firms usually employ them first to assist in adapting to international regulations on quality and to acquire certain quality standards. This has happened particularly in textile and other manufacturing industries such as those in Valencia. They employ large KIS companies familiar with international norms and with the necessary prestige and experience.

The problem of pricing is important in the organization of KIS markets. The influx of multinational KIS companies has not resulted in price reduction, at least for the more sophisticated KIS services. In fact, there has been a rise in prevailing reference prices such as the hourly rate of consultants. In recent years, the fees for a senior consultant from a major company have fluctuated between Pta 25,000 and Pta 40,000 per hour (ECU 150–240). For small consultancies, however, the situation varies enormously. Some very specialized firms with great professional prestige have fees comparable to those charged by large consultancies, although they normally charge less. Many others operate at significantly reduced rates. It is common for consultants to work for Pta 4,000 per hour (ECU 25), although prices are negotiated for each individual case.

KIS activities are also affected by proximity to production factors, especially human resources, to clients, including large companies and the public sector, and the flow of international information, through market contacts. At the domestic level, living conditions, including environmental or residential quality, also influence preferred KIS location. New communication developments have supported two trends. On the one hand, they allow certain KIS head offices to distance themselves from urban centres where their clients are concentrated so that they can chose a suitable location for optimum working conditions. On the other hand, some more routine tasks can be carried out in branch offices where cost criteria can determine location. Differences also seem to reflect the size of the company. Large consultancies usually have headquarters in more prestigious buildings in the new urban developments, for instance in Madrid where most consultancy firms are located around Castellana Street. This does not mean that they do not have other offices determined by more general criteria. SMEs seek strategic locations relatively near their clients.

Interaction in regional policies: three case studies

These cases reveal something of how KIS interaction is influenced by regional policies. The Madrid case study (see Box 11.2) reflects the influence of public sector clients through policy formulation and quality evaluation. This has been reflected in the participation of private KIS agents in the design of policies. The Valencia case study summarizes the distinctive characteristics of the Valencia regional development 'model', compared to other national and European models. The interaction of KIS supply and demand is presented, as well as the instruments used to define the policies. Finally, the Basque Country case study indicates three means of interaction between KIS demand and supply: benchmarking practices through the 'Gazelle companies'; the organization of 'knowledge clusters'; and the evaluation of technological centres by the users. The main conclusions of these studies include:

1   The participation of private KIS in public policies is important if KIS and technology policies are to be adapted to local needs. A too-centralized model reduces potential user interest in policies, and may waste resources. The best example of this participation model is represented by the Valencia regional government. It is probably one of the most successful in Spain, and perhaps in Europe, involving private and public agents who identify, define and execute policies promoting innovation and advanced services in a dynamic SME context. The actions carried out by the Basque SPRI on Gazelle enterprises and knowledge clusters are also excellent examples of the fruitful involvement of private KIS in public policies.
2   The business environment in which companies operate must be taken into account in regional policies. These must be adapted to the economic, industrial, social and urban characteristics of the region. Public policies can facilitate, but not determine, regional dynamism. Therefore, the main effort should be addressed to promoting the market conditions in which KIS and technological developments can be performed on a private basis. Both horizontal and vertical policies are required for this purpose.

3   The role of regional development agencies is important but not critical in the success of KIS and technology policies. In Valencia and the Basque Country the role of IMPIVA and SPRI in developing policies is important not because they lead all policies but because they integrate private and public efforts on a local institutional and markets basis. Madrid is distinctive since its strong KIS and R&D markets confer a different role on the regional agency.

4   The evaluation of services provided by public policies is not easy. Surveys may not be fully representative and the results are affected by other than specific policy factors. The best guarantee of success is continuity of action, the increasing real participation of current and potential users, and the involvement of private KIS companies as a complement to public or semi-public services. For all these success factors, a degree of political independence and professional management is required in the public regional institutions.

Box 11.2 Spain: regional interaction and KIS-related policies

---

Madrid

The measures regarding innovation policy carried out in the Madrid region are directed towards promoting innovation in SMEs, since they face most difficulties in the process of innovation. Support for technological innovation is particularly required in medium-sized companies, with 50–99 employees, as well as activities requiring high-technological intensity. On the other hand, companies with a low R&D intensity and small firms are less aware of this type of support. Support is centred on four basic requirements: credits for the acquisition of equipment; training programmes for businessmen and employees; advice on the initiation of projects; and subsidies for the development of R&D and the orientation of projects.

There is active participation from the beneficiaries of innovation policies in Madrid. Indeed, they must participate in the risks involved and thus must adopt an attitude 'mentally favourable to innovation'. Subsidies are thus the most extensive means of support. In addition, a network of intermediate agents, such as commercial societies, associations and foundations, actively participate in innovation policies, bringing them close to interested companies.

As for the evaluation of the innovation policy, 65 per cent of businessmen in 1994 were not aware of the support offered by the Madrid region for R&D, and about 61 per cent did not know about the actions of IMADE. Among those which knew about IMADE, just over 50 per cent considered that its actions are either negative or only slightly positive, compared with 47 per cent who thought that they are quite or even very positive. Despite this, most companies receiving support from the regional development agency have been satisfied with the results, support- ing the need to persevere with the promotion of regional inno-vation policy.

Valencia

The main features of the Valencia model reflect the region's peculiarities. The interaction with private economic agents arises, on the one hand through territorial and functional decentralization and, on the other hand, through the participation of these agents in the design and application of policies.

Territorial decentralization follows a scheme of industrial localization in the region, focused on placing technological institutes and enterprise inno-vation centres in areas with the highest concentration of firms in relevant sectors. These institutes and centres are themselves organized in a network, fostering significant interaction between them. Thus they not only support SMEs through the services they offer, but also keep in touch with the feed-back from policy through their proximity to client firms. Economic and private agents participate as shareholders in the non-profit industrial associ-ations which fund the technological institutes. In fact, industrial firms hold the majority of the votes on their boards. This structure encourages SME participation through annual fees and the payment for the use of services, in addition to the access to public funds which were quite substantial when the institutes were created.

Some figures help evaluate the policies developed by the technological institutes. In 1995, 53 per cent of the enterprises in Valencia with more than twenty employees participated as shareholders in the region's technological institutes. Of the shareholding enterprises, 85 per cent used their services. The number of times these firms used the services can be estimated from figures of work carried out by the Valencia technological institutes: 130,000 laboratory tests; 84,430 information actions; 5,283 cases of tech-nology advice; training actions for 4,244 employees; and 275 R&D projects.

The Basque Country

1 Knowledge clusters

'Knowledge clusters' have been promoted by the Basque Depart-ment of Industry, Agriculture and Fishing through the Competitive-ness Programme which commits institutions and economic agents to higher rates of competitiveness. The clusters embrace all those agents related to business management at different stages of the creation, adaptation and spread of management techniques, including regional institutions, universities, industrial and service companies and consultancies, through their participation as partners in knowledge clusters, established as associations. The clusters' goal is the promotion, support, improvement and transfer of management knowledge through permanent cooperation and communication with the various agencies and other institutions and enterprises to improve competitiveness. The clusters carry out actions, among others, related to the creation of meeting forums, research projects and studies, the promotion of investment in management expertise, the establishment of international networks and meetings, training in management for technologists, promoting Internet use for working networks, specific actions toward the needs of SMEs and the design of management tools. In the design of all these actions, the cluster's KIS partners play an important role as an intermediate interface between public initiative and private users.

2 Benchmarking – Gazelle companies

The study 'Gazelle Companies in the Basque Country', carried out by SPRI, analysed the growth of 4,000 companies in the Basque Country from 1991–94, of which 89 were identified as 'Gazelle' companies. The term 'Gazelle' was coined in 1994 by Cognetics, a financial consultancy in Massachusetts, to define companies that had doubled their size in the previous four years, or maintained an annual rate of growth above 15 per cent for four consecutive years. The research identified the key factors that made them the most competitive in their sectors, in order to benchmark other enterprises.

Gazelle companies belong to three types of sector: high technology, services, and sectors that are undertaking structural changes. They have been the only types of company making a positive contribution to employment growth in recent years in the Basque Country, mainly in highly qualified employment. The business culture of Basque Gazelles is focused on anticipating market trends, exploring opportunities and reinforcing internal strengths, maintaining tight control over costs, adjusting margins to compete successfully and devoting special attention to quality and customer satisfaction. In general, the most extensive type of innovation is systems innovation, followed by process innovation, new products, technology (with heavy investment in production tools) and finally the introduction of new raw materials. The experience of Gazelle companies encourages enterprises to employ KIS to help achieve benchmarking standards and to exploit the resulting competitive advantage.

3 Evaluation of Basque Country Technological Centres by innovative enterprises

Evaluation of the actions and programmes developed by the technological centres of the Basque Country, in cooperation with innovative enterprise has shown that 56.5 per cent of these enterprises had used the services offered by the centres, although this increased with the size of firm. The main conclusion was that the enterprises value highly the technological results offered by the programmes and the adaptation of these technologies to the needs of the enterprise. They were less positive about the economic impacts, including the balance between project incomes and expenses, and the effects on competitiveness. However, the smallest companies were more satisfied with the increase of competitiveness than average, while the largest companies were less negative in the evaluation of the profit–cost balance of the programmes.

Conclusion

KIS have recently developed in Spain, following wider socio-economic change and growth, and the expansion of advanced services in the 1980s and the second half of the 1990s. The opening-up of new markets, the interrelation of economies in the European Union, and the new demands of international competition have encouraged the development of KIS in Spain. Changes in production systems, the introduction of new technologies and processes of specialization or externalization have had a strong impact on Spanish companies, hence increasing demand for new and knowledge-intensive services.

The use of KIS has nevertheless been strongly linked to the largest companies and major urban areas. Spanish SMEs and peripheral regions have not actively participated in R&D, innovation processes or the use of KIS. Moreover, the state has not given adequate support to alleviate the situation. The tentative policies aimed in this direction have not always received sufficient financial backing or achieved the degree of success hoped for. KIS-related policies have nevertheless been radically reorientated in recent years. Their current involvement in promoting innovation is associated with initiatives to support SMEs and the multiple activities of regional development agencies, some of which are very active in the promotion of KIS and are well integrated in the surrounding production community. Although these policies may not always have been completely successful, there is evidently a growing interest in KIS, and more realistic and effective initiatives are being promoted within a framework of complementary private and public actions. These include, for example, the engagement of private enterprises in public policies in the case of IMPIVA, to coordinate, complement and profit from existing synergies. This is required in an economy which increasingly uses KIS. There is, however, still much to be discovered in the consultancy world, which seems to hide more than it reveals to researchers and decision makers. Many aspects of the relationship between innovation, competitiveness, supply–demand interaction and KIS are likely to form the basis of continuing research in the near future.

Note

This chapter includes some of the main results of a study carried out in Spain for the KISINN project, financed within the European Union Framework IV Programme of Research. The Spanish part of the project was directed by the professors Cuadrado-Roura and Rubalcaba-Bermejo, authors of this chapter, with the collaboration of Laura Nuñez and David Gago, research assistants of the Services Industries Research Laboratory (Servilab), a research centre linked to the University of Alcalá and the Madrid Chamber of Commerce and Industry. The development of the project also counted on a 'Policy Panel', including senior representatives of the Ministry of Industry and Energy (MINER), business associations of the sector (AEC, SEDISI, Ernst & Young) and agencies for the regional development of Madrid (IMADE), the Basque Country (SPRI) and Valencia (IMPIVA). The authors are grateful for the interest shown by all participants in the project.

References

AEC (1996) *La consultoría en España en 1996*, Madrid: AEC.
—— (1996–99) Reports, Madrid: AEC.
—— (2000) *La consultoría en España en 1999*, Madrid: AEC.
Cuadrado, J.R. and Del Río, C. (1993) *Los servicios en España*, Madrid: Editorial Pirámide.
Cuadrado, J.R., Mancha, T. and Garrido, R. (1998) *Convergencia regional en España: hechos, tendencias, perspectivas*, Madrid: Fundación Argentaria.
Cuadrado, J.R. and Rubalcaba, L. (2000) *Los servicios a empresas en la industria española*, Madrid: Instituto de Estudios Económicos.
FIES (Fundación Fondo para la Investigación Económica y Social) (1997) 'Las comunidades autonomas en 1997', *Cuadernos de Información Económica* (March–April): 132–3.
Fuentes Quinta, E. and Barea Tejeiro, J. (1997) 'El déficit público de la democracia española', *Papeles de Economía Española* 68: 86–191.
INE (1993) *Encuesta de servicios de consultoría 1993*, Madrid: INE.
—— (1995) *Encuesta de servicios informáticos*, Madrid: INE.
—— (1997) *DIRCE (Directorio de Empresas)*, Madrid: INE.
Koedijk, K. and Kremers, J. (1997) 'Apertura de mercados, regulación y crecimiento en Europa', *Colección de Estudios de la Caixa* 11: 67–95.
Mañas, E. (1992) 'La demanda de servicios a empresas según tamaño, actividad y localización', *Papeles de Economía Española* 50: 307–11.
Martín, C. (1988) 'Fundamentos económicos de la política tecnológica', *Economía Industrial* (January): 69–78.
MINER (1994) 'Informe sobre la industria española 1994', *Industria y Política Industrial*, Vol. I, Madrid: MINER, Secretaría General Técnica Subdirección General de Estudios.
—— (1995) *Encuesta de servicios a empresas*, Madrid: MINER, Subdirección General de Estudios.

Ministry of Economy (1997) *Monthly Report*, Madrid: Ministry of Economy.

Molero, J. and Buesa, M. (1995) 'Innovación y cambio tecnológico', in García Delago, J.L. (ed.) *Lecciones de Economía Española*, Madrid: Civitas, 143–72.

Myro, R. (1996) 'La competitividad en las regiones industriales españolas', *Cuadernos de Información Económica* (March–April): 133–47.

OECD (2000a) *Main Science and Technology Indicators 2000*, Paris: OECD.

—— (2000b) *Annual Report*, Paris: OECD.

Parejo Gamir, J.A., Calvo Bernadino, A. and Paul Gutierrez, J. (1995) *La política económica de las reformas estructurales*, Madrid: Editorial Centro de Estudios Ramón Areces, S.A.

Quasar (1998) *Informe de evaluacion de los fondos estructurales en la regiones Objectivo 1*, Madrid: Quasar (mimeo).

Raymond, J.L. (1995) 'Crecimiento económico, factor residual y convergencia en los países de la Europea Comunitaria', *Papeles de Economía Española* 63: 93–110.

Rubalcaba, L. (1997) *Los servicios a empresas en Europa: crecimiento y asimetrías*, Madrid: University of Alcalá.

—— (1999) *Business Services in European Industry: Growth, Employment and Competitiveness*, Brussels and Luxemburg: European Commission.

Rubalcaba, L., Gago, D., Ortiz, A. and Cuadrado, J.R. (1998) *Crecimiento y geografía de los servicios a empresas en el contexto de la nueva sociedad servindustrial: el caso de la Comunidad de Madrid*, Madrid: Consejería de Hacienda, Comunidad de Madrid.

Tamames, R. (1995) 'La economía española 1975–1995', *Colección Grandes Temas* 45: 75–95.

Velarde, J. (1996) *Los años perdidos: críticas sobre la política económica española de 1982 a 1995*, Madrid: Ediciones Eilea.

12 Knowledge-intensive services and innovation

The diversity of processes and policies

Peter Wood

Innovation and consultancy activity across Europe take place in diverse economic and political environments. Regional variations are set within different national milieux of innovative change, which strongly influence the contribution of individual economic agencies, not least the knowledge-intensive services (KIS). These national conditions reflect inherited factor endowments, economic and political structures, processes and rates of economic change, business cultures, education and training regimes, and patterns of state involvement and regulation. EU policies also increasingly exert a differential impact. The eight country chapters of this book have explored these influences. Their common concern has been the role of KIS activity in economic development, especially in relation to innovation. This chapter reviews the variety of conditions encompassed by these cases, and their implications for innovation policy at European, national and regional levels.

It should first be recognized that the drives and goals of business innovation are broadly similar throughout Europe. Most generally, they serve the need to sustain long-term returns on commercial capital in a highly competitive and rapidly changing global economy. This need is widely supported by national governments and is commonly used to justify greater European economic integration. In recent decades it has been supported by more liberal business regulation, although this still varies between countries and is generally more active than in the US. Second, the most reliable sources of innovation-based profit are not high-risk new technologies, but market-orientated applications of developing technologies, for example to serve consumer, logistical, communications and media demand. These are likely to be based more on incremental and synthetic innovations than 'breakthrough', discontinuous technical innovations. Thus, many changes have arisen from the extension and refinement of new information and computer technologies (ICT) in applications ranging from retailing, banking and tourism, to traffic management and education. These have created major new markets, transforming the productivity of other commercial and public sector functions by increasing their information-processing and communication capacities. Such developments are often associated with equally important innovations in organizational, management, human resources and marketing methods. Other technical innovations affecting many sectors include those in new materials, biotechnological processes and engineering developments, for example supporting production efficiency or environmental improvements.

It is also generally true that the longer-term, non-commercial promotion of competitive innovation in Europe is the responsibility of state or EU agencies. This takes place primarily through investment in education and science, health care, and the planning and administration of defence and internal security. The public sector is also responsible for basic investment in infrastructure, sometimes in collaboration with large private companies, including a great deal of exploratory research and development. In recent years, the restructuring of public management and administration itself has created major markets for ICT-based innovation. This includes the privatization and outsourcing of services, as well as transfers of responsibility and powers between national, regional and EU agencies. Innovation also continuously redefines employment needs, both creating and destroying jobs. Its public acceptability thus often requires government to alleviate such social impacts, including those in different localities and regions.

In spite of some basic similarities, however, there are marked contrasts in innovative focus between the eight countries examined in Part II of this book. The most obvious contrast is between the more and less industrially developed northern and southern countries. In Germany, France, The Netherlands, much of Italy outside the Mezzogiorno, and the UK, patterns of technical and organizational innovation are dominated by the transformation from old to new forms of industrial-based production, and to leaner forms of economic organization. These changes have brought significant sectoral and regional decline. This has to some extent been compensated for by new technical-based centres of growth, but these often require new skills and new locations. Such economies are also moving towards more flexible, 'post-industrial', service-dominated production systems and regulatory cultures, although perhaps at different rates. The manufacturing sector employs fewer workers than in the past, but its development of new products and process technologies is still important for macro-economic competitiveness. This also depends on the quality of service functions, however, both to support production and to act as market mediators.

For Spain, Portugal, Greece and southern Italy, equally significant transformations are moving them towards new, competitive technical and organizational cultures. This generally involves market liberalization and the exploitation of associated trade advantages. These advantages may be based on labour availability, the rapid growth of consumer demand, and key sectors such as tourism, the media, trade, agriculture or public investment. There may also be distinctive niches in international markets, enabling established language or trading relations to be exploited, for example in Latin America or the Balkans. The key agents of change include government, as well as multinational investors and major indigenous firms. The large small–medium enterprise (SME) sector, however, is closely tied to old styles of production. In general the EU Community Support Framework and other programmes are directed towards supporting innovative change in these countries, reflecting the strategy promoted by the 1996 European Commission *Green Paper on Innovation*. The dominant competitive drivers of change nevertheless remain major multinational and indigenous firms in key industrial or commercial sectors.

National characteristics and European themes

Each 'half' of Europe represented by our sample of countries is internally very diverse, but both halves are attempting to develop in similar directions, towards innovation-orientated globally competitive production.

The Netherlands

Within the 'northern' countries, The Netherlands is a highly integrated, open, and largely trade and service-based economy, exploiting technical, marketing, financial and trading expertise (Chapter 6). The main economic priority is to adapt human resources to support wider technical, organizational and trading relationships with its European hinterland. There is strong institutional support for some industrial, trade and civil engineering sectors, but also concern, expressed in the 1995 national White Paper, over the difficulties of sustaining innovativeness in an increasingly knowledge-intensive technical environment. A conventional view of this knowledge infrastructure appears generally to be accepted, however, with an emphasis on technological links between the private sector and universities and research institutes, and technical R&D by major companies. Local and regional initiatives encourage 'knowledge clusters', advising small to medium-sized manufacturing firms on technological matters through public–private partnerships. This policy emphasis has nevertheless been questioned by Dutch-based research suggesting that external technical knowledge has little influence on the innovativeness of firms (Brouwer and Kleinknecht 1995).

Dutch research also shows, however, that services are at least as effective as manufacturing firms in delivering innovation to markets (Brouwer and Kleinknecht 1996). Nevertheless, less policy consideration seems to have been given to the distinctive nature of service innovation. This also sustains the international competitiveness of other parts of the economy. The Netherlands is self-evidently a service- and information-dominated economy, within which KIS have played a growing but neglected role. The authors of Chapter 6 comment that, 'Innovation is, after all, a strategy to compete'. One significant outcome of such strategy in The Netherlands, pursued by its major commercial players, has been the growth of KIS. We still need to know more about why this is the case.

Germany

German preoccupations are with sustaining manufacturing-based technological strengths, especially in engineering, vehicles and chemicals production. These are based on dominant manufacturing firms linked to active small, and especially medium-sized firm sectors. As well as incorporating new ICT or materials technologies, traditional corporate and management structures are having to respond to intensifying competition. This includes the challenges posed by low east German and Eastern European production costs. Dominant business practice sustains the relative self-sufficiency of manufacturing firms in technical and other service support. This is nevertheless matched by a strong technical consultancy sector, compared to management or marketing consultancy, and long-established public or semi-public 'intermediary agencies', primarily offering technical advice to SMEs. Even so, computer, management, market research and accountancy consultancies have grown strongly. Indigenous consultancy firms have not developed the international markets of their American, British or French counterparts, and this introversion has been encouraged by recent high demand in eastern Germany. Consultancy markets in Germany thus remain distinctively regionally segmented, reflecting the diversity of economic specialization, but have also been actively penetrated by international firms. The non-commercial 'intermediary' business support agencies also operate at the regional level, and their past success has demonstrated the value of the German tradition of public–private partnership at regional and city levels.

The German experience reviewed in Chapter 5 confirms why KIS innovativeness is difficult to identify and measure. Not only is it dependent on specific client–consultancy relations, but it is also embedded within wider 'systems of innovation', driven by regional, sectoral and large and small firm interactions. German evidence for the importance of the quality of client–consultancy relations in determining KIS innovativeness accords closely with that from other countries, including France (Chapter 4) and the UK (Chapter 7). More common in Germany than elsewhere, however, is the integral role of the intermediary agencies in systems of innovation. Their role has become more comprehensive in recent decades, including sustaining coordination and contact systems, as they have needed to differentiate themselves more clearly from private consultancy firms. This experience is also being transferred to eastern regions. These agencies continued to promote innovativeness in spite of the changing context of the 1990s. More radical forms of institutional, social and organizational change may require still further adaptation, and in Germany as elsewhere these changes largely arise from international pressures on major firms. This has especially supported the growth of KIS. As most other chapters demonstrate, however, German regional intermediary agencies continue to provide a model in developing regional and local innovation policies, especially for SMEs, adapted to the conditions of many other countries.

France

French industrial policy generally focuses on state-supported high-technology sectors, for example in computing and aerospace, and also places strong emphasis on infrastructure-led development. Attempts are also being made to transform the technical base of the older industrial regions by linking it to university and corporate research and development (R&D) strengths. Commercial consultancy involvement in these programmes has probably developed more naturally in France than in either Germany or the UK. In this context, the public sector acts as a facilitator, linking users and suppliers of KIS within a wider framework of industrial and labour market policy. Particularly impressive in France is the range of policies that recognize the potential value of KIS in regional SME support, as detailed in Chapter 4. They have been employed especially to help firms in innovative 'seedbed' developments, and to improve the information environment more generally to support innovativeness.

Emphasis is also given to the coordination of these policies at the local level, although the details of how this is done, and no doubt its effectiveness, vary widely. Innovation policy is associated with key regional and urban development strategies, in and around such regional centres as Lyon, Toulouse and Lille. As is often the case in continental Europe, the contributions of local chambers of commerce are important in this process. The headquarters of the large consultancy firms are focused in and around the capital, but a diversity of spatial networking strategies has also been adopted to reflect government-sponsored development programmes around the country. The consultancy sector includes numerous small firms, however, and their regional and local networking is often linked to semi-public intermediary agencies that actively promote links between consultancy supply and demand.

French practice appears to recognize a broadening view of innovation. The role of innovation policy is not only to stimulate the development of technological hardware itself, although it is clearly a significant priority. It also aims to act as an animator in the creation of networks as a breeding ground for organizational learning, including that associated with technical innovation. Many such networks seem to perform better when organized on a territorial basis, with linkages to similar networks in other regions and metropolises. The innovative impacts of consultancy within this interaction system depend on the types of consultancy and client. The availability of technological consultancies employed in support of R&D is regarded as particularly significant.

The United Kingdom

The emphasis in the United Kingdom during the 1980s and 1990s was on national rather than regional institutional involvement in innovation policies. These policies revolved around the competitive promotion of industrial restructuring and entrepreneurship, labour market flexibility, privatization, market deregulation, inward investment and, until the late 1990s, public sector withdrawal. Regional policies have been in abeyance, except in Scotland, Wales and Northern Ireland. However, some regional promotional agencies, publicly supported but largely led by the private sector, have remained active. Sector-based trade organizations and chambers of commerce have been employed as conduits for the promotion of national innovation policy, although in the UK these have traditionally been less proactive in this respect than they have elsewhere in Europe. Some priority has been given to policy support for SMEs, including regional and urban development components. The quality of education and training policies has been regularly reviewed, for example to encourage computing skills in school pupils, and to teach management methods to university undergraduates. More recently, the need to support risk taking, and to encourage geographical clusters of innovative SMEs, has been promoted by the policy debate.

Meanwhile, the growth of commercial KIS in the UK has been one of the few consistently positive economic trends since the early 1980s. It now employs an increasing share of skilled technical and management labour and entrepreneurial effort, and recruits large numbers of graduates. Computer and various other business consultancies have had a major impact, as has the routine outsourcing of specialist business functions. British business culture has thus been more positively directed towards adaptation to change. This has influenced much manufacturing, but has especially thrived in relation to the larger and more dynamic service sector, as well as the privatized utilities and public agencies. Consultancies have primarily influenced approaches to innovativeness in the corporate, rather than the SME sector, especially in the already more prosperous southern regions.

In spite of the growing involvement of commercial KIS in mainstream economic development, the explicit adaptation of their experience to innovation policies has remained tentative, and selectively directed towards SMEs and peripheral regions. For SMEs, much of their innovative direction often depends on customers and suppliers, with business planning advice coming from their accountants and banks. More strategic initiatives, such as those that relate to new technologies or markets, are promoted to some by government advisory schemes such as the 'Business Link' programme. The involvement of strategically orientated consultancies has been confined to short-term pilot studies, or small-scale spin-offs from public sector advisory programmes. It remains to be seen how far the establishment of regional development agencies in 1998–99, and their encouragement of innovative business clusters, will affect patterns of business support, especially in more peripheral regions. The faltering growth of KIS employment in Scotland and the North in the mid-1990s, described in Chapter 7, suggests a reinforced concentration of consultancy services in the South, which may weaken the innovative potential of other regions.

Italy

Italian experience links that of northern and southern Europe. It reflects Italy's threefold regional division and the polarization between a dominant monopoly/oligopoly sector and the SME sector, elements of which are embedded within flexible regional production systems. Industrially, Italy is characterized by a low presence of multinational firms compared with other industrial countries, and a greater dominance of state enterprises and SMEs. Innovation also lags, with a relatively weak electronics/IT sector, but stronger machine tools and regionally based innovation systems. The national policy approach developed in earlier decades towards the economically backward South was effectively withdrawn in the 1990s. EU support and priorities have replaced this, to emphasize the development of an innovation-orientated economic infrastructure. This aims to encourage inter-organizational links to foster technical and managerial innovation. Meanwhile, national policies have encouraged rationalization and diversification of large firms, including the financial and other services. There has also been a crisis of state enterprise, with moves towards privatization, and the restructuring of public services.

The complexity of innovation-related policies in Italy, originating from EU, national and local institutions, is detailed in Chapter 8. Perhaps most significant is the commitment of public and private regional agencies, especially in the North, to the promotion of local competitiveness and innovation. Service support seems to be integral to these strategies, often through business 'service centres'. The potential contribution of KIS has also been recognized, with many programmes experimenting in supporting their use. Innovation is thus promoted in a great diversity of local and regional socio-institutional contexts. However, the move during the 1990s towards a greater 'localism' in policy has encouraged inequality. The better-established communities of the North are able to promote new developments more successfully than those in the South. KIS developments in Italy, as elsewhere, closely mirror this wider economic variability. Consultancy use has also been stimulated by EU-sponsored programmes and regulation, whose policy focus is the development of knowledge-based human resources and interlinked groupings of SMEs. In practice, the implementation of many of these expertise-support policies in the North is driven by commercial opportunities, while in the South they largely depend on public sector initiatives.

Such diversity might be interpreted a strength, if it reflected adaptation to a wide range of local economic trajectories. As commonly in the past, however, there seems to have been very little attempt in Italy to evaluate alternative policies. The contributions of various approaches are therefore likely to remain unclear. In the case of the Mezzogiorno, KIS can support development only as part of a much broader shift of policy towards promoting capacities for mutual institutional support. These are already much more firmly embedded in the North, and EU and other support is unlikely to overcome the long-established advantages of more prosperous regions, now reinforced by the development of advanced KIS.

Greece

Like the other three countries that entered the EU during the 1980s, Greece has since undergone very rapid economic development, one outcome of which is the growth of KIS. There are still, however, chronic contrasts between modern and traditional elements in the economy, and between Athens and Salonica (Thessaloniki), the dominant centres of modernization, and much of the rest of the country. The Greek debate over the role of service developments in economic restructuring has a wider relevance in southern Europe. In Greece as elsewhere, the state is active in promoting innovation. EU policies and programmes have also been instrumental in establishing the norms, standards and requirements of development. Much consultancy activity, especially by Greek firms, has been directed towards immediate business problems arising from technical and regulatory change, such as computing innovation, legal advice, auditing and quality certification. This has done little to challenge the traditional short-term entrepreneurial culture. KIS activities supporting longer-term business strategy, and a more innovative, collaborative culture based on organizational and social change, are less in evidence. Nevertheless, some consultancies are involved in the restructuring of major industrial firms, arising from communications developments in shipping and maritime developments, for example, in tourism, or in supporting Greek investment in the Balkans and Eastern Europe.

The impact of consultancy on innovation, as for any other influence, is undoubtedly constrained by the small size of the market and its general dependence on outside sources of technical development. The main economic needs are to adapt outside innovation to Greek conditions, and to support Greek business in taking advantage of international developments. Such adaptive innovation, both technical and organizational, must be what is primarily required by any small, rapidly developing nation. Consultancy offers expertise in promoting such processes that is unlikely to be matched by academic or public sector agencies. A high proportion of the Greek consultancy market in the 1990s was in the public sector, supporting 'adaptive' structural innovation, for example arising from privatization. The private sector market was also heavily orientated towards adaptive change, in the banking sector, support for mergers and takeovers, or the expansion of foreign service firms in the Greek market.

Chapter 9 also emphasizes problems in Greek policy application, including its lack of consistency and continuity. More specifically, it points to a danger that the consultancy market may be distorted by the direction of much effort towards EU project preparation, rather than innovation programmes themselves. This 'project logic' creates substantial fictitious demand, vulnerable to policy changes in the future. For these and other reasons, in Greece as in

Portugal, the quality of consultancy work is more a matter of concern than in the more commercially orientated, competitive markets of Germany, France or the UK.

As elsewhere, intermediate institutions are being experimented with in peripheral regions to encourage networking between local firms, consultancies and innovative institutions such as universities, science parks and innovation centres. This may hardly compensate for the huge concentration of innovative KIS into core regions. There is also scope in Greece to support the professional and commercial strength of established consultancies in key fields such as shipping, public works, environmental impact assessment and retail studies. Their comparative advantages may then be exploited in outside markets, as has already occurred in the Balkans and Eastern Europe.

Portugal

The same pattern of rapid KIS growth accompanying economic transformation that we saw in Greece has occurred in Portugal. Inward investment has expanded while the traditional manufacturing and agricultural sectors have struggled, especially the SMEs. Much new capital has been directed towards various elements of the service sector, including retailing, the media, tourism, and business and financial services. The workforce is young and increasingly well qualified. Demand for consultancy has, once more, been encouraged by government privatization and public sector outsourcing programmes, as well as by the requirements of EU projects. As in southern Italy, these often relate to innovation policies, emphasizing infrastructure and communications support, promotional programmes, and the development of technical and management competence.

Portugal also displays characteristic contrasts in regional development. Four territorial profiles as models of innovation are presented in Chapter 10 which might well also be adapted to other southern European cases. First, the Lisbon metropolitan area is the main focus of R&D, advertising and marketing, the media and information technology. Like Athens in Greece, it heavily dominates the national scene. Second, the northern coastal industrial districts, based on traditional textiles, clothing and shoe industries, require support to foster specialization and keep down costs. Third, the mass tourism region of the Algarve is dominated by the real estate, construction and travel/accommodation services. Finally, KIS developments are severely limited by the small scale of regional markets in rural areas. Case studies in the chapter suggest the possible contribution of KIS in these different regional contexts, as well as in seven sectoral situations, such as agriculture, various manufacturing sectors and in the services.

Demand is not only constrained by small-scale markets, but also by the attitudes of firms to employing outside advice. Chapter 10 offers some revealing insights into the variety of client needs and the types of consultancy clients might call upon. Portugal illustrates a more general policy dilemma. Innovation-support policies tend to focus on more proactive firms, especially in the SME economy. But they must also seek to widen demand for support services, including KIS, through the more general promotion of modernization. This requires programmes that are not just directed towards already innovative firms. Significant questions are also raised in Portugal, as in Greece, about the quality of advice available from the limited range of consultancies and the lack of client experience in employing them, especially in conditions of rapid growth. More active public and professional engagement and supervision may be required than in established national KIS markets. This must not inhibit the dynamism of effective client–consultancy relationships, however, or the ability of clients to become self-reliant through experience.

Spain

In Spain, sources of innovation capital are again dominated by public enterprise and multinational firms. The major innovative sectors are electronics and computing, engineering, and the electrical and basic metal industries. Indigenous firms remain generally cautious, emphasizing cost-cutting and incremental changes rather than significant strategic innovation. Innovation policy needs therefore to support the competitive restructuring of traditional industrial sectors, the promotion of R&D, support for SMEs and the improvement of the economic infrastructure, including IT.

Spanish KIS are described in Chapter 11 as 'paradigmatic activities' for the wider organizational changes facing Spanish industry. They promote greater organizational and labour market flexibility, emphasize quality, develop economies of scale through specialization and encourage greater awareness of international market developments. The more progressive KIS segments primarily serve the major companies and the public sector in and around the major urban areas. Demand is thus very polarized. As in Portugal, most SMEs in Spain employ outside support only for routine, cost-cutting purposes and, as in Greece, their potentially innovative contribution in the wider context of change is poorly developed.

The location of KIS in Spain reflects the economic characteristics of the different regions on a grander scale than in these other countries. The most advanced consultancies and major companies have headquarters mainly in Madrid and Barcelona, with other provincial offices established as opportunities arise. Once again, there are no policies directed towards the development or use of consultancies, but many aspects of economic policy, from the EU to the local level, might engage SMEs with consultancy services. Chapter 11 indicates the variety of such policies and, as does Chapter 8 on Italy, exemplifies the local institutional diversity that has emerged. The recent interest of larger consultancy firms in SME clients, and in collaboration with progressive small to medium-sized consultancy firms, could challenge the polarity of KIS impacts between large organizations and SMEs. As the market matures, the tendency for experts employed by large firms to break away and establish small consultancies may also grow, encouraging integration between large and small firms' use of consultancies.

Europe north and south

Generally, in southern countries the significance of IT and computer software in driving innovative change is now universal, overlapping into consumer and public sector markets. Such trends are often led by these sectors, rather than manufacturing. Higher-level management and technical skills, often imported, are focused in multinational companies, major state infrastructure and specialist export sectors/regions. All of these are supported by international consultancy firms. The potential for developing indigenous innovative expertise largely rests in specific industrial, regional and trading resources, local market knowledge, and adaptation of these to international markets. The involvement of major consultancy firms and smaller spin-off firms in the dissemination of technical and professional experience in the future would seem to be integral to these processes.

Regional variability is inherent in economic change throughout Europe. The complexity of innovation-based change is probably greatest in the northern countries. Here, it depends on interdependent trends in the division of labour between manufacturing and private and public services, forms of organization and regional patterns. For example, much of the dynamism of the UK economy remains concentrated in its southern regions; Germany's east–west divide has shown little sign of diminishing; innovation in France is tied to emerging regional and national networks, based around Paris, Lyon and other urban conglomerations. Even in The Netherlands there is a contrast between the concentration of industrial innovation in the South East compared with the service-dominated cities of the Randstad.

Each country is also distinctive in the regulatory regime supporting innovation processes. The most distinctive features of high-income economies are the greater development of small to medium-sized consultancy companies, their involvement with corporate clients, and the orientation of a higher proportion of consultancy supply towards strategic rather than routine operational advice. As the economies of southern Europe develop, consultancy markets are likely to resemble northern European practice, although there may be important problems of maintaining quality during any transitional period. In both situations, however, the main difficulty created by the growth of the commercial KIS market is to strengthen the range of effective innovation-orientated support to SMEs and peripheral regions.

Although KIS have grown at high rates across the EU, low levels of demand and supply in some regions of both northern and southern countries often constrain their contribution to change. The traditional slowness of many commercial and public institutions to adapt to change is reflected in attitudes to consultancies. This perhaps offers them a particular role in fostering change.

Unfortunately, polarization remains inherent in KIS provision. To compensate, regional centres of 'self-help', run by private, public or non-profit agencies, have multiplied across the EU. These are attempts to counter the geographical and sectoral inequality that appears to characterize knowledge-based economies. If these inequalities intensify as specialist consultancies extend their influence, new questions will be raised about the effectiveness of such intervention in widely differing sectoral and regional circumstances.

Policy, innovation and the role of consultancies

Across Europe, state involvement in promoting technical innovation and associated changes in business practice take many forms. These include:

1 Undertaking or subsidizing basic research and development, guided by science and technology policies, in universities, research institutes and private companies.
2 Collaborating with private companies, as the dominant client, in developing particular innovations, for example in defence, health care or communications.
3 Encouraging the commercial applications of developed inventions by private firms, for example through tax breaks, investment support, financial and other advice, marketing assistance or training.
4 Encouraging the innovativeness of private firms in particular sectors or size segments, especially SMEs, through measures such as financial and management advisory services and fiscal incentives.
5 Supporting business development more broadly to encourage competitive technical, organizational or marketing adaptation in specific sectors, firm size segments or regions/areas.

At the national scale, the dominant innovation policies are primarily directed, not towards non-innovative or peripheral activities, but towards key sectors and dominant firms where success is most likely to be achieved. They may, however, be linked with more comprehensive change programmes to stimulate the wider innovative environment, both nationally and in particular regions or localities. At any scale, the success of innovation-orientated programmes also depends on the wider business milieu, based in social and cultural norms. These influence more general economic policies, traditions of education and training, infrastructure investment patterns, and market intervention or regulation conventions. Transferring such conditions to less innovative economic sectors or regions is thus not a simple, or purely economically based, task. Innovation is inherently a sectorally and regionally uneven process, supported by much innovation policy. Those 'remedial' policies designed to spread innovativeness more widely face strong countervailing processes, many driven by other, and often more powerful, aspects of policy itself.

Any innovative impacts of consultancy have probably reinforced the position of the corporate sector, and intensified the challenges faced by less innovative, more policy-dependent activities. There are also innovative small firms, of course, including those in specialist regions, such as southern Germany, Emilia Romagna or southern France. These are associated with supportive technical and business KIS activity. Once more, most policies in these regions generally support established success. Where policies are most needed, especially among SMEs or in peripheral or developing regions, consultancy use is commonly least developed.

The direction of remedial innovation policies during the 1990s suggests that the inherent divorce between them and consultancy activity has been a growing handicap, although largely unrecognized. For example, policies traditionally sponsor or encourage technical R&D, including the adoption of ICT. Nowadays, more than this is required (see Chapter 1), including flexible training, skills development or organizational adaptation. In the past, these changes might have been regarded as the responsibility of firms themselves. In practice, successful organizations increasingly employ outside consultancies to help. Non-commercial intermediary agencies may make some contribution, but their aggregate activities remain small and specialized compared to the general range and level of advice available from the innovation-orientated economy, especially by the KIS sector. Outside the corporate–public sector nexus supporting the core processes of innovation, any individual measures are likely to have marginal impact. In many cases, this is all that is intended, primarily to support SMEs or particular communities. In comparison, the growth of consultancy over the past twenty years has been a major trend affecting the environment of change that is still to be fully recognized in innovation policy throughout Europe.

The 1994 White Paper, Growth, Competitiveness, Employment

The questions for industrial policy makers raised by the rapid modern development of specialist business-related consultancy, and especially the international dimension of European KIS, are exemplified by the European Commission 1994 White Paper on *Growth, Competitiveness, Employment* (European Commission 1994). These themes have been echoed in similar national documents published since. The specific areas of policy where KIS might exert a significant impact may be illustrated by examining some of the major aims of the White Paper (relevant section numbers are indicated):

1   *Exploiting the competitive advantages associated with the shift to a knowledge-based economy* (section 2.3(b)): KIS are a rapidly growing resource of human knowledge and expertise affecting especially the organizational capacity of business. They potentially contribute to the quality of training, efficiency of industrial organization, capacity for continuous

improvements in production processes, the exploitation of R&D, the fluidity of market conditions and the availability of competitive service infrastructure.

2   *Supporting the 'supply side' in reducing the time lag of change* (section 2.3(d)): KIS promote this by encouraging continual structural adjustment in EU industry (including supporting privatization programmes) and the development of 'clusters' of competitive activities, drawing on the regional diversity of the EU. At the regional level, KIS may provide the matrix of such clusters, linking and augmenting the skills of other participants.

3   *Developing frameworks for inter-company cooperation* (section 4.3(a)(i)): KIS advise on these, contributing experience inherent in their own mode of operation.

4   *Developing entrepreneurial ability to respond flexibly to new demands* (section 5.2(b)): KIS develop new services for the 'information society', based on their role as codifiers of knowledge.

5   *Supporting training* (section 5.3): KIS encourage the diffusion of best practice in the application of ICT, including that of training.

The various KIS specialisms have very different effects. In exploring the role of existing and potential policy arenas, however, a primary distinction can be made at European, national and regional levels, between KIS activities that serve the technical needs of specific sectors, and thus are most germane to the sector-based innovation policies; those supporting the dissemination and application of IT and communications innovation, affecting IT and communications policy; and those supporting more general managerial innovation in all sectors, impinging on many aspects of industrial and regional development policy. In practice, these activities have become increasingly overlapping and interdependent.

It is clear that KIS are implicated in many facets of European growth, competitiveness and employment. Market evidence suggests that their main impacts are on the adaptive and innovative capacities of international/national corporations, and state-driven initiatives supporting R&D, trade, privatization, infrastructure, regulation and public sector reform. The growth of KIS suggests that many organizations are successfully reorganizing the expertise requirements of market and technical changes without regard to industrial or innovation policy intervention.

The effective use of consultancies is nevertheless still a relatively new management challenge, which even large private and public sector clients have yet to master fully. Anti-consultancy business cultures still sometimes prevail and consultancies themselves often adopt a short-term and opportunistic view of their role. There are still significant barriers to tapping the international expertise available through consultancy, even in the corporate and public sectors. In-house recruitment, training and loyalties often remain the preferred basis for change and innovation.

The policy significance of consultancies is perhaps easier to demonstrate where markets for their services are less developed. It has been argued above that their growth is associated with inherent polarizing tendencies, adversely affecting small firms and peripheral regions. The influence of consultancies therefore raises special problems in relation to section 2.6 of the European Commission White Paper, 'Supporting the development of SMEs'. 'Consultancy-type' provision is already commonplace for routine financial and legal inputs among such clients. Many consultancies are themselves dynamic SMEs. However, the orientation of consultancy markets towards corporate needs, and SME lack of capacity and experience in employing consultancies, discourage more strategic use. This is a primary justification for the continuing role of non-commercial intermediary agencies. As we have seen, however, these agencies are also challenged by the growth of consultancy, because of the pace and range of change they promote, and the model they established of the collaborative relationships required to benefit clients.

Consultancies and innovation policy

Innovation policies with at least some KIS component began to complement more traditional measures of loans and investment subsidies in the 1980s. They have focused on information and advisory support to SMEs, to enhance their technological, marketing or training abilities, and on elements of regional development, for example associated with technology parks or other types of seedbed investment. These measures nevertheless appear to be largely *ad hoc*, small scale, and marginal in relation to the wider needs of business information availability. In most cases consultancy activity is strictly regulated by public or semi-public development agencies, which often carry out advisory work themselves. These appear to have been most successful in the technical sphere in Germany. More recently they have broadened their remit there, and have also been promoted in France, The Netherlands, northern Italy and the UK. They often rely on a responsive mode of working, however, targeted towards firms with established innovative potential.

The involvement of consultancies remains generally limited to that of small specialist firms acting as the agents of public bodies at a local level. Bessant and Rush (1995) have reviewed policies for supporting technology transfer to SMEs in the UK, including a critique of their use of consultancies. They suggest in general that many innovation policies still reflect traditional linear models of innovation, rather than the interactive processes identified by modern research (see Chapter 1). However, they tend to neglect the dynamism and growing interactive influence of consultancies in the wider economy. Their identification of six problems in incorporating consultancies into innovation policy (ibid.: 112–13) nevertheless raises some key issues.

1 The effective matching of consultancies and users

This process has been the focus of the developing market for consultancy over the past twenty years. Bessant and Rush argue that, 'large firms know how to make effective use of consultancies' (ibid.: 112). In fact, the UK evidence cited in Chapter 7 suggests that this is too sanguine a view for some large firms. Nevertheless, matching SME clients and suitable types of consultancy has become the dominant concern of policy agencies. The main need highlighted by Bessant and Rush is for the training of users and consultants, and the exchange of experience about effective user–consultancy relations.

2 The need to develop quality control of consultants engaged in policy initiatives, and for project monitoring, based on acceptable standards and certification

Bessant and Rush suggest government action over this. In practice, in the northern industrialized countries at least, the rapid growth of consultancy in recent years has led to the development of a competitive market, with standards established through performance and reputation. Various forms of quasi-professional certification have also emerged, including membership of trade organizations and the adoption of codes of practice. Government quality control has had little impact, except where public programmes select particular consultancies. The same may not be the case in the southern European countries, such as Greece and Portugal, where the common problems of ensuring quality of consultancy provision have been highlighted in Chapters 9 and 10. Even here, accreditation and quality audit procedures instituted at the national level, especially by national and EU agencies in association with major companies, are likely to be the most effective basis for quality control, rather than detailed intervention in client–consultancy exchanges by public sector regulators. The effective use of consultancies might also become an integral part of any programmes intended to improve management quality more generally in corporate, state and SME sector agencies.

3 Where there is public intervention, problems of detailed project management need to be addressed

This is likely to be inherently difficult where, as we have seen, intensive, programme-specific interchange is fundamental to success. These difficulties are compounded in innovation processes which require complex inter-organizational arrangements. The inter-firm basis of much modern innovation is emphasized by Bessant and Rush themselves, with large and small firms employing consultancy inputs at various levels and stages of project development. In such a context, public monitoring of individual projects is unlikely to be effective, and might even be counter-productive. At the very least, decentralized and expert forms of liaison are required.

4 *The need to promote and subsidize effective long-term development of consultancy use through possible outreach programmes, especially to SMEs and in peripheral regions*

Sustained mutual experience appears to be the key to success in the client–consultancy relationship, often through repeat projects. The involvement of the same personnel also aids the process, especially for small businesses. Localized consultancy firms or intermediary agencies may be able to provide this continuity, perhaps building on more routine forms of support. Difficulties arise over more occasional, strategically significant business development programmes, when such familiarity is inherently rarer. Forms of collaboration with clients and outside consultancies may have to be developed by local agencies on behalf of SMEs to draw on wider experience of such programmes.

5 *Extending the period over which subsidized access to consultancy is sustained*

Bessant and Rush argue that consultancy advice for larger and experienced firms may be expected to become self-financing. The wider social benefits of raising managerial and technical capability may nevertheless justify continuing subsidies to smaller and inexperienced clients. Without this, they suggest, the net effects of growing consultancy influence will favour firms able to support commercial relationships with consultancies. As this book has argued, this effect is already demonstrable across European economies at large. The proposed solution nevertheless raises the prospect of clients developing 'subsidy dependency', discouraging the development of independent expertise in engaging consultancies. Nevertheless, the principle of sustaining process consultancy, to enable innovation to be delivered, remains important.

6 *The use of consultancies in guiding potential beneficiaries of innovation policies to appropriate support, through 'one-stop' access*

As we shall see below, this might well be an important role for consultancies, if only to overcome confusion arising from the multiplicity of public policies and agencies.

Bessant and Rush regard consultancy as a detailed element of wider policy support for innovation. Any adoption of such 'best practice' must also take account of wider policy shifts in recent years towards supporting the economic and social environment of innovation, rather than just aid to individual firms or even the physical infrastructure of development. For example, the EU Community Support Framework, reflecting the *Green Paper on Innovation* (European Commission 1996: see Chapter 1), has emphasized inter-firm and inter-organizational relationships. It also addresses the social and political context of business success, including the availability of venture capital, information and advisory services, promotional agencies, and support for management quality improvements.

In many countries, as well as in Brussels, these changes became the orthodoxy of the 1990s. Publicly supported intermediary agencies, such as the Steinbeis Foundation/RKW in Baden-Württemberg/Germany, Business Link in the UK, or the regional innovation centres in The Netherlands, and the many agencies based on local authority–chambers of commerce collaboration in France and Italy, are now generally favoured as integrators of these various types of initiative, linking area development policies to the needs of individual firms.

The need for such agencies to adapt to the pace of organizational and technical innovation, as private consultancies must do to survive, has led some, for example in Germany and The Netherlands, to recast their activities. A cautious and even suspicious attitude still prevails, however, towards the quality and validity of private consultancy itself, especially for inexperienced SMEs. Such traditional attitudes to consultancies are perhaps understandable, but increasingly out of date. Policies to encourage organizational and technical adaptation to global competition must tap the experience and dynamism of consultancy markets. As in best commercial practice, a key condition for this is a clear definition of the division of client–consultancy skills and responsibilities between public and private sector agencies.

In spite of recent changes of policy approach and in the range of agency activity, the pace of commercial change still raises questions about the provision of information and expertise by other non-commercial agencies. Internationalization especially strengthens potential consultancy influence beyond what regional or even nationally based agencies can offer. Innovation policy must address the intensity of international competition, the factors influencing rates of innovation dissemination, and the capacity of firms to absorb and use new knowledge, especially through organizational and human resources development. In the mainstream economy, consultancy firms are now active agents in each of these processes, sharing mutual learning capacities between European, national and regional levels of innovation experience. A growing interdependence between different levels in the global learning process is also increasingly influenced, and even structured by KIS (see Chapter 3).

Client–consultancy interaction in remedial innovation policy

How might market processes be encouraged to support innovation among SMEs, traditional sectors of the economy and peripheral regions? A demand-based and process-focused approach is required, addressing problems of market responsiveness, the sectoral and local networking of experience, the assessment of alternative outcomes, the development of client–consultancy trust, and appropriate critical feedback of results. Good corporate managements control consultancy quality and delivery during often sustained periods of process consultancy. In contrast, in spite of the advice of Bessant and Rush (1995), the limited policies supporting access to consultancy impose severe limits on both time and financial resources. More extended interchange with clients is usually reserved for the intermediary agencies themselves.

For many consultancies, experience with policy-sponsored projects is also generally regarded as onerous and unprofitable. Public agencies are seen as obstructing sustained and productive client–consultancy relations. Such schemes draw mainly on small, specialized consultancies, often engaged by officials with limited commercial experience. They commonly appear more interested in the regulation of inputs than in the process of exchange and the quality of outputs.

Perhaps a more fundamental problem lies in the scale of demand, and the relevance of the integrating skills of major consultancies to the needs of the SMEs, declining sectors and peripheral locations. These conditions are especially limiting where the scope of innovation policy is drawn too narrowly, for example focusing on technical innovation alone, on manufacturing SMEs or on individual firms. Regionally based development strategies should aim to overcome demand threshold problems. In IT and computer software development, for example, organizational and technical innovation in banking, retailing, public administration, transport, health care and other community activities, as well as in manufacturing, are often interdependent. Together they present a volume and range of needs that may justify the selective engagement, especially in strategic roles, of specialist consultancies. Often the required expertise is generic, supporting responses to global competitive change, such as market analysis. The public sector itself might take a lead. The EU, like other public bodies, employs major consultancy firms to monitor and audit programmes. It seldom encourages their more positive involvement to benefit local implementation across a range of development policies.

Partnerships between non-commercial intermediary agencies and consultancies could enable the latter to be incorporated into policy implementation, as is already often the case in national sector-based policies. Various forms of consultancy collaboration might be developed, including perhaps the exchange of seconded staff between intermediary agencies. If suspicion of consultancies remains a barrier to such developments, commercial experience suggests that this can be overcome through the active and professional management of agency–consultancy relations. Public sector agencies have, after all, gained much experience as clients of consultancies in recent years. As in all cases, the key to success is a clear division of labour in relation to the different skills brought to bear. In any case, non-commercial agencies also attract suspicion from SMEs, and very few of the programmes they operate are properly evaluated. One of the advantages of employing consultancy expertise, often cited by private companies, is that its costs, control and impacts are more transparent than when internal staff are employed. The effectiveness of any policies may remain difficult to establish, but the culture of much consultancy is nowadays inherently aligned to project specification and delivery.

Consultancy involvement in innovation policies has other implications. Training of agency and SME clients in the use of consultancies may form part of improvement and updating more generally in the quality of management. Consultancies do not substitute for client technical or R&D deficiencies, but they may help to identify and overcome these through process innovation. Locally based consultancies should also be encouraged to improve their services, especially by fostering awareness of, and even links to national/international consultancies.

The growth of consultancies over the past twenty years has made them an integral part of modern economic organization. The dominant sectors of the European economies, including the public sector itself, increasingly rely upon them as agents and supporters of change. Policies intended to foster and encourage change thus cannot ignore consultancy influence. Policies directed towards supporting a 'learning economy' commonly depend on non-commercial advisory agencies. In the mainstream economy, however, consultancies have flourished because of their flexibility and adaptability. Public agencies generally find this more difficult, especially in relation to globally driven commercial change. New arrangements are required in which such agencies collaborate with global, national and local consultancies to support programmes of sector and community economic development. These should draw on the expertise which only consultancies can bring to bear on the effective adaptation of enterprise to global change, and the market delivery of innovation.

References

Bessant, J. and Rush, H. (1995) 'Building bridges for innovation: the role of consultants in technology transfer', *Research Policy* 24: 97–114.

Brouwer, E. and Kleinknecht, A.H. (1995) 'An innovation survey in services: the experience with the CIS questionnaire in The Netherlands', *STI Review* 16, Paris: OECD, 141–8.

—— (1996) 'Determinants of innovation: a micro-econometric analysis of three alternative innovation output indicators', in Kleinknecht, A.H. (ed.) *Determinants of Innovation and Diffusion*, London and Basingstoke: Macmillan Press, 99–124.

European Commission (1994) *Growth, Competitiveness, Employment*, White Paper, Brussels: European Commission.

—— (1996) *Green Paper on Innovation* (*Bulletin of the European Union*, Supplement 5/95), Brussels: European Commission.

Index